Electronic Thin-Film Reliability

Thin films are widely used in the electronic device industry. As the trend for miniaturization of electronic devices moves into the nanoscale domain, the reliability of thin films becomes an increasing concern. Building on the author's previous book, *Electronic Thin Film Science* by Tu, Mayer, and Feldman, and based on a graduate course at UCLA given by the author, this new book focuses on reliability science and the processing of thin films. Early chapters address fundamental topics in thin-film processes and reliability, including deposition, surface energy, and atomic diffusion, before moving on to systematically explain irreversible processes in interconnect and packaging technologies. Describing electromigration, thermomigration, and stress-migration, with a closing chapter dedicated to failure analysis, the reader will come away with a complete theoretical and practical understanding of electronic thin-film reliability. Kept mathematically simple, with real-world examples, this book is ideal for graduate students, researchers, and practitioners.

King-Ning Tu is a Professor in the Department of Materials Science and Engineering at the University of California at Los Angeles. Since receiving his Ph.D. in Applied Physics from Harvard University in 1968, he has gained 25 years' experience at IBM T.J. Watson Research Center as a Research Staff Member in the Physical Science Department. He is a Fellow of the American Physical Society, the Metallurgical Society, the Material Research Society, and an Overseas Fellow of Churchill College, Cambridge. He is also an academician of Academia Sinica of the Republic of China. Professor Tu has published over 450 journal papers, authored a book (*Solder Joint Technology*, 2007) and co-authored a textbook (*Electronic Thin Film Science*, 1992).

Electronic Thin-Film Reliability

KING-NING TU

University of California at Los Angeles

CAMBRIDGE
UNIVERSITY PRESS

University Printing House, Cambridge CB2 8BS, United Kingdom

One Liberty Plaza, 20th Floor, New York, NY 10006, USA

477 Williamstown Road, Port Melbourne, VIC 3207, Australia

314-321, 3rd Floor, Plot 3, Splendor Forum, Jasola District Centre, New Delhi - 110025, India

103 Penang Road, #05-06/07, Visioncrest Commercial, Singapore 238467

Cambridge University Press is part of the University of Cambridge.

It furthers the University's mission by disseminating knowledge in the pursuit of
education, learning and research at the highest international levels of excellence.

www.cambridge.org
Information on this title: www.cambridge.org/9780521516136

First published 2011

A catalogue record for this publication is available from the British Library

Library of Congress Cataloging in Publication data
Tu, K. N. (King-Ning), 1937–
    Electronic thin-film reliability / King-Ning Tu.
        p.   cm.
    Includes bibliographical references and index.
    ISBN 978-0-521-51613-6
    1. Thin films–Textbooks.   2. Reliability (Engineering)–Textbooks   I. Title.
    TA418.9.T45T82 2010
    621.3815'2–dc22   2010033855

ISBN 978-0-521-51613-6 Hardback

Dedicated to my wife, Ching Chiao Tu

Contents

Preface   *page* xv

1   Thin-film applications to microelectronic technology   **1**

   1.1   Introduction   1
   1.2   Metal-oxide-semiconductor field-effect-transistor (MOSFET) devices   1
      1.2.1   Self-aligned silicide (salicide) contacts and gate   5
   1.3   Thin-film under-bump-metallization in flip-chip technology   7
   1.4   Why do we seldom encounter reliability failure in our computers?   11
   1.5   Trend and transition from micro to nano electronic technology   11
   1.6   Impact on microelectronics as we approach the end of Moore's law   12
   References   12

2   Thin-film deposition   **14**

   2.1   Introduction   14
   2.2   Flux equation in thin-film deposition   15
   2.3   Thin-film deposition rate   17
   2.4   Ideal gas law   17
   2.5   Kinetic energy of gas molecules   19
   2.6   Thermal equilibrium flux on a surface   20
   2.7   Effect of ultrahigh vacuum on the purity of the deposited film   20
   2.8   Frequency of collision of gas molecules   21
   2.9   Boltzmann's velocity distribution function and ideal gas law   22
   2.10   Maxwell's velocity distribution function and kinetic energy of
       gas molecules   24
   2.11   Parameters of nucleation and growth that affect the
       microstructure of thin films   26
   References   28
   Problems   28

3   Surface energies   **30**

   3.1   Introduction   30
   3.2   Pair potential energy, bond energy, and binding energy   31

3.3	Short-range interaction and quasi-chemical assumption	33
3.4	Surface energy and latent heat	35
3.5	Surface tension	36
3.6	Liquid surface energy measurement by capillary effect	38
3.7	Solid surface energy measurement by zero creep	41
3.8	Surface energy systematics	44
3.9	Magnitudes of surface energies	46
	3.9.1 Thermodynamic approach	46
	3.9.2 Mechanical approach	46
	3.9.3 Atomic approach	49
3.10	Surface structure	51
	3.10.1 Crystallography and notation	51
	3.10.2 Directions and planes	54
	3.10.3 Surface reconstruction	54
	References	56
	Problems	57
4	**Atomic diffusion in solids**	**60**
4.1	Introduction	60
4.2	Jump frequency and diffusional flux	61
4.3	Fick's first law (flux equation)	64
4.4	Diffusivity	65
4.5	Fick's second law (continuity equation)	66
	4.5.1 Derivation of the continuity equation	69
4.6	A solution of the diffusion equation	71
4.7	Diffusion coefficient	73
4.8	Calculation of the diffusion coefficient	74
4.9	Parameters in the diffusion coefficient	76
	4.9.1 Atomic vibrational frequency	76
	4.9.2 Activation enthalpy	79
	4.9.3 The pre-exponential factor	81
	References	83
	Problems	84
5	**Applications of the diffusion equation**	**86**
5.1	Introduction	86
5.2	Application of Fick's first law (flux equation)	87
	5.2.1 Zener's growth model of a planar precipitate	87
	5.2.2 Kidson's analysis of planar growth in layered thin films	89
5.3	Applications of Fick's second law (diffusion equation)	93
	5.3.1 Effect of diffusion on composition homogenization	93
	5.3.2 Interdiffusion in a bulk diffusion couple	95
5.4	Analysis of growth of a solid precipitate	108

	5.4.1	Ham's model of growth of a spherical precipitate (C_r is constant)	109
	5.4.2	Mean-field consideration	112
	5.4.3	Growth of a spherical nanoparticle by ripening	113
	References		116
	Problems		117

6   **Elastic stress and strain in thin films**   **118**

	6.1	Introduction	118
	6.2	Elastic stress–strain relationship	120
	6.3	Strain energy	123
	6.4	Biaxial stress in thin films	124
	6.5	Stoney's equation of biaxial stress in thin films	127
	6.6	Measurement of thermal stress in Al thin films	131
	6.7	Application of Stoney's equation to thermal expansion measurement	133
	6.8	Anharmonicity and thermal expansion	134
	6.9	The origin of intrinsic stress in thin films	134
	6.10	Elastic energy of a misfit dislocation	135
	References		138
	Problems		138

7   **Surface kinetic processes on thin films**   **141**

	7.1	Introduction	141
	7.2	Adatoms on a surface	143
	7.3	Equilibrium vapor pressure above a surface	145
	7.4	Surface diffusion	146
	7.5	Step-mediated growth in homoepitaxy	149
	7.6	Deposition and growth of an amorphous thin film	152
	7.7	Growth modes of homoepitaxy	153
	7.8	Homogeneous nucleation of a surface disc	155
	7.9	Mass transport on a patterned surface	159
		7.9.1 Early stage of diffusion on a patterned surface	159
		7.9.2 Later stage of mass transport on a patterned structure	161
	7.10	Ripening of a hemispherical particle on a surface	163
	References		167
	Problems		167

8   **Interdiffusion and reaction in thin films**   **170**

	8.1	Introduction	170
	8.2	Silicide formation	172
		8.2.1 Sequential Ni silicide formation	172
		8.2.2 First phase in silicide formation	178
	8.3	Kinetics of interfacial-reaction-controlled growth in thin-film reactions	180
	8.4	Kinetics of competitive growth of two-layered phases	185

8.5	Marker analysis in intermetallic compound formation	186
8.6	Reaction of a monolayer of metal and a Si wafer	189
References		189
Problems		190

9   Grain-boundary diffusion   **192**

9.1	Introduction	192
9.2	Comparison of grain-boundary and bulk diffusion	194
9.3	Fisher's analysis of grain-boundary diffusion	197
	9.3.1 Penetration depth	200
	9.3.2 Sectioning	200
9.4	Whipple's analysis of grain-boundary diffusion	202
9.5	Diffusion in small-angle grain boundaries	206
9.6	Diffusion-induced grain-boundary motion	207
References		209
Problems		210

10   Irreversible processes in interconnect and packaging technology   **212**

10.1	Introduction	212
10.2	Flux equations	214
10.3	Entropy generation	216
	10.3.1 Heat conduction	217
	10.3.2 Atomic diffusion	218
	10.3.3 Electrical conduction	218
10.4	Conjugate forces with varying temperature	220
	10.4.1 Atomic diffusion	221
	10.4.2 Electrical conduction	222
10.5	Joule heating	222
10.6	Electromigration, thermomigration, and stress-migration	223
10.7	Irreversible processes in electromigration	225
	10.7.1 Electromigration and creep in Al strips	226
10.8	Irreversible processes in thermomigation	229
	10.8.1 Thermomigration in unpowered composite solder joints	229
10.9	Irreversible processes in thermo-electric effects	232
	10.9.1 Thomson effect and Seebeck effect	233
	10.9.2 Peltier effect	235
References		235
Problems		235

11   Electromigration in metals   **237**

11.1	Introduction	237
11.2	Ohm's law	242

11.3	Electromigration in metallic interconnects	243
11.4	Electron wind force of electromigration	246
11.5	Calculation of the effective charge number	249
11.6	Effect of back stress and measurement of critical length, critical product, and effective charge number	251
11.7	Why is there back stress in an Al interconnect?	252
11.8	Measurement of back stress induced by electromigration	254
11.9	Current crowding	256
11.10	Current density gradient force of electromigration	259
11.11	Electromigration in an anisotropic conductor of beta-Sn	261
11.12	Electromigration of a grain boundary in anisotropic conductor	264
11.13	AC electromigration	266
	References	267
	Problems	268

12	**Electromigration-induced failure in Al and Cu interconnects**	**270**
12.1	Introduction	270
12.2	Electromigration-induced failure due to atomic flux divergence	271
12.3	Electromigration-induced failure due to electric current crowding	271
	12.3.1 Void formation in the low-current density region	272
12.4	Electromigration-induced failure in Al interconnects	276
	12.4.1 Effect of microstructure in Al on electromigration	276
	12.4.2 Wear-out failure mode in multilayered Al lines and W vias	277
	12.4.3 Solute effect of Cu on electromigration in Al	277
	12.4.4 Mean-time-to-failure in Al interconnects	277
12.5	Electromigration-induced failure in Cu interconnects	279
	12.5.1 Effect of microstructure on electromigration	281
	12.5.2 Effect of solute on electromigration	282
	12.5.3 Effect of stress on electromigration	285
	12.5.4 Effect of nanotwins on electromigration	286
	References	287
	Problems	288

13	**Thermomigration**	**289**
13.1	Introduction	289
13.2	Thermomigration in flip-chip solder joints of SnPb	291
	13.2.1 Thermomigration in unpowered composite solder joints	291
	13.2.2 In-situ observation of thermomigration	292
	13.2.3 Random states of phase separation in the two-phase eutectic structure	293
	13.2.4 Thermomigration in unpowered eutectic SnPb solder joints	295
13.3	Analysis of thermomigration	298
	13.3.1 Driving force of thermomigration	299

		13.3.2	Thermomigration in eutectic two-phase alloys	301
13.4			Thermomigration under DC or AC stressing in flip-chip solder joints	302
13.5			Thermomigration in Pb-free flip-chip solder joints	303
13.6			Thermomigration and creep in Pb-free flip-chip solder joints	304
References				306
Problems				307

14   Stress migration in thin films   309

14.1		Introduction	309
14.2		Chemical potential in a stressed solid	311
14.3		Diffusional creep (Nabarro–Herring equation)	313
14.4		Void growth in Al interconnects driven by tensile stress	317
14.5		Whisker growth in Sn/Cu thin films driven by compressive stress	319
	14.5.1	Morphology of spontaneous Sn whisker growth	319
	14.5.2	Stress generation (driving force) in Sn whisker growth	323
	14.5.3	Effect of surface Sn oxide on stress-gradient generation	325
	14.5.4	Measurement of stress distribution by synchrotron radiation micro-diffraction	328
	14.5.5	Stress relaxation by creep: broken oxide model in Sn whisker growth	332
References			334
Problems			335

15   Reliability science and analysis   336

15.1		Introduction	336
15.2		Constant volume and non-constant volume processes	337
15.3		Effect of lattice shift on divergence of mass flux in irreversible processes	338
	15.3.1	Initial distribution of current density, temperature, and chemical potential in a device structure before operation	338
	15.3.2	Change of the distributions during device operation	340
	15.3.3	Effect of lattice shift on divergence of mass flux	341
15.4		Physical analysis of electromigration failure in flip-chip solder joints	341
	15.4.1	Distribution of current density in a pair of joints	342
	15.4.2	Distribution of temperature in a pair of joints	343
	15.4.3	Effect of current crowding on pancake-type void growth	346
15.5		Statistical analysis of electromigration failure in flip-chip solder joints	350
	15.5.1	Time-to-failure and Weibull distribution	353
	15.5.2	To calculate the parameters in Black's MTTF equation	355
	15.5.3	Modification of Black's equation for flip-chip solder joints	357
	15.5.4	Weibull distribution function and JMA theory of phase transformations	359

15.5.5   Physical analysis of statistical distribution of failure   360
15.6   Simulation   361
References   361
Problems   362

Appendix A: A brief review of thermodynamic functions   363
Appendix B: Defect concentration in solids   366
Appendix C: Derivation of Huntington's electron wind force   368
Appendix D: Elastic constants tables and conversions   373
Appendix E: Terrace size distribution in Si MBE   380
Appendix F: Interdiffusion coefficient   385
Appendix G: Tables of physical properties   388
Index   392

Preface

The book is intended as a textbook for first and second year graduate students in the Department of Materials Science and Engineering. It can also be used as a reference book for self-study by engineers in the microelectronic industry. The early chapters in this book evolve from *Electronic Thin Film Science*, by K. N. Tu, J. W. Mayer, and L. C. Feldman, and published by MacMillan in 1993. The contents of this book have been taught in a graduate class on "Thin film materials science" at UCLA for over 15 years.

The emphasis in thin-film research is twofold: (1) to invent or to process new thin-film materials having useful functions in applications, and (2) to improve the reliability of functional thin films in large-scale applications, for example, in consumer electronic products. To achieve these goals, on the basis of the discipline of thin-film materials science, requires the study of correlation among structure—properties—processing—performance—reliability of thin films. There are textbooks on the processing of thin films, such as deposition methods by sputtering, electroplating, and MBE growth. There are also textbooks on characterization techniques such as SEM, TEM, RBS, XPS, UPS, Auger, and STM, etc. However, there is no textbook on thin-film reliability science.

When a technology is mature and in mass production, and has widespread application, reliability issues become crucial. As the trend of miniaturization of electronic devices moves into the nanoscale region, the reliability concern of nanotechnology will become serious in the near future. Reliability of nanotechnology may depend on the experience and understanding of reliability in microelectronic technology.

To have a reliable device, it is important to include the concept of reliability into the design and processing of materials in making the device. Thus, there is a strong link between the processing and reliability. It is the goal of this book to combine the science therein, but the emphasis will be on reliability.

What is reliability science? Typically, we tend to assume that the microstructure in a device is stable in its lifetime of usage. Unfortunately, this is not true. In most electronic applications, we must apply an electric field or an electric current. Under a high current density, electromigration occurs in the microstructure and induces circuit failure due to opening by void formation or shorting by extrusion. The high current density causes joule heating and the temperature rise will lead to thermal stress in the device between different materials having different thermal expansion coefficients. The stress gradient and temperature gradient, in addition to electromigration, may induce atomic diffusion and lead to microstructure change and phase transformations. Under a stress gradient or temperature gradient, it means that pressure or temperature is not constant. What is unique in these changes is that they occur in the domain of non-equilibrium thermodynamics or irreversible processes. The basic science that is needed in order to develop an understanding of these reliability problems and to find a way to prevent them from occurring is reliability science.

At the start of the book, the fundamental subjects needed in thin-film processes and reliability such as deposition, surface energy, atomic diffusion, and elastic stress–strain in thin films will be reviewed. The essence of irreversible processes and entropy production will be covered in Chapter 10. This is followed by chapters on electromigration, thermomigration, and stress-migration, with a few examples of reliability failure. The final chapter will discuss failure analysis on the basis of both physical and statistical analyses. Appendixes A–G cover some of the very basic and useful topics and data related to this book.

It is worth mentioning that electromigration itself does not necessarily lead to microstructure failure. Only when there is atomic flux divergence in the microstructure may failure occur. Furthermore, even flux divergence is not enough to cause failure. We require a non-constant volume process in which lattice shift does not occur. In the absence of lattice shift, the non-constant volume change can lead to the generation of extra lattice sites which account for void and hillock or whisker formation.

In preparing the book, I have been helped greatly by students in my classes and in my research group. In particular, I would like to thank Miss Hsin-Ping Chen, Miss Tian Tian, and Mr. Daechul Choi at UCLA for typing part of the text, and revising figures and references. Mr. Choi proofread the book. I would also like to thank Professor Andriy M. Gusak at Cherkasy National University, Ukraine, for a review of Chapter 10 and Chapter 15. Appendix C on the derivation of electron windforce was taken from the lecture notes of Prof. Gusak. Research funding support on reliability study from NSF, SRC, National Semiconductor Corporation, Hitachi (Japan), Seoul Technopark (South Korea), and Advanced Semiconductor Engineering (Taiwan, ROC) is acknowledged.

King-Ning Tu, October 2009.

1 Thin-film applications to microelectronic technology

1.1 Introduction

Layered thin-film structures are used in microelectronic, opto-electronic, flat panel display, and electronic packaging technologies. A few examples are given below. Very large-scale integration (VLSI) of circuits on computer chips are made of multilayers of interconnects of thin metal films patterned into submicron-wide lines and vias. Semiconductor transistor devices rely on the growth of epitaxial thin layers on semiconductor substrates, such as the growth of a thin layer of p-type Si on a substrate of $n+$-type Si [1–3]. The gate of the transistor device is formed by the growth of a thin layer of oxide on the semiconductor. Solid-state lasers are made by sandwiching thin layers of light-emitting semiconductors between layers of a different semiconductor. In electronic and optical systems, the active device elements lie within the top few microns of the surface; this is the province of thin-film technology. Thin films bridge the gap between monolayer (or nanoscale structures) and bulk structures. They span thicknesses ranging from a few nanometers to a few microns. This book deals with the science of processing and reliability of thin films as they apply to electronic technology and devices [4]. To begin, this chapter describes the application of thin films to modern advanced technologies with examples.

1.2 Metal-oxide-semiconductor field-effect-transistor (MOSFET) devices

Advances in layered thin-film technology have been pivotal to the evolution of integrated circuits and opto-electronics. Today, we can fabricate hundreds of millions of transistors on a piece of Si chip the size of a fingernail. These transistors must be interconnected by thin-film lines to form circuits in order to function together. The basic circuit in a memory device is very simple. It consists of a transistor and a capacitor. A schematic cross-section of such a field-effect transistor is shown in Fig. 1.1, consisting of a transistor junction of $n+/p/n+$ type and a gate with a thin gate oxide over the p-type channel. The conductor on the gate is a bilayer structure consisting of a silicide and a heavily doped poly-Si, where the silicide is a metallic compound of metal and silicon. The $n+$ regions in the transistor junction are the source and drain regions and are connected by silicide contacts to the "word" line. Hence, silicide is used as a gate contact as well as source and drain contacts. There is a "bit" line connecting the source contact to the capacitor. The

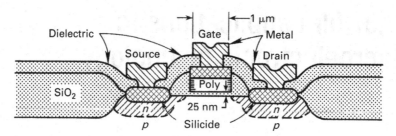

Fig. 1.1   Cross-section of a FET consisting of a transistor junction of $n+/p/n+$-type and a gate with a
thin gate oxide over the p-type channel.

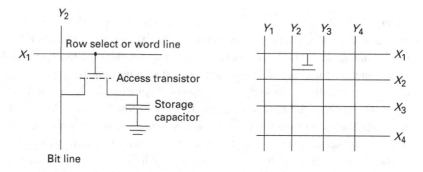

Fig. 1.2   Schematic diagram of an array of two-dimensional integrated circuits of MOSFET. Used with
permission from *VLSI Technology*, S. M. Sze (1988), p. 494.

capacitor serves as a memory unit of either "1" (when the capacitor is full of charges) or
"0" (when the capacitor is empty or stores no charge). The metal-oxide-semiconductor
(MOS) field-effect transistor (FET) serves as a control (or gate) to allow the capacitor to
discharge or not to discharge so that we can read or detect the two states of the capacitor,
either full or empty [1–3].

Fig. 1.2 depicts an array of two-dimensional integrated circuits of MOSFET. In the x-
coordinate, we have x_1, x_2, x_3, x_4, and so on, and in the y-coordinate, we have y_1, y_2, y_3, y_4,
and so on. At each coordination point of (x, y), for example, take (x_1, y_2), we build a
memory unit consisting of an FET and a capacitor. To operate the memory unit, a turn-
on voltage is applied from the "word" line to open the gate. It attracts electrons to the
p-type region and forms an inversion layer below the gate oxide. The inversion layer
with electrons now electrically connects the two $n+$ regions. If the capacitor is full of
stored charges, it will discharge so that a signal pulse can be detected at the end of the
"bit" line. When this happens, we have identified a memory bit of "1" at the point (x_2, y_3).
On the other hand, if the capacitor has no stored charges, there will be no discharge and
no signal will be detected when we open the gate; then we have a memory bit of "0" at
the point (x_2, y_3). The two-dimensional circuit integration as shown in Fig. 1.2 enables
us to operate and detect every coordinate point on the two-dimensional integration of
circuits, so it is called random access memory (RAM). Often, we use dynamic random

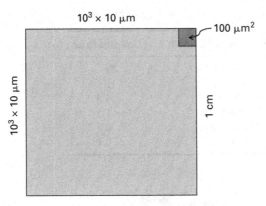

Fig. 1.3   Schematic diagram of a Si chip of size of 1 cm × 1 cm, divided into $10^3 \times 10^3 = 10^6$ small squares, so that each squares has an area 10 μm × 10 μm.

access memory (DRAM) to describe the device because the capacitors leak, since they are interconnected with lines, so we have to recharge them frequently and the recharge is a dynamic process. In certain devices, when the gate is isolated, we can use it as a floating gate.

In Fig. 1.3, we depict a Si chip of size 1 cm × 1 cm, and we divide it into $10^3 \times 10^3 = 10^6$ small squares, so that each of the squares has an area 10 μm × 10 μm. In each square, or cell area, if we can fabricate a FET and a capacitor, we have made a chip which has one million memory units. Needless to say, we should be able to interconnect them with their bit lines and word lines. In addition, we should also be able electrically to connect the chip to the outside circuit when we want to use it. The latter is a function of electronic packaging technology.

Next, we divide the 10 μm × 10 μm area into four smaller areas, i.e. cells about 5 μm × 5 μm. If we can shrink and build a FET and a capacitor in the smaller area, we will have a chip which has four million units of memory. This is the principle behind the miniaturization of the Si microelectronics industry in the last quarter century, or the essence of progress as suggested by Moore's law. The advance of one generation means the increase of a factor of four in circuit density on a chip. It goes from 1, 4, 16, 64, 256, to 1024 and so on. The industry started with about a one-thousand memory unit in the late 1960s and has advanced to about one billion memory units per chip today. Table 1.1 lists the dimensional changes of cells in several generations of devices. As the cell size becomes smaller, the feature size of the transistor, capacitor, and interconnects elements in the cell should also become smaller. There is a scaling law behind the shrinkage, which affects the electrical behavior of the transistor as well as the interconnect.

The VLSI of circuits is achieved by interconnecting all the transistors together using multilayers of Al or Cu thin-film interconnects. The process and reliability of multilayered thin-film interconnect structures are crucial to device applications. Today, there are eight or more layers of interconnects built on the transistors. Fig. 1.4 shows a scanning electron microscopic (SEM) image of a two-level Al interconnect on a Si surface after

Table 1.1. Dimensional changes of cells in several generations of devices

Cell density	Lithographic line width (μm)	Cell area (μm^2)
1 Mb	1	33
4 Mb	0.7	11
16 Mb	0.5	4.5
64 Mb	0.35 (deep UV)	1.5
256 Mb	0.25 (X-ray)	0.5
1 Gb	0.18	0.15

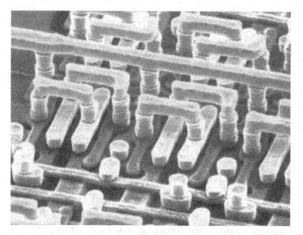

Fig. 1.4   SEM image of a two-level Al interconnect on a Si surface after the insulating dielectric has been etched away. The width of the Al lines is 0.5 μm and the spacing between them is 0.5 μm, so the pitch is 1 μm.

the insulating dielectric has been etched away. The width of the Al lines is 0.5 μm and the spacing between them is 0.5 μm, so that the pitch is 1 μm. On an area 1 cm $\times$ 1 cm, we can have 10^4 lines and each of them has a length of 1 cm, so the total length of interconnects in such a layer is 100 m. When we build eight such layers on a chip the size of a fingernail, the total length of interconnect is over 1 km, if we include those interlevel vias, i.e. the interconnects between layers. Fig. 1.5(a) shows a SEM image of an eight-level Cu interconnect structure taken after the interlevel dielectric was etched away. Fig. 1.5(b) shows a cross-sectional transmission electron microscopic image of a six-level Cu interconnect structure built on a Si surface. Here, the width of the narrowest interconnect via is 0.25 μm. The alignment of the vias between layers is a very challenging issue in device manufacturing.

The production cost of making the layered metallization structure is now more than half the cost of production of the whole Si wafer. In the interconnect metallization, silicide of C-54 $TiSi_2$, $CoSi_2$ or $NiSi$ has been used as contacts to gate as well as contacts to source and drain of the FETs. Tungsten has been used as interlayer vias in the Al interconnect technology. It is also used as the first-level vias in Cu interconnect technology. To use

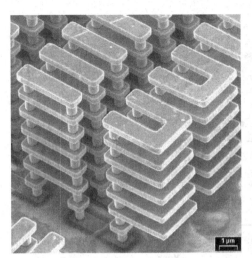

Fig. 1.5 (a) SEM image of an eight-level Cu interconnect structure taken after the interlevel dielectric was etched away. (b) A cross-sectional transmission electron microscopic image of a six-level Cu interconnect structure built on a Si surface. Here, the width of the narrowest interconnect via is 0.25 μm.

Cu in interconnect technology, the Cu must be coated with a very thin adhesion and diffusion barrier layer of Ta, or TiN. The processing, properties, and reliability of these thin films are relevant to the success of the technology.

Clearly, if the size of a cell is 1 μm $\times$ 1 μm, its circuit elements must be smaller than 1 μm. In the trend of miniaturization, not only the lateral dimension will become smaller; the vertical dimension such as the gate oxide thickness must be thinner too. No doubt we cannot keep shrinking the size in order to make smaller and smaller devices. The trend of miniaturization or the progress in scaling down device dimension has been modeled by Moore's law, which stated that the on-chip circuit density will double every 18 months. Fig. 1.6 shows a schematic curve of circuit density of memory and logic units on a chip in a central processing unit plotted against year. It is a log-linear plot to follow Moore's law.

1.2.1 Self-aligned silicide (salicide) contacts and gate

A critical dimension in the MOSFET device discussed above is the gate width or the distance between the source and drain contacts. The gate width is called the "feature size" of the device. Today, it is of nanoscale down to 45 nm and soon to be 33 nm and beyond. In fabrication, if the gate and contacts are made of different materials, it will require two different processing steps or two lithographic steps to manufacture them, so a high-precision alignment is required and it is difficult to control the feature size in nanoscale. The salicide process was invented to overcome this crucial step. Fig. 1.7 shows the schematic diagrams of the salicide process. Both gate and source/drain contacts are made of C-54 $TiSi_2$. Over the gate oxide is a layer of heavily doped poly-Si, and

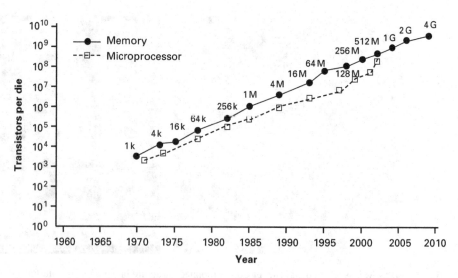

Fig. 1.6   Curve showing the circuit density of memory and logic units on a chip in a central processing unit plotted against the year. It is a log-linear plot and follows Moore's law.

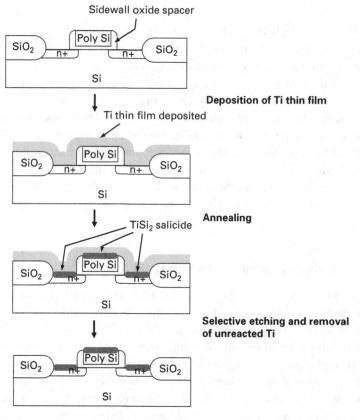

Fig. 1.7   Schematic diagrams showing the salicide process.

over the source/drain contact regions is a layer of heavily doped single crystal $n+$-Si. They are separated by two sidewalls of SiO_2. The spacing between the sidewalls and the thickness of the sidewall determine the lateral dimension of the feature size. When a Ti thin film is deposited and annealed, C-54 $TiSi_2$ forms on the gate contact and the source/drain contacts, simultaneously. But the Ti on the sidewall oxide does not react with the oxide to form silicide and the unreacted Ti can be selectively etched away, so that electrical insulation between the gate and the contacts can be achieved in one lithographic step. This "salicide" process avoids the high-precision alignment issue and enables the self-alignment of the gate and the contacts in production.

In the thin-film literature, there are many publications on silicide formation by the reaction between thin metal film and Si. Since there are hundreds of millions of silicide contacts and gates on a Si chip and they should have the same microstructure and electrical properties, a controlled salicide formation has been a very important processing step in VLSI device fabrication. The kinetics of silicide formation will be covered in Chapter 8.

1.3 Thin-film under-bump-metallization in flip-chip technology

How to connect the on-chip VLSI circuits to external circuits is the major function of electronic packaging technology [5–6]. To provide external electrical leads to all these on-chip interconnect wires, we may need several thousands of input/output (I/O) electrical contacts on the surface of the chip in a central processing unit. At present, the only practical and reliable way to provide such a high density of I/O contacts on the chip surface is to use an area array of tiny solder balls. We can have 50 μm diameter solder balls with a spacing of 50 μm between them, so the pitch is 100 μm. We place 100 of them along a length of 1 cm or 10 000 of them on an area of 1 cm^2. Typically the diameter of solder balls used today is about 100 μm, and the processing of solder balls in the electronic packaging industry is called "bumping technology." Because of the use of so small a size and large a number of solder balls, the International Technology Roadmap for Semiconductors (ITRS) has, since 1999, identified "solder joint in flip-chip technology" as an important subject of study concerning its yield in manufacturing and its reliability in application.

What is flip-chip technology? It is a technology to provide a large number of electrical connections between a Si chip and a packaging substrate using solder joints. The Si chip is flipped face down so that its circuits of very large-scale integration face the substrate. The electrical connections are achieved by an area array of solder bumps between the chip and its substrate. Fig. 1.8 shows an area array of solder balls on a chip surface. To join the chip to a substrate, the chip will be flipped over, so that the VLSI side of the chip is upside down, facing the substrate.

Flip-chip technology has been used for over 30 years in making mainframe computers. It originated from the "controlled collapse chip connection" or "C-4" technology in packaging chips on ceramic modules in the 1960s. Generally speaking, the advantages of flip-chip technology are: smaller packaging size, large I/O lead count, and higher

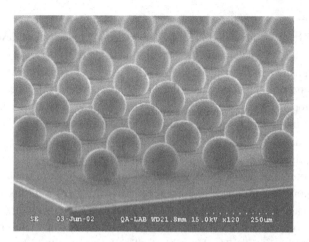

Fig. 1.8   SEM image of an area array of solder balls on a chip surface.

performance and reliability. Now it is used widely in consumer products such as in chip-size packaging, where the packaging substrate is of nearly the same size as the chip. The small packaging size is needed in handheld devices, where the form factor is important. In handheld terminals or computers, the demand for higher performance and greater functionality will require a large number of electrical I/O lead counts in the area array. The higher performance is because the solder bumps in the central part of the chip allow the device to operate at lower voltage and higher speed. Besides, flip-chip solder bumping is the only existing technology that can provide the reliability needed. We shall discuss reliability issues such as electromigration and stress-migration in later chapters.

At first, when VLSI chip technology was developed, a packaging technology of a high density of wiring and interconnection was required. This led to the development of multilevel metal–ceramic modules and multi-chip modules for mainframe computers. In a multilevel metal–ceramic module, many levels of Mo wire were buried in the ceramic substrate. Each of these modules could carry up to a hundred pieces of Si chip. Several of these ceramic modules were joined to a large printed circuit board and resulted in the two-level packaging scheme for mainframe computers shown in Fig. 1.9. It consisted of a first-level packaging of chip to ceramic module and a second-level packaging of ceramic module to polymer printed circuit board.

A schematic diagram of the cross-section of the first-level flip-chip C-4 solder joint is shown in Fig. 1.10. In the first-level packaging, the under-bump-metallization (UBM) on the chip side is a tri-layer thin film of Cr/Cu/Au. Actually, in the tri-layer the Cr/Cu has a phased-in microstructure for the purpose of improving the adhesion between the Cr and Cu and strengthening its resistance against solder reaction which may leach out the Cu so that the phase-in Cr/Cu, formed by the co-deposition of Cr and Cu having a composition gradient, can last several reflows in solder reaction. A "reflow" means that the solder is experiencing a temperature slightly above its melting point so that it melts. It is in the molten state that the solder will react with Cu to form the metallic bond or the formation of intermetallic compound (IMC) in the solder joint. On the substrate side

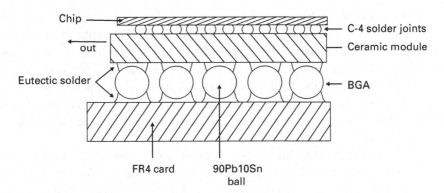

Fig. 1.9 Schematic diagram of the two-level packaging scheme for mainframe computers. It consists of a first level of chip to ceramic module packaging and a second level of ceramic module to polymer printed circuit board packaging.

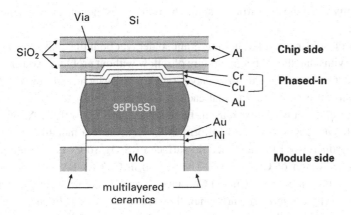

Fig. 1.10 Schematic diagram of the cross-section of a flip-chip solder joint.

of the joint, the metal bond-pad on the ceramic surface is typically Ni/Au. The solder which joins the UBM and the bond-pad is a high-Pb alloy such as 95Pb5Sn or 97Pb3Sn.

Initially, the on-chip solder bumps were deposited by evaporation and patterned by lift-off. Later, they were deposited by selective electroplating. Recently, in a new process of C-4, a template is used to form a two-dimensional array of solder balls, then a chip with a receiving array of UBMs is placed on the template so that all the balls can be transferred to the chip in a reflow, in which the balls react with the UBMs. The advantage of the new process is that no thick photo-resist is needed in selective electroplating, and the template can be used repeatedly to cut cost.

In the C-4 process using selective electroplating, after etching away the photo-resist and the electrical conducting line used for the plating, an array of cylindrical solder columns or bumps remains on the chip surface. The high-Pb bump has a melting point of over 300 °C. During the first reflow (around 350 °C), the column bumps change to ball-shape bumps on the UBM. Since the SiO_2 surface cannot be wetted by molten solder, the base of the molten solder bump is defined by the size of the UBM, thus the molten solder bump balls up or stands up on UBM contact. Therefore, the UBM contact controls the dimensions (height and diameter) of the solder ball when its volume is given. Often, the UBM contact is called "ball-limiting metallization" or BLM. As the BLM controls the height of the fixed volume of a solder ball when it melts, this is the meaning of the word "control" in "controlled collapse chip connection." Without the control, the solder ball will spread on the UBM, and then the gap between the chip and the module is too small.

To join the chip which already has an array of solder balls to a ceramic module, a second reflow is used. During the second reflow, the surface energy of the molten solder balls provides a self-aligning force to position the chip on the module automatically. When the solder melts in order to join the chip to the module, the chip will drop slightly and rotate slightly. The drop and rotation are due to the reduction of surface tension of the molten solder balls, which achieve the alignment between the chip and its module, so it is a controlled collapse process. The word "collapse" means that the chip drops and rotates slightly when the area array of solder balls becomes molten and wets the pads on the module.

The high-Pb solder is a high melting-point solder, yet both the chip and the ceramic module can withstand the high temperature of reflow without problem. Additionally, the high-Pb solder reacts with Cu to form a layer-type Cu_3Sn, which can last several reflows without failure. It is worth noting that each of the metals in the tri-layer of Cr/Cu/Au has been chosen for a particular reason. First, solder does not wet the Al wire, so Cu is selected for its reaction with Sn to form IMCs to achieve a metallic joint. Second, Cu does not adhere well to the dielectric surface of SiO_2, so Cr is selected as a glue layer for the adhesion of Cu to SiO_2. The phased-in Cu/Cr UBM was developed to improve the adhesion between Cu and Cr. Since Cr and Cu are immiscible, their grains form an interlocking microstructure when they are co-deposited. In such a phased-in microstructure, the Cu adheres better to the Cr, and also it will be harder for the Cu to be leached out to form IMC with Sn during reflow. Furthermore, the phase-in microstructure provides a mechanical locking of the IMC. Finally, Au is used as a surface passivation coating to prevent the oxidation or corrosion of Cu. It also serves as a surface finish to enhance solder wetting. In the literature, much study on interdiffusion and reactions in the bilayer thin films of Cr/Cu and Cu/Au has been performed.

In the second-level packaging of ceramic module to polymer board, i.e. to join the ceramic substrate to a polymer printed circuit board, another area array of solder balls is placed on the back side of the ceramic substrate. They are called ball-grid-array (BGA) solder balls, which have a much large diameter than the C-4 solder balls. Typically the BGA solder ball diameter is about 760 μm. They are eutectic SnPb solder with a lower melting point (183 °C), which is reflowed around 220 °C. Sometimes, composite solder balls of high-Pb and eutectic SnPb are used, with the high-Pb as the core of the ball. It is obvious that during this reflow (the third reflow) of the eutectic solder, the high-Pb solder joints in the first-level packaging or the high-Pb core in the composite solder balls will not melt. In certain applications, the high-Pb core in the composite can be replaced by a Cu ball.

Due to the environmental concern of Pb, there are four anti-Pb bills pending in US Congress, including one from Environmental Protection Agency. The European Union has issued a Directive on July 1st, 2006 calling for a ban on Pb-containing solders in all electronic consumer products. While car batteries use up to 88 % of Pb and electronic devices use up to 1 % of Pb in industry applications, the garbage of consumer products is the source of 20 to 30 % of Pb affecting underground water. This is the reason for the ban.

At the moment, no chemical element has been found to replace Pb and to function as well as Pb in solder joints. The most promising Pb-free solders to replace the eutectic SnPb are the eutectic SnAgCu, eutectic SnAg, and eutectic SnCu. These Sn-based solders have a very high concentration of Sn, e.g. the eutectic SnCu has 99.3 wt. % of Sn and the eutectic SnAgCu has about 95 to 96 wt. % of Sn. The very high concentration of Sn in these Pb-free solders has led to a high rate of Cu–Sn reaction and in turn has caused reliability problems in the thin-film Cr/Cu/Au UBM discussed above. No suitable thin-film UBM for Pb-free soldering is available to date, especially when reliability against electromigration is taken into consideration.

1.4 Why do we seldom encounter reliability failure in our computers?

We have mentioned that electromigration is a serious reliability problem. However, we seldom hear that our computers have failed because of electromigration. This is due to the fact that the electronic industry recognized the problem a long time ago and has spent much effort in conducting research into understanding the problem. The mean-time-to-failure (MTTF) of a computer operating under certain accelerated test conditions has been measured, so that the lifetime of the computer in ordinary use can be predicted. At the design stage, reliability has been taken into account so that the operating conditions of a computer, such as the applied electric current density and temperature, are specified. Thus, we will not encounter computer failure in ordinary use.

1.5 Trend and transition from micro to nano electronic technology

Towards the end of Moore's law, the transition from micro to nano electronic technology is expected. The transition can be either top-down or bottom-up. While the bottom-up approach starting from molecular-sized transistors has many merits, the challenge to achieve a large-scale integration of these transistors is not trivial at all. The top-down approach has limits, not only physical limits such as lithography, but also financial limits from the cost of production. The most likely transition will be a hybrid of nano technology and the existing microelectronic MOSFET technology. New applications will come from energy, medical, and biological devices. At the end of Moore's law, the microelectronic industry will not disappear. The growth in the microelectronic industry will just not be exponential and the profit will not be massive. It will continue to exist as the automobile and aerospace industry.

In consumer electronic products, the trend will be towards more portable, wireless, and handheld devices. Handheld terminals will come soon and the replacement of the hard disc by the memory stick will be the beginning. Thin films will still be used widely. Electronic packaging technology will become more important, especially three-dimensional IC and system packaging. Reliability will be of concern.

1.6   Impact on microelectronics as we approach the end of Moore's law

At the moment, however, there are concerns about when will be the end of Moore's law and what will happen then? Actually, there are economical as well as physical limits to Moore's law. Physically, we cannot keep shrinking the thickness of the gate oxide and the width (or the feature size) of the transistor. Economically, the gain in device performance such as speed may not justify the very rapidly increasing cost of miniaturization into the deep submicron or nanoscale region. The impact on the Si microelectronic industry of approaching the end of Moore's law is not that the industry will disappear; rather, that the exponential growth of the industry in the last half century will not continue. Instead, its growth will become normal. The 2009 International Technology Roadmap for Semiconductors published by the Semiconductor Industry Association indicates that the Si technology still has room for progress in the next 10 to 15 years [7].

Before the end of Moore's law, instead of dedicating all our efforts to the shrinking of device dimensions, we should have a new paradigm in expanding the applications of what we have already. The existing complementary metal-oxide-semiconductor (CMOS) technology has great capability and potential in many handheld and wireless consumer product applications, especially if we can combine it with optical devices, microelectrical mechanical system (MEMS) nanosensors for applications in energy and health. Take the example of a cell phone. Most of us carry a cell phone and use it every day. When we hold a cell phone in our hand, with a few sensors it will be able to detect our body temperature, our pulse rate, and blood pressure. Even more advanced diagnosis can be performed. Indeed, when we have a cell phone which can perform as a mobile computer, much more useful functions are with us. In these applications of consumer electronic products, new circuit design and new packaging technology and reliable materials will be required.

References

[1] Yuan Taur and Tak H. Ning, *Fundamentals of Modern VLSI Devices* (Cambridge University Press, Cambridge, 1998).

[2] S. M. Sze (ed.), *VLSI Technology* (McGraw-Hill, New York, 1983).

[3] K. N. Tu, J. W. Mayer, and L. C. Feldman, *Electronic Thin Film Science* (Macmillan, New York, 1992).

[4] M. Ohring, *Reliability and Failure of Electronic Materials and Devices* (Academic Press, San Diego, 1998).

[5] K. N. Tu, *Solder Joint Technology* (Springer, New York, 2007).

[6] K. L. Puttlitz and K. A. Stalter (eds), *Handbook of Lead-Free Solder Technology for Microelectronic Assemblies* (Marcel Dekker, New York, 2004).

[7] International Technology Roadmap for Semiconductors, *Semiconductor Industry Association* (San Jose, 2009). See website http://public.itrs.net/.

2 Thin-film deposition

2.1 Introduction

Thin-film deposition can be regarded as a phase change from a gas phase to a solid phase on a substrate [1, 2]. Typically, we need to know the growth rate, the purity, and the microstructure of the deposited film. The growth rate is controlled by the flux equation, to be discussed in the next section. For a high-purity film, it will require ultrahigh vacuum deposition, to be discussed in Section 2.7. The consideration of microstructure of the film, e.g. whether it is amorphous, polycrystalline, or epitaxial single crystal, will require specification of the deposition conditions or the selection of deposition parameters to be discussed in Section 2.11.

Fig. 2.1 depicts the two key parts in a vacuum chamber in film deposition; they are the target and the substrate. Let us assume that they are of the same material. The target is kept at temperature T_1 and the substrate is kept at temperature T_2, and $T_2 < T_1$. At the equilibrium condition, there are fluxes of atoms departing from and returning to the target surface, as there are on the substrate surface on the basis of the principle of micro-reversibility. These fluxes establish an equilibrium pressure, P_1 and P_2, on the target and the substrate surfaces, respectively, and $P_2 < P_1$. Due to the pressure difference or pressure gradient, it will lead to a flow of gas or mass transport via the gas state from the target to the substrate. The mass transport will produce supersaturation on the substrate surface, so condensation will occur and a thin film will be deposited on the substrate. The condensation involves nucleation and growth of the thin solid film. The mass transport will cause an undersaturation on the target surface, so evaporation will continue. In essence, it is thermal deposition, for example, to heat a tungsten coil by joule heating so that a metal wire of lower melting point around the coil can be evaporated and a film of the metal can be deposited on a cold substrate. Similarly in e-beam evaporation, we use an e-gun to melt and evaporate a metal target in a crucible and collect the vapor on a cold substrate to form a film.

The basic processes involved in the deposition are governed by three equations. First, the gas state is governed by the ideal gas law [3]. Second, the mass transport between the target and the substrate is governed by the flux equation. Third, the evaporation from the target and condensation on the substrate surface is governed by the equilibrium flux equation on these surfaces. We shall discuss these equations in this chapter. The atomic process behind the desorption from a crystalline target surface and the nucleation and growth process in condensation on a substrate surface will be discussed in later chapters.

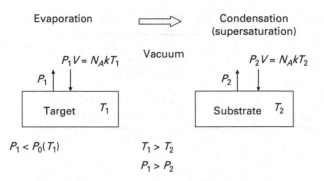

Fig. 2.1 Schematic diagram depicting the transfer of a flux of gas atoms from the target surface to the substrate surface in thin-film deposition in a vacuum chamber.

Many experimental techniques can be used to transfer mass or matter from a target to a substrate [4–6]. There are physical vapor deposition (PVD) techniques such as e-beam evaporation, sputtering, and molecular beam epitaxy. In sputtering, there are DC, RF, magnetron, and reactive sputtering. The last is for depositing chemical compounds in which the compound composition can be maintained and the undesirable effect of different partial pressures of the elements in the compound on deposition will be reduced. There are also chemical vapor deposition (CVD) techniques such as the decomposition of silane for Si thin-film deposition, and selective area deposition of W vias in multi-layered Al interconnects. However, today the fabrication of Cu multilayer interconnect metallization is by process of electrochemical plating. While the CVD of Cu is possible in laboratories, the large-scale production in real device manufacturing is not yet available. Then there are atomic layer deposition (ALD) techniques, for example, in ultra-thin gate oxide formation. These deposition techniques will not be discussed here. Only the basic concepts in thin-film deposition are given in this chapter.

There are four key deposition parameters that will affect the purity and microstructure of the deposited thin film: the rate of deposition, the vacuum level in the deposition chamber, the substrate temperature, and the surface structure of the substrate, as shown in Table 2.1. How these deposition parameters can affect the film purity and microstructure can be explained on the basis of our understanding of nucleation and growth in the phase change from gas to solid. In this chapter, the emphasis is on the behavior of the gas phase as regards thin-film deposition. The nucleation of a surface step and the stepwise growth model of a surface step will be covered in Chapter 7.

2.2 Flux equation in thin-film deposition

Atomic flux, J, is defined as the number of atoms passing through a unit area in a unit time, or the number of atoms depositing on a unit area on a substrate surface in a unit time. The unit of J is the number of atoms/cm^2 s. In Fig. 2.2, we assume a stream of parallel atoms passing through an area A with a constant velocity v. It depicts a flow of air through

Table 2.1. Correlation between deposition parameters and nucleation and growth

Deposition parameter	Nucleation and growth
Rate of deposition	Supersaturation
Vacuum level	Impurity
Substrate temperature	Undercooling and diffusion
Surface structure of the substrate	Surface and interfacial energy

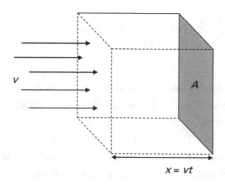

$$x = vt$$

Fig. 2.2   Depiction of a stream of parallel atoms passing through an area A with a constant velocity v. For details, see text.

a window or a jet of water running through a faucet. We envision a volume $V = xA$ in front of the area A, as shown in Fig. 2.2, where $x = vt$ and t is a period of time. Within the volume V, the number of atoms is $N = CV$, where C is the concentration of atoms per unit volume, and C is a constant here, independent of time and distance. The unit of C is the number of atoms/cm^3. In the period t, the total number of atoms that can reach the area A are those within the volume V. Thus, according to the definition of J, we have

$$J = \frac{N}{At} = \frac{CV}{At} = \frac{CvtA}{At} = Cv \tag{2.1}$$

This is a simple yet powerful equation. Sometimes the velocity is called "drift velocity" by writing $J = C < v >$. Note that Fick's first law of diffusion is also a flux equation, and its relationship to Eq. (2.1) will be discussed in Chapter 4. The physical picture of atomic flux presented in the above assumes that atoms in the gas have a uniform velocity and there is no interaction among the atoms because they move parallel to each other. In other words, we have ignored their collision in the gas. To include collision, we must consider a distribution function of velocity, using Boltzmann's or Maxwell's velocity distribution function, so the velocity can be in any direction and have a different magnitude. These distribution functions will be presented later in this chapter.

2.3 Thin-film deposition rate

To apply the flux equation of $J = Cv$ to thin-film deposition, we consider A to be an area on a substrate and J is the atomic flux in a gas phase being deposited onto A. The total number of atoms deposited on A in a period of t is equal to $N = JAt$. So the deposition rate is

$$\frac{dN}{dt} = JA$$

If we take the atomic volume to be Ω, we have the following relationship due to conservation of volume in the growth of a thin film of thickness y on the area of A,

$$V = \Omega JAt = yA$$

so the growth rate or thickening rate of the film is

$$\frac{dy}{dt} = \Omega J = \Omega Cv \tag{2.2}$$

To calculate the growth rate, we need to know Ω, C, and v. We recall that C is the concentration of atoms in the gas phase and we will use the ideal gas law to obtain C; v is velocity of the depositing atoms and we will use the equation of the kinetic energy of gas molecules to obtain v, to be discussed below. The atomic volume, Ω, for most solid elements can be deduced from the crystalline unit cell of the element. Take aluminum as an example; when we deposit Al thin film on a substrate, the Al film has the face-centered cubic (fcc) lattice with a lattice parameter $a = 0.405$ nm (4.05 Å), which can be measured by X-ray diffraction. Therefore, we can obtain the atomic volume from the unit cell volume. There are four Al atoms per unit cell. Thus, the atomic volume of Al can be calculated as $\Omega = (0.405 \text{ nm})^3/4 = 0.0166 \text{ nm}^3$. A discussion of the calculation of Ω is also given in Section 3.10.1.

Since the densest packing direction in the fcc lattice is along <110>, the equilibrium interatomic distance in Al is $a/\sqrt{2} = 0.29$ nm. Using a hard-sphere model, we can calculate the atomic volume to be $\frac{4}{3}\pi a_0^3 = \Omega = 0.0122 \text{ nm}^3$, where a_0 is equal to half of the interatomic distance. Note that this value is about 30% smaller than that calculated from the unit-cell volume. This is because the hard-sphere model does not include the interstitial volume between the atoms in the unit cell. The number of Al atoms per unit volume is given by $1/\Omega = C$, or there are about 6×10^{22} Al atoms/cm^3. We have $C\Omega = 1$ for pure solid elements.

2.4 Ideal gas law

Electronic thin films are almost invariably deposited in vacuum systems. All vacuum systems have a finite background pressure which defines the purity of a film grown in the vacuum system. In this section we calculate the impinging flux of impurity atoms from the background gas pressure. This impinging flux will then be compared to the flux of deliberately deposited atoms. It will show why we have to have an ultrahigh vacuum in order to obtain a high-purity thin film.

The basic relation describing vacuum is the ideal gas law,

$$pV = RT = N_A kT \tag{2.3}$$

where p is pressure (newton/m^2 or N/m^2); V is molar volume or the volume of one mole of gas, 22.4×10^3 cm^3 or 22.4 l; R is the gas constant, 8.31 J/K-mole; N_A is the number of molecules in one mole of gas, i.e. Avogadro's number $= 6.02 \times 10^{23}$ molecule/mole; k is Boltzmann's constant $= 1.38 \times 10^{-23}$ J/K; T is the absolute temperature in kelvin (K $= °$C $+ 273.16$).

Other commonly used units of pressure are:
1 torr $= 1$ mm Hg $= 1333$ dyne/cm^2 $= 133.3$ N/m^2
1 atmosphere $= 760$ torr $= 1.013 \times 10^6$ dyne/cm^2 $= 1.013 \times 10^5$ N/m^2
1 pascal (Pa) $= 1$ N/m^2 $= 7.5 \times 10^{-3}$ torr

Using the ideal gas law, we derive the relation between gas density and pressure. This relation is then used to estimate the flux of gas atoms impinging on a substrate surface. At 1 atmosphere and $0°$C, the molar volume of an ideal gas is 22.4×10^3 cm^3 or 22.4 l. This quantity is obtained by using the ideal gas law,

$$V = \frac{N_A kT}{p}$$

$$= \frac{6.02 \times 10^{23} \text{ (molecules/mole)} \times 1.38 \times 10^{-23} (\text{J/K}) \times 273(\text{K})}{1.013 \times 10^5 \text{ (N/m}^2)}$$

$$= 22.4 \times 10^3 \text{ cm}^3$$

The gas concentration or density at $25°$C (298 K) is

$$n = \frac{N_A}{V} = \frac{6.02 \times 10^{23}}{24.4 \times 10^3} = 2.46 \times 10^{19} \text{molecules/cm}^3$$

where n (or C) is the concentration or the number of molecules (or atoms) per unit volume. Since

$$n = \frac{N_A}{V} = \frac{p}{kT}$$

we see that n is directly proportional to p at a given temperature. This is an important relation in vacuum technology, since it controls the purity of the deposited film. At 1 torr and $25°$C, we have $n = n_1$,

$$n_1 = \frac{2.46 \times 10^{19}}{760} = 3.24 \times 10^{16} \text{ molecules/cm}^3$$

In industrial processes, a vacuum of 10^{-7} torr is commonly obtained by using mechanical pumps; in this vacuum the background vapor density is

$$n_1 \times 10^{-7} = 3.24 \times 10^9 \text{ molecules/cm}^3$$

In an ultrahigh vacuum system, a vacuum of 10^{-11} torr is achieved and we have

$$n_1 \times 10^{-11} = 3.24 \times 10^5 \text{ molecules/cm}^3$$

The flux of atoms impinging on a solid surface is given by the product of particle concentration and velocity, or $J = Cv$. We need to know the velocity of gas atoms or molecules, to be given below.

2.5 Kinetic energy of gas molecules

To calculate the velocity, we take the mean kinetic energy of a molecule in an ideal gas to be

$$\overline{E_k} = \frac{3}{2}kT = \frac{1}{2}mv_a^2 \tag{2.4}$$

$$v_a = \left(\frac{3kT}{m}\right)^{1/2} = \left(\frac{3RT}{M}\right)^{1/2}$$

where m is the mass of a molecule, M is the molar weight, and v_a is the root mean square velocity (see Section 2.10). We take $M = 28$ gram/mole for nitrogen gas and obtain

$$v_a = \left(\frac{3 \times 8.31(\text{J/K-mole}) \times 10^7(\text{erg/J}) \times 298\,\text{K}}{28(\text{g/mole})}\right)^{1/2}$$

$$= (26.1 \times 10^8 \text{erg/g})^{1/2} = 5.2 \times 10^4 \text{ cm/s}$$

The magnitude of v_a is of the order of magnitude of the speed of sound in air.

The rate of molecules or atoms impinging on a surface per unit area per unit time has been derived in Section 2.1 by considering that only those molecules within a distance $v_a t$ from the surface can hit the surface in the time t. The rate of impingement per unit area per unit time (or atomic flux) is $J = nv_a$, which is Eq. (2.1) except that we replace C by n because of the use of ideal gas law, and v by v_a. The flux of atoms impinging on a surface is equal to the product of concentration (atom/cm^3) times velocity (cm/s) and has the unit of atom/cm^2 s. In the above derivation, we did not consider the velocity distribution of the particles, nor the concentration gradient of the particles. If we take the velocity distribution of an ideal gas, e.g. the Maxwell distribution of velocities, shown in Section 2.10, a more complete derivation will show that the flux is given by

$$J = \frac{1}{4}n\bar{v}$$ (2.5)

where $\bar{v}$ is the mean velocity (see Section 2.10).

2.6 Thermal equilibrium flux on a surface

Owing to desorption and absorption, a solid free surface is at equilibrium with the partial pressure on the surface due to its own gaseous atoms which depart from and arrive at the surface. We define such fluxes as the thermal equilibrium flux. If the pressure on the solid surface is defined as the equilibrium pressure p, we have an equilibrium flux on the surface (see Section 2.10 for the mean velocity),

$$J = \frac{1}{4}n\bar{v} = n\sqrt{\frac{kT}{2\pi m}} = \frac{p}{kT}\sqrt{\frac{kT}{2\pi m}} = p\sqrt{\frac{1}{2\pi mkT}}$$ (2.6)

Knowing the equilibrium pressure of a metal, we obtain the equilibrium flux at a given temperature. For example, it is known that metals such as Cd and Zn have a very high equilibrium pressure. In other words, they evaporate easily. Therefore, these metals should be avoided in a deposition chamber. In thin-film deposition, the deposition flux must be greater than the equilibrium flux, otherwise desorption occurs instead of deposition.

2.7 Effect of ultrahigh vacuum on the purity of the deposited film

Knowing the concentration and velocity of molecules in air, we can estimate the flux of air molecules impinging on a surface when exposed to one atmosphere of air,

$$J_c = \frac{1}{4}(2.36 \times 10^{19} \times 5.2 \times 10^4) = 3.2 \times 10^{23} \text{ molecules/cm}^2 \text{ s}$$

Most solids have atomic densities $n = 1/\Omega = 5$ to 9×10^{22} atom/cm^3, meaning that there are about 10^{15} atom/cm^2 in a monolayer (ML) using the estimate that one ML $= n^{2/3}$. Using this relation, we can estimate the time for a ML of air atoms to be collected on a substrate surface when we expose it to one atmospheric pressure of air:

$$t = \frac{10^{15}}{3.2 \times 10^{23}} = 3.1 \times 10^{-9} \text{ s}$$

Therefore, if we expose a substrate to air, we will have collected a very thick layer of air molecules in no time, assuming that the gas sticks to the surface, so any film deposited

Table 2.2. The time needed to deposit one monolayer of background gas on a surface under various vacuum conditions

p (torr)	C (atom/cm^3)	$J = Cv$ (atom/cm^2 s)	Time for monolayer deposition (s)
760 (atmospheric pressure)	2.46×10^{19}	1.2×10^{24}	10^{-9}
1	3.24×10^{16}	1.5×10^{21}	10^{-6}
10^{-7}	3.24×10^{9}	1.5×10^{14}	10
10^{-11} (UHV)	3.24×10^{5}	1.5×10^{10}	10^{5}

in air is not pure at all. On the other hand, in an ultrahigh vacuum of 10^{-11} torr, we have

$$J_c = \frac{1}{4}(3.24 \times 10^5 \times 5.2 \times 10^4) = 4.2 \times 10^9 \text{ molecules/cm}^2 \text{ s}$$

and it will now take

$$t = \frac{10^{15}}{4.2 \times 10^9} = 2.4 \times 10^5 \text{ s}$$

which is about three days, for the residual gas to form a monolayer of deposit on the surface. Therefore, we can have a high-purity film in an ultrahigh vacuum when the deposition time is short: several minutes. For epitaxial growth rates of one monolayer (about 10^{15} atom/cm^2 s), we require the rate of impingement of residue air atoms to be at least 10^{-4} times the deposition rate in order to maintain high-purity films. This rate is equivalent to a vacuum better than 10^{-10} torr. Table 2.2 lists the time needed to deposit one monolayer of background gas on a surface under various vacuum conditions.

2.8 Frequency of collision of gas molecules

In a gas, molecules collide with each other. Therefore, they do not have the same velocity and do not move in the same direction. This leads to a distribution of velocity. To see the effect of collision, we assume that a molecule has a diameter of d and an average velocity v. Fig. 2.3 shows the cross-section of collision to be πd^2, depicted by the broken circle, which means that any other molecule passing through this cross-section will cause a collision. Since the molecule is moving with an average velocity v, it further means that any molecule in the volume $V = \pi d^2 vt$ will collide with the moving molecule in a period t. Thus, we can calculate the collision frequency if the velocity and the concentration of molecules in the volume are given.

We take the velocity to be $v = 5 \times 10^4$ cm/s and the concentration of molecules in air in the standard condition to be $n = 2.46 \times 10^{19}$ molecules/cm^3, and $d = 0.3$ nm. The frequency of collision or the number of collisions per second is equal to the number of

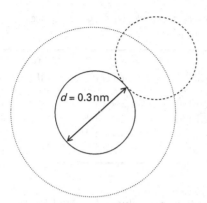

Fig. 2.3   Schematic diagram of the cross-section of collision, depicted by the dotted circle. Any other molecule (depicted by dashed circle) passing through this cross-section will cause a collision.

molecules in the volume $\pi d^2 v$,

$$\text{frequency of collisions} = \pi \times (3 \times 10^{-8})^2 \times 5 \times 10^4 \times 2.46 \times 10^{19} \cong 4 \times 10^9/\text{s}$$

Clearly, the frequency of collision is very high. As a consequence, the velocity of molecules in air will frequently change in direction and in magnitude.

Then, in an ultrahigh vacuum of 10^{-11} torr, the concentration of molecules is $n = 3.24 \times 10^5$ molecule/cm^3. The frequency of collision becomes

$$\text{frequency of collisions} = 5 \times 10^{-5}/\text{s}$$

which means just a few collisions per day. If we conduct film deposition in UHV, molecules tend to fly directly to the substrate without collision. It is a direct beam deposition, as depicted in Fig. 2.2. This is the case with MBE.

2.9   Boltzmann's velocity distribution function and ideal gas law

In Section 2.4 we used the ideal gas law to calculate the concentration of gas phase as a function of pressure, and in Section 2.5 we used the equation of the kinetic energy of molecules to calculate their velocity. Here and in the next section, we shall use velocity distribution functions to derive the ideal gas law and the equation of the kinetic energy of molecules.

We consider an assembly of gas particles at low pressure and they are non-interacting chemically. They have the same mass m, but different velocity v. In Cartesian coordinates x, y, z, the velocity has components of v_x, v_y, v_z. In Boltzmann's law of velocity distribution, we consider only the kinetic energy of the particles, i.e. $E = \frac{1}{2}mv^2$. In the distribution, the probability of finding a particle whose x-component of velocity lies between v_x and $v_x + dv_x$ is given as

$$P(v_x) = B \exp\left(-\frac{mv_x^2}{2kT}\right) \qquad (2.7)$$

and

$$\int_{-\infty}^{\infty} P(v_x)dv_x = 1$$

We have

$$B = \left(\frac{m}{2\pi kT}\right)^{1/2}$$

If we plot $P(v_x)$ against v_x, it is a bell-shape curve as shown in Fig. 2.4. In the following, we use the distribution function to derive the ideal gas law.

Now we consider gas particles in an elastic wall container, as shown in Fig. 2.5. A particle of momentum mv hitting the wall will bounce back elastically. There is no loss

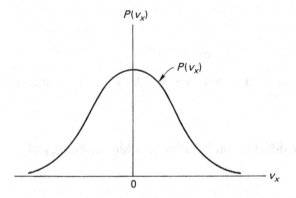

Fig. 2.4   Plot of Boltzmann's velocity distribution function $P(v_x)$ against v_x, a bell-shape curve.

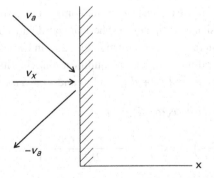

Fig. 2.5   Gas particles in a container of elastic wall. A particle of momentum mv hitting the wall will bounce back elastically.

of energy and the momentum change is $2mv_x$, normal to the surface of the wall. The number of particles hitting an area A of the wall in a unit time is J_xA. The force on the area A is the rate of change of momentum per unit time. Therefore, the pressure is

$$p = \frac{F}{A} = 2mv_xJ_x = 2mv_xnv_x = 2mn \int_0^\infty v_x^2 P(v_x)dv_x$$

$$= 2mn \left(\frac{m}{2\pi kT}\right)^{1/2} \int_0^\infty v_x^2 \exp\left(-\frac{mv_x^2}{2kT}\right) dv_x$$

(2.8)

Using the integration

$$\int_0^\infty x^2 \exp(-\alpha x^2)dx = \frac{1}{4}\sqrt{\frac{\pi}{\alpha^3}}$$

(2.9)

we obtains

$$p = 2mn \left(\frac{m}{2\pi kT}\right)^{1/2} \left(\frac{2kT}{m}\right)^{3/2} \frac{1}{4}\sqrt{\pi}$$

$$= nkT = N_A kT/V$$

or

$$pV = N_A kT$$

which is the ideal gas law that we have used to calculate the concentration in the flux equation.

2.10   Maxwell's velocity distribution function and kinetic energy of gas molecules

For a very large number of particles, the velocity distribution is expected to have spherical symmetry. It is inconvenient to express the spherical symmetry in Cartesian coordinates as in Boltzmann's velocity distribution function. Also, we would like to have a distribution function in terms of the velocity itself instead of its components. We need to perform a transformation from Cartesian coordinates to spherical coordinates in order to obtain Maxwell's velocity distribution function, as shown in Fig. 2.6. In Cartesian coordinates, we have the probability of finding a particle in equilibrium having its velocity components between v_x and $v_x + dv_x$, v_y and $v_y + dv_y$, and v_z and $v_z + dv_z$, as

$$P(v_x, v_y, v_z)dv_x dv_y dv_z = P(v_x)dv_x P(v_y)dv_y P(v_z)dv_z$$

$$= \left(\frac{m}{2\pi kT}\right)^{3/2} \exp\left[-\frac{m(v_x^2 + v_y^2 + v_z^2)}{2kT}\right] dv_x dv_y dv_z$$

$$= \left(\frac{m}{2\pi kT}\right)^{3/2} \exp\left[-\frac{mv^2}{2kT}\right] dv_x dv_y dv_z$$

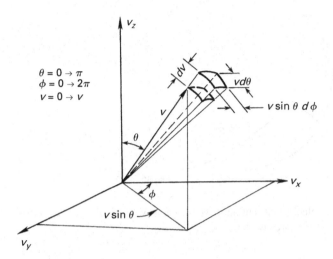

$\theta = 0 \to \pi$
$\phi = 0 \to 2\pi$
$v = 0 \to v$

Fig. 2.6   Transformation from Cartesian coordinates to spherical coordinates to obtain Maxwell's velocity distribution function.

where $v = (v_x^2 + v_y^2 + v_z^2)^{1/2}$ is the velocity of the particle. We perform a coordinate transformation from (v_x, v_y, v_z) to (v, θ, ϕ) as shown in Fig. 2.6, and we have

$$dv_x dv_y dv_z = (dv)(vd\theta)(v\sin\theta d\phi) = v^2 \sin\theta dv\, d\theta\, d\phi$$

Thus, we obtain

$$P(v,\theta,\phi)dv\, d\theta\, d\phi = \left(\frac{m}{2\pi kT}\right)^{3/2} v^2 \exp\left(-\frac{mv^2}{2kT}\right) \sin\theta dv\, d\theta\, d\phi \qquad (2.10)$$

For a large number of particles, the distribution of v has a spherical symmetry, so the variable v is independent of θ and ϕ, since

$$\int_0^{2\pi} d\phi \int_0^{\pi} \sin\theta d\theta = 4\pi$$

We have Maxwell's velocity distribution function as

$$P(v)dv = 4\pi \left(\frac{m}{2\pi kT}\right)^{3/2} v^2 \exp\left(-\frac{mv^2}{2kT}\right) dv \qquad (2.11)$$

Figure 2.7 is a plot of Maxwell's velocity distribution function against velocity. Two curves at two temperatures, T_1 and T_2, are given, where $T_2 > T_1$. Using this distribution function, we calculate the "mean velocity", $\bar{v}$, to be

$$\bar{v} = \int_0^{\infty} vP(v)dv = \left(\frac{8kT}{\pi m}\right)^{1/2} \qquad (2.12)$$

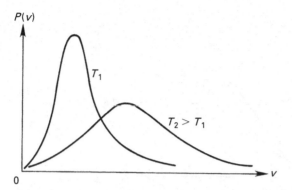

Fig. 2.7   Plot of Maxwell's velocity distribution function against velocity. Two curves at two temperatures, T_1 and T_2, are given, where $T_2 > T_1$.

which has been used to calculate the equilibrium flux on a solid surface in Section 2.6. Also, we obtain the "mean square velocity," $\overline{v^2}$, which is equal to

$$\overline{v^2} = \int_0^\infty v^2 P(v)dv = \frac{3kT}{m}$$

or

$$\frac{1}{2}m\overline{v^2} = \frac{3}{2}kT \tag{2.13}$$

It is the kinetic energy equation for gas particles that we have used to calculate the velocity of a gas particle in the flux equation. Note that the "root mean square velocity" is $\sqrt{\overline{v^2}}$, as in Section 2.5.

2.11   Parameters of nucleation and growth that affect the microstructure of thin films

In the above, we have discussed the three equations that control the gas molecules to be deposited on a solid surface: the ideal gas law, Eq. (2.3); the flux equation, Eq. (2.1); and the thermal equilibrium flux equation on a surface, Eq. (2.6). On condensation on the solid surface, the microstructure of the deposited thin film will be controlled by a set of parameters to be discussed below and which will be explained in the following chapters.

The microstructure of a thin film, for example, can be an epitaxial single crystal film on a single crystal substrate, such as an epitaxial n-type Si film grown on a p-type Si wafer; or it can be polycrystalline, such as an Al thin film grown on an amorphous quartz substrate. The polycrystalline film may have texture. Furthermore, we can also deposit amorphous thin films. Generally speaking, there are four key deposition parameters that will affect the microstructure of a thin film, as listed in Table 2.1: the deposition rate, the vacuum level in the deposition chamber, the substrate temperature, and the surface structure of the substrate. How these deposition parameters can affect the microstructure of a thin film will be explained on the basis of our understanding of nucleation and growth in phase change from gas to solid.

The basic concept in nucleation is that it has to overcome the surface energy barrier of nucleation. Thus, a certain amount of undercooling is required below the equilibrium phase transition temperature in order for the nucleation event to occur. In other words, a certain amount of supersaturation is needed for a nucleation event to occur in vapor phase condensation. Homogeneous nucleation is rare, but heterogeneous nucleation is common. This is because the event of nucleation in thin-film deposition requires supersaturation to overcome the nucleation barrier, and heterogeneity can greatly reduce the nucleation barrier. All four deposition parameters mentioned in Table 2.1 affect the nucleation of the deposited film. The deposition rate is used to control the supersaturation, so the deposition flux must be larger than the equilibrium flux, which was discussed in Section 2.6. The substrate temperature will affect the magnitude of the equilibrium flux as well as the diffusivity of atoms on the substrate surface. The substrate temperature can be regarded as an indication of the degree of undercooling in nucleation. The larger the undercooling, the higher the nucleation rate. Then, the vacuum level is used to control the impurity or heterogeneity in the deposition chamber. Also, the substrate surface structure (or surface energy) will affect the misfit or homogeneous and heterogeneous nucleation of film on the substrate surface. On the growth of the film after nucleation, it will mostly depend on the surface diffusivity, surface epitaxial steps, and rate of deposition.

For example, if we want to deposit an epitaxial film of Si on a Si wafer, we can regard it as an epitaxial growth process, without nucleation. Thus, we should reduce all the factors which may enhance nucleation. Therefore, the substrate surface should be tapered, typically with a 7° cut so that surface steps are abundant and no nucleation of surface steps is required. The nucleation of a surface step will be analyzed in Section 7.7. The other deposition parameters will be selected to reduce nucleation too. The deposition flux must be slightly larger than the equilibrium flux so that no large supersaturation exists during the deposition. A high substrate temperature is used to reduce undercooling as well as to enhance the diffusion of adatoms to go to the steps. An ultrahigh vacuum is required so that no impurity can enhance heterogeneous nucleation. All these requirements lead to the use of MBE deposition for the growth of semiconductor films of Si on Si or the growth of superlattices.

If we want to deposit an amorphous Si film, we also need to suppress the nucleation of crystalline Si grains, but the selection of deposition parameters will be completely different from those in epitaxial growth discussed above. We will use a glassy surface such as a quartz substrate, so that no epitaxial growth between crystalline phases can occur. A low substrate temperature is used so that surface diffusion of Si adatoms cannot occur after the gas atoms have impinged on the substrate surface, nor can atomic rearrangement occur on the surface for the formation of a critical nucleus of crystalline Si. Also, a high rate of deposition may jam the atoms into an amorphous structure before atomic rearrangement can take place. Ultrahigh vacuum is not required. Actually, a small amount of impurities such as hydrogen may even help to maintain the stability of the amorphous structure. However, for a high-purity amorphous film, UHV is still preferred.

Table 2.3. Deposition conditions and the corresponding microstructure of the deposited thin film

Thin-film deposition condition	Epitaxial single crystal film	Textured film	Polycrystalline film	Amorphous film
Rate of deposition	Very low rate $\sim$0.1 nm/s	Medium rate $\sim$1 nm/s	Medium rate $\sim$1 nm/s	High rate $\sim$10 nm/s
Vacuum level	Ultrahigh vacuum	10^{-7} torr	10^{-7} torr	10^{-5} torr
Substrate temperature	High temperature	Room temperature	Room temperature	Low temperature
Substrate structure	Single crystal and lattice matching with the film	Crystalline substrate	Oxidized substrate	Glassy substrate

If we want to deposit a polycrystalline Si thin film with large grain size in applications to solar-cell or flat-panel-display technology, a large surface area substrate is required. We need to reduce the nucleation rate but enhance the growth rate, especially in the lateral direction which is parallel to the substrate surface. However, grain growth in thin films tends to be limited by the film thickness. Generally speaking, it is hard to obtain a grain size much larger than the film thickness. This is because when a columnar grain distribution is obtained, it tends to have vertical grain boundaries that are normal to the substrate surface. When the triple points of grain boundaries achieve the local equilibrium, there is no driving force for grain growth or for the migration of a vertical grain boundary, except for abnormal grain growth. The growth of large-grain polycrystalline Si thin films on a large substrate surface has been a very challenging problem, when the constraints of production costs are taken into account.

In Table 2.3, a list of deposition conditions and the corresponding microstructures of the deposited films are given. These serve as a useful guide as to how to select the deposition parameters in order to obtain thin film with the desirable microstructure. Typically, one should know the deposition conditions before asking for a thin-film deposition.

References

[1] J. L. Vossen and W. Kern, *Thin Film Processes* (Academic Press, New York, 1978).

[2] M. Ohring, *The Materials Science of Thin Films* (Academic Press, Boston, 1992).

[3] J. C. Slater, *Introduction to Chemical Physics* (McGraw-Hill, New York, 1939).

[4] L. Maissel and R. Glang (eds), *Handbook of Thin Film Technology* (McGraw-Hill, New York, 1970).

[5] L. Eckertova, *Physics of Thin Films* (Plenum Press, New York, 1986).

[6] J. W. Mayer and S. S. Lau, *Electronic Materials Science* (Macmillan, New York, 1989).

Problems

2.1　Why do we need an ultrahigh vacuum in order to obtain a high-purity thin film?

2.2　If we want a textured metallic thin film, what kinds of deposition conditions will be needed?

2.3 Why is it difficult to deposit a very large-grain Si thin film on a glassy substrate for solar-cell applications?

2.4 In depositions at very low temperature, if we assume no surface diffusion, what is the growth rate of the film?

2.5 The equilibrium vapor pressure above a silicon surface at 1123 K corresponds to 6.9×10^{-9} Pa. At equilibrium the fluxes of atoms leaving and returning to the surface are equal.
(a) What is the flux of atoms/m^2 s?
(b) What is the number of Si atoms/m^3 in the vapor?
(c) What is the velocity of the Si atoms in m/s?

3   Surface energies

3.1   Introduction

Surface energy is an underlying concept in understanding thin-film process. By definition, thin film has a very large surface-to-volume ratio. Surface energy controls the nucleation as well as the heterogeneous epitaxial growth processes. It also plays a key role in many applications of thin films, for example, in MEMS devices. Generally speaking, surface energy is the extra energy expended to create a surface, so surface energy is positive. It is important to know that metals have high surface energies and oxides have low surface energies, so a native oxide can grow on a metal.

The surface energy determines whether or not one material wets another and forms a uniform adherent layer as in heterogeneous epitaxial growth. A material with a very low surface energy will tend to wet a material with a higher surface energy so that epitaxial growth is possible. On the other hand, if the material to be deposited has a higher surface energy than the substrate surface, it tends to form clusters ("ball up") on the low-surface-energy substrate. The epitaxial growth of a superlattice structure of $ABABAB$ requires that in addition to a good lattice parameter match between A and B, the surface energy of A and B should be nearly the same. There is a well-known wetting principle that if A wets B, B will not wet A but ball up on A. In order to have A wetting B and B wetting A in growing the superlattice, their surface energies should be the same. For this reason, we can have superlattices consisting of two very similar semiconductor films or two oxide films. It will be very difficult to grow a superlattice of dissimilar materials, for example, a superlattice of silicide and silicon such as $CoSi_2$ and Si.

Waterproofing is a good example of manipulating surface energies. Organic materials tend to have low surface energies, so a car is waxed with an organic substance and water droplets form on a waxed surface. On the other hand, Benjamin Franklin, who was looking at a large pond of water, noticed the surface roughness of the water. When he put a spoonful of oil on the surface of the pond, the pond became very smooth. One cubic centimeter of the material (about one spoonful) over a large pond forms a surface film of about 1 nanometer thickness. It wets the water surface and dramatically changes the surface properties.

Surface energy is a positive quantity because energy is added to create a surface. In nature, a liquid tends to ball up to reduce its surface area, and crystals tend to facet in order to expose those surfaces of the lowest energy. When we break a solid, two new surfaces are created and bonds between atoms are broken. It is clear that the surface energy is related to the bond energy and to the number of bonds broken in creating the surface. This in turn is related to the binding energy of the material. Below, we shall discuss the relationships between pair potential energy, bond energy, binding energy, surface energy, as well as latent heats [1–9].

3.2 Pair potential energy, bond energy, and binding energy

The binding energy is defined as the energy needed to transform one mole of solid or liquid into gas at a low pressure. In the transformation, all the bonds between atoms are broken. The binding energy is nearly the same in magnitude as the energy of sublimation (transforming a solid to gas) or the energy of evaporation (transforming a liquid to gas), except that the latter two are generally measured at one atmosphere pressure rather than at a low pressure. The latent melting heat is the difference between the energy of sublimation and that of evaporation. These energies are related to a fundamental energy in materials, that is, the interatomic potential energy or pair potential energy between two atoms. The bond energy is defined as the pair potential energy between two atoms when they are at their equilibrium positions.

A schematic curve of interatomic potential energy ϕ as a function of interatomic distance r is shown in Fig. 3.1(a). The minimum potential energy, $-\varepsilon_b$, or the bond energy corresponds to the equilibrium interatomic separation, a_0. For crystalline solids, the equilibrium interatomic distance can be measured by X-ray diffraction. We estimate a_0 from the lattice parameter, a. For example, aluminum forms an fcc lattice with a lattice parameter $a = 0.405$ nm (4.05Å). Since the densest packing direction is along <110>, the equilibrium interatomic distance in Al is $a/\sqrt{2} = 0.29$ nm.

An atom in a solid, displaced from its equilibrium position, experiences a restoring force given by

$$F = -\frac{d\phi}{dr} \tag{3.1}$$

A schematic curve of the force as a function of distance is shown in Fig. 3.1(b). For displacements of the order of 0.1% of the interatomic spacing, the displacement is proportional to the force. Recall that the elastic limit is 0.2%. In solids, this linear displacement is the origin of Hooke's law where strain is directly proportional to stress. In this bottom region of the potential curve as shown in Fig. 3.1(a), we can approximate the potential energy by a parabolic function,

$$\phi = \frac{1}{2}kr^2 \tag{3.2}$$

where k is the force constant. We have $F = -kr$. Clearly, the force is linearly proportional to displacement. Furthermore, we have the equation of motion of $F = ma$ as

$$m\frac{d^2r}{dt^2} = -kr \tag{3.3}$$

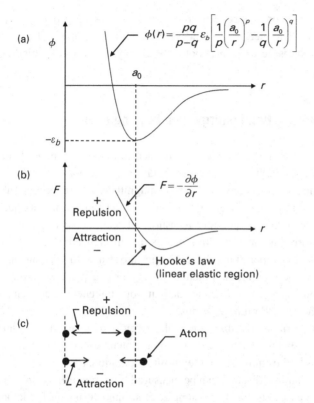

Fig. 3.1 (a) Schematic curve of interatomic potential energy ϕ as a function of interatomic distance r. The minimum potential energy, $-\varepsilon_b$, or the bond energy corresponds to the equilibrium interatomic separation, a_0. (b) Schematic curve of the restoring force as a function of distance. (c) The direction of the force between two atoms. The force is defined to be positive or negative according to whether it increases or decreases the interatomic distance.

where m is mass of the atom. It is an equation of simple harmonic motion and a simple solution of the equation of motion is

$$r = \cos \omega t \tag{3.4}$$

and $\omega = \sqrt{\frac{k}{m}}$. Since $k = \frac{\partial^2 \phi}{\partial r^2}$, we can calculate k when ϕ is given. Then we obtain ω by knowing k and m. Since $\omega t = 2\pi$, we obtain the atomic vibrational frequency,

$$\nu = \frac{1}{t} = \frac{1}{2\pi}\sqrt{\frac{k}{m}} \tag{3.5}$$

which will be related to atomic jump frequency in Chapter 4 on atomic diffusion.

The direction of the force is indicated in Fig. 3.1(c). In the figure, an atom is placed at the origin and another at the equilibrium separation. If the atoms are displaced towards each other, a repulsive force acts to increase the interatomic distance in order to push them back to their equilibrium positions. The force is defined as positive or negative according to whether it increases or decreases the interatomic distance. The internal repulsive force is positive and internal attractive force is negative. On the other hand, if an external force is applied, the applied tensile force tends to pull atoms apart, so the external tensile force is positive and the external compressive force is negative.

3.3 Short-range interaction and quasi-chemical assumption

While the shape of the interatomic potential energy curve (or the force) controls many of the physical properties of an aggregate of atoms, such as bulk modulus and thermal expansion, we shall consider here only whether the potential is short range or long range and how the surface energy is affected by it. The short-range interaction approximation is important, since it is the basis of the quasi-chemical approach in thermodynamics. In most cases, the potential can in general be represented by

$$\phi(r) = \frac{pq}{p-q} \varepsilon_b \left[\frac{1}{p} \left(\frac{a_0}{r} \right)^p - \frac{1}{q} \left(\frac{a_0}{r} \right)^q \right] \tag{3.6}$$

where ε_b is the minimum potential energy and p and q are numbers whose value depends on the shape of the potential. For short-range interactions in some simple solids, such as a frozen inert gas, solid Ar, the Lennard-Jones potential ($p = 12$ and $q = 6$) applies, and Eq. (3.6) is reduced to

$$\phi(r) = \varepsilon_b \left[\left(\frac{a_0}{r} \right)^{12} - 2 \left(\frac{a_0}{r} \right)^6 \right] \tag{3.7}$$

At the equilibrium position, $r = a_0$, the potential is at its minimum, $\phi(a_0) = -\varepsilon_b$. The attractive interaction is due to the van der Waals force because each atom is electrically neutral. The interaction is of short range because of the power-law dependence of the potential. When the interatomic separation is twice the equilibrium distance, i.e. $r = 2a_0$, the potential energy is reduced by a factor of about 32, or only 3% of ε_b:

$$\phi(r = 2a_0) = -\frac{\varepsilon_b}{32} \tag{3.8}$$

For this reason, we can ignore the interaction energy beyond nearest neighbors and approximate the binding energy by taking only the bonds between the nearest neighbors. This simple estimate shows why we are interested in the short-range interactions here. Note that the repulsive interaction is of even shorter range.

To apply the short-range interaction to quasi-chemical approximation, recall that in thermodynamics, the ideal solution is defined to have enthalpy $\Delta H = \varepsilon_{AA} + \varepsilon_{BB} - 2\varepsilon_{AB}$ $= 0$, and entropy $\Delta S =$ ideal mixing, where the ε_{AA}, ε_{BB}, ε_{AB} are quasi-chemical bond energy in pairs of A–A, B–B, and A–B atoms, respectively. Hence, only the nearest bonds are counted in the approximation.

We define n_c to be the coordination number, i.e. the number of nearest neighbors of an atom in a liquid or solid. By using the short-range interaction or quasi-chemical approach, we obtain the binding energy for one mole by counting only the nearest-neighbor bonds,

$$E_b = \frac{1}{2} n_c N_A \varepsilon_b \qquad (3.9)$$

where N_A is Avogadro's number and the factor of $1/2$ arises because we have counted each bond twice in the produce $n_c N_A$. The maximum number of rigid spheres that can be brought into contact with another sphere of the same radius is 12 in a crystalline solid. The coordination number, n_c, equals 12 for such close-packed hexagonal (hcp) and fcc structures. For the more open, covalently bonded diamond structure of silicon, $n_c = 4$, i.e. there are four nearest neighbors. For body-centered cubic (bcc) structures such as iron, $n_c = 8$.

For comparison, we consider a long-range interaction in ionic crystals where the attractive interaction between a positive ion and a negative ion is Coulombic. The Coulomb potential is proportional to $1/r$ and it is long range, since it decays very slowly with increasing r. For an ionic crystal, we take the exponential parameters $p = 12$ (repulsive interaction remains the same) and $q = 1$ in Eq. (3.6). The profile of the potential is shown in Fig. 3.2.

For metals, the cohesion is due to the interaction of the regularly arranged positive ions with the "electron sea" wherein the electrons move freely. In the free-electron model, the positive ions are shielded by the free electrons from interacting with each other; electrical neutrality is achieved locally and the attractive interaction between atoms is again of short range. The parameter q in the pair interaction potential of metals can be taken as close to six. For this reason, we can use the nearest-neighbor interaction approximation to estimate the binding energy and surface energies of metals.

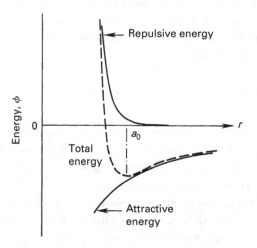

Fig. 3.2   Pair potential curve for an ionic crystal; we take the exponential parameters $p = 12$ (repulsive interaction remains the same) and $q = 1$ in Eq. (3.6).

In covalent solids such as semiconductors and inorganic materials where electrons are shared between neighboring positive ions, the screening effect again leads to short-range attractive interaction and the chemical bond picture is prevailing. Nevertheless, the bonds are more directional in covalent solids.

Using the nearest-neighbor interaction approximation, we can compare the magnitudes among the heat (energy) of sublimation, evaporation, melting and crystallization, and surface energies. The heat of sublimation ΔE_s at low pressures (i.e. the binding energy), where all the bonds are broken,

$$\Delta E_s = \frac{1}{2} n_c N_A \varepsilon_b \tag{3.10}$$

depends on the coordination number n_c. In its molten state, the metal atoms may have about 11 nearest neighbors instead of 12, so the heat of evaporation will be about 10% less than the heat of sublimation because there are approximately 10% fewer bonds to be broken in evaporation. Consequently, the heat of melting or of crystallization is only approximately 10% of the heat of sublimation.

3.4 Surface energy and latent heat

To evaluate surface energy from the point of view of nearest-neighbor bonds, we first define N_s to be the number of atoms per unit area and E_s/A to be the surface energy per unit area [10–16]. Across an arbitrary atomic plane, each atom has on average $n_c/2$ nearest neighbors on each side, so the number of bonds to be broken per unit area when we cleave along the plane is $\frac{1}{2} n_c N_s$. Since we create two surfaces in cleaving, the surface energy per unit area is $n_c N_s \varepsilon_b / 4$.

The ratio of surface energy/atom, E_s/AN_s, to latent heat of sublimation/atom, $\Delta E_s/N_A$, is

$$\frac{E_s/AN_S}{\Delta E_s/N_A} = \frac{^{1}/_{4} n_c \varepsilon_b}{^{1}/_{2} n_c \varepsilon_b} = \frac{1}{2} \tag{3.11}$$

The argument of $n_c N_s/2$ broken bonds is oversimplified for crystalline solids since it ignores the packing configuration of atoms in a crystal. Crystals tend to show faceted surfaces, indicating that they have different surface energies on different atomic planes. On (111) surface of fcc metals, each atom has three broken bonds instead of six, so the (111) surface has the lowest surface energy in fcc metals. From the measured values, Table 3.1, we can calculate the ratio of surface energy per atom to latent heat per atom, and also the interatomic potential energies.

The latent heat of Au is 60 kcal/mole, or 2.6 eV/atom. To convert the surface energy per cm² to units of eV/atom, we use the lattice parameter of Au, 0.4078 nm, to calculate a value of 1.39×10^{15} atom/cm² in the (111) plane. If we assume that the measured

Table 3.1. Relationship between solid–vapor surface energy and latent heat of evaporation.

Metal	*Solid–vapor surface energy (erg/cm^2)	*Latent heat of evaporation (kcal/mole)	Ratio of surface energy per atom to latent heat per atom	Interatomic potential energy (eV/atom)	$^\Delta$Cohesive energy (kcal/mole)
Copper	1700	73.3	0.22	0.58	80.4
Silver	1200	82	0.15	0.65	68
Gold	1400	60	0.24	0.47	87.96

* Data from B. Chalmer, *Physical Metallurgy* (Wiley, New York, 1959). Similar surface energies are given in Table 3.2.
$^\Delta$ Data from C. Kittel, *Introduction to Solid State Physics*, 6th edn (Wiley, New York, 1986), 55.

surface energy of 1400 erg/cm^2 is for the (111) surface of Au, we obtain

$$1400 \text{ erg/cm}^2 = \frac{1400 \times (1/1.6) \times 10^{12} \text{ eV/cm}^2}{1.39 \times 10^{15} \text{ atom/cm}^2} = 0.636 \text{ eV/atom}$$

Thus, the ratio of $0.636/2.6 = 0.24$, as given in Table 3.1. We use the (111) plane which has only 3 broken bonds out of 12 nearest neighbors. Hence, we expect the ratio to be 3/12 or 0.25, which agrees well with the calculated value of 0.24.

To calculate the interatomic potential energy ε_b from the measured latent heat, we rearrange Eq. (3.9),

$$\varepsilon_b = \frac{2\Delta E_s}{n_c N_A} \tag{3.12}$$

Since the heat of evaporation is between liquid and gas, we shall take the coordination number n_c to be 11 (following the same argument used for the molten state). We have for Au,

$$\varepsilon_b = \frac{2 \times 60}{11 \times 23} = 0.47 \text{ eV/atom}$$

In the last column of Table 3.1, we list the calculated theoretical cohesive energies at 0 K and low pressure. This is not the same as the value obtained from latent heat, since the latter is measured at the melting point and at 1 atom. Knowing the bond energy ε_b and interatomic distance a_0 of Au, as well as the fcc crystal structure, many physical properties of Au can be calculated and simulated.

3.5   Surface tension

Surface properties of a liquid can be described by a thermodynamic variable, the surface tension, γ. The basic definition is that the reversible work, dW, on the material upon increasing its surface area dA is

$$dW = \gamma \, dA \tag{3.13}$$

The surface tension has dimensions of work/area or erg/cm^2. Three quantities enter the scientific literature in this connection: surface energy, surface tension, and surface stress. All have the same unit, energy/area, or force/length. The interrelationship of these quantities arises because solids can change their "surface energy" in two ways, either by increasing the physical area, as may occur from a cleave, or by changing the arrangement of atoms on a surface, as in a surface reconstruction. The former case simply involves creating (forming) more surface area; the latter case involves the detailed arrangements of atoms on a solid surface area and may be thought of as the work involved in stretching a surface. The different processes are related through the surface stress tensor. A diagonal element of the stress tensor may be written in the form $S = \gamma + d\gamma/d\varepsilon$ where the first quantity γ is the surface energy and the second quantity is the change in the surface energy with a variation in strain. In liquids, no elastic deformation is possible, so $d\gamma/d\varepsilon = 0$. Then the surface energy is equal to the surface tension in liquids.

In this book we use the following notation and concepts. The phrases "surface tension" and "surface energy" are used interchangeably and scalar quantity is denoted by γ. We shall occasionally refer to the total surface energy defined below. The concept of "surface tension" implies a force. For example, the shape of a droplet on a planar surface is an equilibrium configuration set up by the balance of surface forces acting on the drop. The force is a vector quantity, acting in the surface plane, with the magnitude of the surface tension, γ. The total surface energy E_s is energy, while surface tension is an energy/area. By definition, for area A

$$E_s = \gamma A \tag{3.14}$$

The surface tension can also be thought of as a force/length, since erg/cm^2 is equivalent to dyne/cm. In this sense the force $\left| \vec{F} \right|$ along a line of length l is

$$\left| \vec{F} \right| = \gamma l \tag{3.15}$$

In the following discussions we speak of surface tension as the scalar quantity γ, the surface energy/area; it actually becomes a vector quantity acting in the surface plane when considering phenomena such as the balance of forces acting on a liquid column under capillary action.

The relationship between surface energy and surface tension can also be shown directly by the example of stretching a soap film. In Fig. 3.3, we show a soap film covering the rectangular area formed by a U-shaped wire and a straight wire. If we pull the straight wire by a distance d, the work done in changing the shape is

$$W = 2F_s l d \tag{3.16}$$

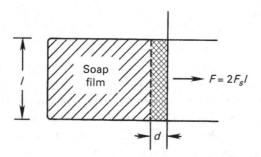

Fig. 3.3   Schematic diagram of a soap film covering the rectangular area formed by a U-shaped wire and a straight wire.

where F_s is the force per unit length and l is the width of the soap film. The factor 2 arises because the film has two surfaces. Since the increase in area is $2ld$, the increase in surface energy is $2ld\gamma$, which should equal the work done,

$$2ld\gamma = 2F_sld \tag{3.17}$$

Therefore, the surface energy per unit area and the surface tension per unit length have the same magnitude.

   A review of surface energy per area units shows

$$\frac{erg}{cm^2} = \frac{dyne}{cm} = \frac{10^{-3} \, J}{m^2} = \frac{10^{-3} \, N}{m}$$

Another physical unit is eV/atom. Since $1 \, eV = 1.6 \times 10^{-12}$ erg and there are approximately 10^{15} atom/cm^2 on a typical surface, surface tension is on average approximately 1000 erg/cm^2,

$$\gamma = \frac{1000 \, erg}{cm^2} \times \frac{cm^2}{10^{15} \, atom} \times \frac{1 \, eV}{1.6 \times 10^{-12} \, erg} = 0.6 \, eV/atom$$

The value of 0.6 eV/atom is in the order of magnitude of the bonding energy of an atom in a solid: it takes roughly that much energy to take an atom out of the surface.

3.6   Liquid surface energy measurement by capillary effect

Surface energies are often measured in the liquid state by taking the material up to its melting point, and watching either how droplets form or how a meniscus forms in interaction with a solid wall. In Fig. 3.4, we consider the rise of a liquid column in a capillary tube of diameter $2r$ to reach the equilibrium height h. The driving force of the

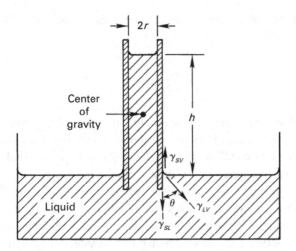

Fig. 3.4 Schematic diagram showing the rise of a liquid column in a capillary tube of diameter $2r$ to reach the equilibrium height h.

rise is the reduction of surface energy of the inside wall of the tube. The rise, however, increases the potential energy of the liquid column. The change in total energy of the process is

$$\Delta E = \rho V g \frac{h}{2} - (\gamma_{SV} - \gamma_{SL})\, 2\pi r h \tag{3.18}$$

where ρ and $V (= \pi r^2 h)$ are the density and volume of the liquid column, g is the gravitational constant, and γ_{SV} and γ_{SL} are the surface energy per unit area of the wall unwetted (surface-to-vapor) and wetted (surface-to-liquid) by the liquid. The first term on the right-hand side of Eq. (3.18) is the potential energy of the liquid column. The mass of the column is $\rho V g$ and the center of gravity of the mass is at $h/2$. The second term is due to the change in surface energy of the inside wall of the tube by the wetting of the liquid. At equilibrium, we have

$$\frac{dE}{dh} = 0$$

or

$$\rho g \pi r^2 h - 2\pi r\,(\gamma_{SV} - \gamma_{SL}) = 0$$

$$\gamma_{SV} - \gamma_{SL} = \frac{\rho g h r}{2} \tag{3.19}$$

At the edge of the liquid column as shown in Fig. 3.4, the surface tensions (energies) are balanced,

$$\gamma_{SV} - \gamma_{SL} = \gamma_{LV} \cos\theta \tag{3.20}$$

where γ_{LV} is the liquid-to-vapor surface energy, and θ is the contact angle. Combining the last two equations, we have

$$h = \frac{2\gamma_{LV}\cos\theta}{\rho r g} \tag{3.21}$$

This relates the liquid surface energy to measurable quantities of the liquid column (r, h, θ, and ρ). The measurement of θ deserves further discussion, but we shall first illustrate an application of the capillary effect in electronic packaging technology.

The capillary effect has been used to fill Cu-plated holes in thick multilayer printed circuit boards with molten solder for mechanical and electrical connection. The board has multilayers of embedded wires, and holes are drilled through the board and plated with Cu for interconnection of the wires. Metallic pins are inserted into the top side of the holes for contacts to external circuits. Then, the bottom side of the board is dipped into molten solder or passing over a solder fountain to enable the low-surface-energy molten solder to fill the holes and to solder the pins. The success of the process depends on the capillary effect, which in turn depends on the surface energy of the molten solder, its contact angle to the Cu, and the aspect ratio (height-to-diameter) of the hole. To define the process, we need to measure the contact angle.

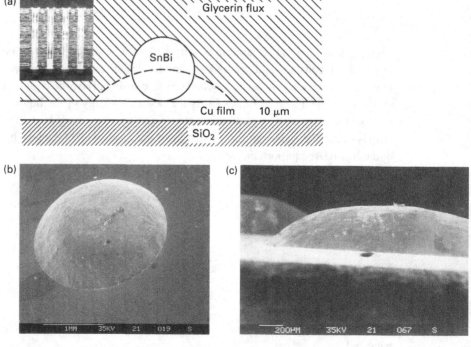

Fig. 3.5   (a) Schematic diagram of a bead of SnBi solder placed on a Cu surface and immersed in glycerin flux. (b) and (c) SEMs of the top and side views respectively of the wetting of the solder on a Cu surface; the contact angle $\theta = 40°$.

Fig. 3.5(a) shows a schematic diagram of a bead of SnBi solder placed on a Cu surface and immersed in glycerin flux. The flux removes surface oxides. Upon heating to 137 °C, the melting point of the solder, the SnBi solder spreads out to wet the Cu (broken curve in Fig. 3.5(a)). After equilibrium is reached, the temperature is lowered to solidify the solder, and the contact angle can be measured. Scanning electron micrographs of the top and side views of the wetting of the solder on a Cu surface are shown in Fig. 3.5(b) and 3.5(c), respectively; and the contact angle $\theta = 40°$.

To calculate h in Eq. (3.21), we assume that $\lambda_{LV} = 250$ erg/cm^2, $\rho \cong 10$ g/cm^3, the gravitational constant $g = 980$ dyne/g, and the hole diameter $= 0.5$ mm. We obtain $h = 1.6$ cm, which should be greater than the thickness of the board. The capillary effect can pull the molten solder all the way through the hole. In practice, to protect the Cu surface from oxidation and to enhance the capillary action, the Cu surface is coated with a thin layer of immersion Sn. The contact angle of the SnBi solder to the Sn surface is zero. This can be shown experimentally by replacing the Cu layer in Fig. 3.5(a) with a Sn layer.

On the other hand, we see in Eq. (3.15) that if $\gamma_{SL} > \gamma_{SV}$, the contact angle θ will be greater than 90°, which means that the liquid will ball up and will not wet the solid surface. In this case, $\cos \theta$ is negative and it gives a negative value to the height h in Eq. (3.21). For example, when we insert a glass tube into mercury, we see the negative capillary effect as the column of mercury goes below the surrounding mercury level.

3.7   Solid surface energy measurement by zero creep

We can extend the capillary technique of measuring liquid surface energies to measuring solid surface energies by turning the arrangement in Fig. 3.4 upside down (i.e. by hanging a wire from the ceiling and measuring its rate of elongation or *creep rate* under its own weight). Zero creep means the strain rate is zero; that is, at zero creep the weight of the wire is balanced by the surface tension of the wire surface, and thus the surface tension of the wire can be determined by knowing its weight.

If we take a glass fiber or a wire made of metallic glass (amorphous alloy), there are no grain boundaries in the wire and the analysis is quite similar to that in the last section. In Fig. 3.6(a), we consider a glassy wire of diameter $2r$ and length l hanging down. In order to reduce surface energy the wire shortens in length, but the weight of the wire balances this tendency. We assume that the wire shortens a small length dl and that the diameter is increased by dr in order to reach equilibrium. Since the volume of the wire must remain the same, we have

$$\pi r^2 l = \pi (r + dr)^2 (l + dl) \tag{3.22}$$

By ignoring the higher order terms, we obtain

$$dr = -\frac{r}{2l} dl \tag{3.23}$$

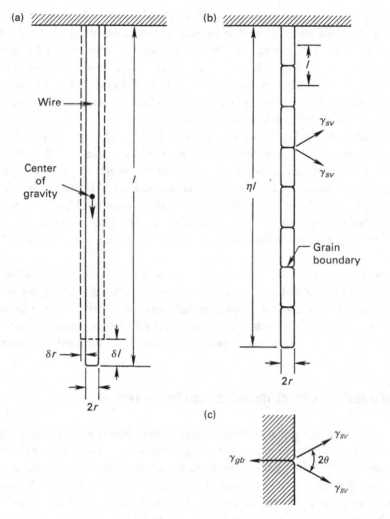

Fig. 3.6   (a) Schematic diagram of a glassy wire of diameter $2r$ and length l hanging down. In order to reduce surface energy the wire shortens in length, but the weight of the wire balances this tendency. (b) Schematic diagram of a wire having a bamboo-type grain structure; we assume that there are η number of grains in the wire. (c) Equilibrium at a triple point where a grain boundary meets the surface.

where dl is negative (decrease in length) and dr is positive (increase in radius). To calculate the energy change of the shrinkage, we start with the total energy E of the wire,

$$E = \rho V g(l/2) - 2\pi r l \gamma_{SV} \tag{3.24}$$

where ρ and $V (= \pi r^2 l)$ are the density and the volume of the wire (the factor of $1/2$ comes in because the center of gravity of the wire is at $l/2$) and γ_{SV} is the surface energy per unit area of the wire. Then,

$$dE = \frac{1}{2}\rho g \pi (2l^2 r \, dr + 2r^2 l \, dl) - 2\pi \gamma_{SV}(r \, dl + l \, dr) \tag{3.25}$$

By substituting Eq. (3.23) into Eq. (3.25), we have

$$dE = \left(\frac{1}{2}\pi r^2 l \rho g - \pi r \gamma_{SV}\right) dl \tag{3.26}$$

At equilibrium, $dE/dh = 0$, we have

$$l = \frac{2\gamma_{SV}}{\rho r g} \tag{3.27}$$

which has the same form as Eq. (3.21). It shows that at zero creep we can determine γ_{SV} by measuring l and r, which gives the weight of the wire when its density is known. Consider the density of Au, 19.3 g/cm^3, and surface energy, 1400 erg/cm^2, the length of an Au wire of diameter of 0.02 cm at zero creep can be calculated to be approximately 14.8 cm with weight of 0.1g. If we perform the creep experiment in an ultrahigh vacuum environment and if we can increase the temperature without introducing crystallization of the glassy wire, we may be able to determine the surface energies of a glass in the solid state.

When the wire crystallizes, it develops grain boundaries and we must take the grain-boundary energy into account. Actually, this is common for crystalline wires. In Fig. 3.6(b), we consider a wire having a bamboo-type grain structure and we assume that there are η number of grains in the wire. The total energy of the wire is

$$E = \eta \rho \vartheta g \frac{\eta l}{2} - \eta 2\pi r l \gamma_{SV} - \eta \pi r^2 \gamma_{gb} \tag{3.28}$$

where $\vartheta (= \pi r^2 l)$ is the volume of a single grain and γ_{gb} is the grain-boundary energy per unit area. The energy change upon a small change of wire length is

$$dE = \frac{\eta^2 \pi \rho g}{2} r^2 l \, dl - \eta \pi r \gamma_{SV} \, dl + \eta \pi \frac{r^2}{l} \gamma_{gb} dl \tag{3.29}$$

At equilibrium, we have

$$\frac{dE}{dh} = 0$$

$$\frac{\eta \pi \rho g}{2} r^2 l - \pi r \gamma_{SV} + \frac{\pi r^2}{l} \gamma_{gb} = 0 \tag{3.30}$$

To relate γ_{SV} and γ_{gb}, we consider a joint where a grain boundary meets the wire surface (see Fig. 3.6(c)). If we assume that the two surface tension vectors are equal, we have

$$\gamma_{gb} = 2\gamma_{SV} \cos \theta \tag{3.31}$$

By substituting this relationship into Eq. (3.27), we obtain

$$\gamma_{SV} = \frac{\rho V g}{2\pi r \left(1 - (2r \cos \theta)/l\right)} \tag{3.32}$$

Note that if the grains are long (i.e. $l \gg r$), we can drop the second term in the denominator, and Eq. (3.32) becomes the same as Eq. (3.27).

In zero-creep experiments, the major portion of the wire is often replaced by a weight so that a short wire can be used. In this case, Eq. (3.32) is still applicable except that the term $2\pi r$ in the denominator is replaced by πr. The factor of 2 comes from the fact that the center of gravity is located at half the length of the wire when no external weight is used. When we use a weight, potential energy is measured at the full length of the wire from the ceiling.

3.8 Surface energy systematics

We now consider the systematics of surface energies. Fig. 3.7 is organized with respect to atomic number and gives the surface tension values of liquid materials at their melting point. Table 3.2 gives the surface tensions of some liquid halides, oxides, and sulfides, as well as polymers. The oxides, halides, and sulfides have lower surface tensions than most of the metals. Under normal evaporation conditions, many metals will ball up upon deposition on an oxide or a halide. Hence, a glue layer with a strong adhesion to the

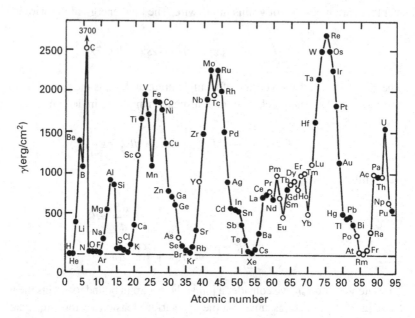

Fig. 3.7 Surface energies plotted against atomic number. The surface tension values of liquid materials are given at their melting point.

Table 3.2. Surface tension of selected solids and liquids*

Material	γ (erg/cm^2)	T (°C)
W (solid)	2900	1727
Nb (solid)	2100	2250
Au (solid)	1410	1027
Ag (solid)	1140	907
Ag (liquid)	879	1100
Fe (solid)	2150	1400
Fe (liquid)	1880	1535
Pt (solid)	2340	1311
Cu (solid)	1670	1047
Cu (liquid)	1300	1535
Ni (solid)	1850	1250
Hg (liquid)	487	16.5
LiF (solid)	340	−195
NaCl (solid)	227	25
KCl (solid)	110	25
MgO (solid)	1200	25
CaF$_2$ (solid)	450	−195
BaF$_2$ (solid)	280	−195
He (liquid)	0.308	−270.5
Na (liquid)	9.71	−195
Xenon (liquid)	18.6	−110
Ethanol (liquid)	22.75	20
Water (liquid)	72.75	20
Benzene (liquid)	28.88	20
n-Octane (liquid)	21.80	20
Carbon tetrachloride (liquid)	26.95	20
Bromine (liquid)	41.5	20
Acetic acid (liquid)	27.8	20
Benzaldehyde (liquid)	15.5	20
Nitrobenzene (liquid)	25.2	20
Perfluoropentane (liquid)	18.6	−110

* From G. A. Somorjai, *Chemistry in Two Dimensions* (Cornell, Ithaca, NY, 1981) [12]

oxide or the halide will be needed. On deposition of Cu on the SiO$_2$ surface, a thin film of Cr, Ti, or Ta is widely used as the glue layer.

It is also interesting to consider semiconductor/insulator problems. Silicon tends to form a thin oxide. The formation and control of oxide layers on Si is one of the key features in the fabrication of integrated circuits. On the other hand, silicon deposited on an insulator would tend to ball up because it has a higher surface energy than that of oxides. Hence, the formation of a heterostructure is always difficult, particularly if the heterostructure involves a superlattice of different types of material. A superlattice requires forming the structure twice (i.e. first material *B* on material *A*, and then *A* on *B*). For one interface the surface energy balance will be unfavorable. It is easier to grow superlattices if both materials are semiconductors, because many semiconductors have similar surface energies. GaAs and GaAlAs, for example, are not very different in this regard; therefore, superlattices of them can be formed.

3.9   Magnitudes of surface energies

The magnitude of the surface tension (or surface energy per unit area) for many materials used in device technology is about 1000 erg/cm^2. In the following sections, we consider three different kinds of approach – thermodynamic, mechanical, and atomic – to describe surface energy and the magnitude of surface tension.

3.9.1   Thermodynamic approach

From the point of view of thermodynamics, there are two important relations:

$$\frac{d\gamma}{dT} = \frac{-S_s}{A} \tag{3.33}$$

and

$$E_s = \gamma A - T\frac{d\gamma}{dT}A \tag{3.34}$$

where S_s is the entropy of the surface, A is the area, γ is the surface tension, and T is the temperature. These relations allow the surface tension at room temperature to be estimated from tabulated values of $d\gamma/dT$, which are assumed to be independent of temperature.

To estimate the surface energy for silicon, using Table 3.3 we find that

$$\gamma = 730 \text{ erg/cm}^2 \text{ at the melting point}$$

$$\frac{d\gamma}{dT} = -0.1 \text{ erg/cm}^2 \,^\circ\text{C}$$

The melting point of silicon T_m (Si) is 1410 °C. Then at room temperature, the surface energy of silicon is given by

$$\gamma_{RT} \text{ (Si)} = 730 + (1410 - 25) \times 0.1 = 869 \text{ erg/cm}^2$$

The surface energy at the melting point is not very different from the surface energy at temperatures of interest. The small change of the surface energy is associated with the entropy contribution, which is small.

3.9.2   Mechanical approach

The mechanical approach uses the mechanical properties of solids to estimate the surface tension. We separate a solid by mechanical force into two pieces which are far apart.

Table 3.3. Surface tension of liquid metals*

Metal	γ_{LV} (erg/cm^2)	$d\gamma_{LV}/dT$ (erg/cm^2 °C)
Al	866	−0.50
Cu	1300	−0.45
Au	1140	−0.52
Fe	1880	−0.43
Ni	1780	−1.20
Si	730	−0.10
Ag	895	−0.30
Ta	2150	−0.25
Ti	1650	−0.26

* From L. E. Murr, *Interfacial Phenomena in Metals and Alloys*
(Addison-Wesley, Reading, MA, 1975) [11]

Two new surfaces are created and twice the surface energy is added to the system. The
surface tension is

$$\gamma = \frac{E\,(\infty)}{2A} = \frac{1}{2}\int_0^{R_F}\frac{F_y}{A}dy \tag{3.35}$$

where $E\,(\infty)$ is the energy required to bring surfaces to infinity; F_y is the applied force
normal to the surfaces created; R_F is the range over which the applied force operates,
and A is the area.

$$\gamma = \frac{Y}{2}\int_0^{R_F}\frac{y}{a}dy = \frac{YR_F^2}{4a} \tag{3.36}$$

where we have expressed the stress F_y/A in terms of Young's modulus Y and the strain as
y/a, where a is the interatomic distance before applying force and y is the displacement
in the y direction. Young's modulus is the material constant that relates the stress or
the applied force/area, F_y/A, to the strain or the fractional change in length in the y
direction, y/a, of the solid; $F_y/A = Yy/a$. If we assume that the force is short range and
that $R_F \approx 10^{-8}$ cm,

$$\gamma = \frac{10^{-16}Y}{4a}$$

As shown in Table 3.4, Y is about the same value for many materials. Using $a = 0.25$ nm
and $Y = 10^{12}$ dyne/cm^2, we obtain the canonical number of 1000 erg/cm^2 for γ.

Since 1 eV $= 1.602 \times 10^{-12}$ erg, we make the following conversion from surface
energy per unit area to surface energy per surface atom:

$$1000\frac{\text{erg}}{\text{cm}^2} = 1000 \times \frac{(1/1.602) \times 10^{12}\text{eV}}{10^{15}(a^2/\text{cm}^2)\text{cm}^2} = 0.6\frac{\text{eV}}{a^2}$$

Table 3.4. Values of Young's modulus and γ_{LV}

Material	Y (dyne/cm^2)	γ_{LV} (erg/cm^2)
Aluminum	6.0×10^{11}	866
Gold	7.8×10^{11}	1410
Iron (cast)	9.1×10^{11}	1880
Tantalum	18.6×10^{11}	2150

where we have taken "a" to be atomic diameter and there are 10^{15} atom/cm^2. For reference, it is known that on a Si (111) surface, each surface atom has one broken bond. Hence, we can assume that the surface energy per atom of Si (111) surface $1\,eV/a^2$ or about 1700 erg/cm^2.

The mechanical approach connects γ to modulus, a bulk property of materials. The strain associated with Young's modulus is a valid concept for small displacements, but we have used it for an enormous displacement. Hooke's law for small displacements, which is the basis for the elasticity theory of solids, is violated here. The force between two atoms is essentially zero when we get to distances R_f the order of 1 Å. The physical meaning of R_f, the dissociation distance in a solid, is that beyond R_f the force needed to separate two atoms decreases. A derivation of R_f will be given in Chapter 6.

The interesting dependence of surface energy γ on Young's modulus Y is given in Table 3.4. In general, surface energies scale approximately with Young's modulus. It is not exact, by factors of six or seven, but it is a way of connecting the bulk properties to the surface properties of the material. To estimate a surface energy, we might first look at Young's modulus, which is tabulated for almost every material known.

Another way to look at the connection between Young's modulus and surface energy is to see that both of them are linearly related to the interatomic potential energy discussed in Section 3.1. Briefly, let us consider the bulk modulus K, which is similar to Young's modulus and is defined as

$$\Delta p = -K \frac{\Delta V}{V} \tag{3.37}$$

where ΔV is the volume change upon a pressure change of Δp in compression.

If the compression is carried out adiabatically (i.e. $dQ = 0$), we can use the first law of thermodynamics at constant entropy to obtain $p = (dE/dV)_s$, and

$$K = -V \left(\frac{dp}{dV} \right) = V \left(\frac{d^2E}{dV^2} \right)_s \tag{3.38}$$

To evaluate the second derivative of E,

$$\frac{dE}{dV} = \frac{dE}{dr} \frac{dr}{dV}$$

$$\frac{d^2E}{dV^2} = \frac{d^2E}{dr^2} \left(\frac{dr}{dV} \right)^2 + \frac{dE}{dr} \frac{d^2r}{dV^2} \tag{3.39}$$

If we take volume $V = N_A r^3$ and binding energy $E(r) = \frac{1}{2} n_c N_A \phi(r)$ where $\phi(r)$ is the interatomic potential function, and if we also assume that $\phi(r)$ obeys the Lennard-Jones potential as given by Eq. (3.7), we obtain

$$\left(\frac{dr}{dV}\right)^2 = \left(\frac{1}{3N_A r^2}\right)^2 = \frac{1}{9N_A a_0 V}\bigg|_{r=a_0}$$

$$\frac{dE}{dr} = \frac{1}{2}n_c N_A \frac{d\phi(r)}{dr}\bigg|_{r=a_0} = 0 \tag{3.40}$$

$$\frac{d^2E}{dr^2} = \frac{1}{2}n_c N_A \frac{d^2\phi(r)}{dr^2}\bigg|_{r=a_0} = \frac{36 n_c N_A \varepsilon_b}{a_0^2}$$

Hence,

$$K = \frac{4 n_c N_A \varepsilon_b}{V} = \frac{8\Delta E_s}{V} \tag{3.41}$$

where ΔE_s is the latent heat of sublimation as given by Eq. (3.11). Thus, there is a linear relation between K and ε_b as shown by Eq. (3.41). We have already shown that surface energy is proportional to ε_b.

3.9.3 Atomic approach

Neither the thermodynamic nor the mechanical approach shows specifically the crystallographic orientation dependence of the surface energy. A third way to think about the calculation of surface energy which reveals this orientation dependence focuses on the interaction between atoms. Consider a bulk array of atoms with a pairwise potential energy ϕ. The potential energy represents the bonding energy of an atom to another atom in the solid. Then, we define

$$\gamma_0 = \sum_{k \neq 1} \frac{\phi_{kl}}{2A} \tag{3.42}$$

where the quantity of γ_0 is the total binding energy/area of a bulklike surface atom, and A is the surface area. For a simple cubic crystal, each atom has six nearest neighbors, twelve second-nearest neighbors, eight third-nearest neighbors, and so on. The energy required to remove an atom from the bulk is then

$$\phi_0 = 6\phi_1 + 12\phi_2 + 8\phi_3 + K \tag{3.43}$$

If bonds are broken, there is an *increase* in the potential energy (a decrease in the binding energy) of the atom. The surface energy is the excess in potential energy over a bulklike atom. The energy ϕ_0 is related to the sublimation energy, ΔE_s, which is equal to $\phi_0/2$. In short-range approximation, we take $\phi_0 = 6\phi_1$.

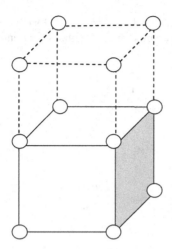

Fig. 3.8   Schematic diagram of a simple cubic cleaved along the (100) plane.

If we take a simple cubic and cleave it along the (100) plane (Fig. 3.8), the bond is broken for only one nearest neighbor, four second-nearest neighbors, four third-nearest neighbors, and so on. This modifies Eq. (3.43) when applied to a surface and it shows that surface energies are not equal to bulk-binding energy ϕ_0. Nearest neighbors are a lattice constant away, second-nearest neighbors are $\sqrt{2}$ lattice constants away. Thus,

$$\gamma_{100} = (\phi_1 + 4\phi_2 + 4\phi_3 + K)/a_0^2 \qquad (3.44)$$

is the difference between bulk bindings and surface bindings and a_0^2 is the area/surface atom.

The ratio R is

$$R = \frac{\gamma_{001}}{\gamma_0} = \frac{\phi_1 + 4\phi_2 + 4\phi_3 + K}{6\phi_1 + 12\phi_2 + 8\phi_3} \simeq \frac{\phi_1}{6\phi_1} \qquad (3.45)$$

It is clear that different crystallographic surfaces have different surface energies. R cannot be evaluated without a specific surface potential. For illustrative purposes we take the Lennard-Jones potential, Eq. (3.7), to calculate R. We can show that $R_{111} < R_{001} < R_{011}$, and so on. When the surface energy is plotted against crystallographic orientation, we have the well-known Wuff plot. For semiconductors, there is no potential as simple as that given by Lennard-Jones. It is assumed that the surface energies are proportional to the number of unpaired bonds/cm^2. For a semiconductor such as Ge, this corresponds to

$$Ge(100) = 1.25 \times 10^{15} \text{bond/cm}^2$$

where there are two dangling bonds per surface atom and to

$$Ge(111) = 0.72 \times 10^{15} \text{bond/cm}^2$$

where there is one dangling bond per surface atom. The (111) surface is the lowest-energy surface of the principal surfaces and, in general, is the lowest surface energy in germanium. The first-nearest-neighbor bonds are by far the dominant contribution because of short-range interaction.

3.10 Surface structure

3.10.1 Crystallography and notation

Here, we summarize the crystal structure of solids so that the number of atoms per cm^2, N_s, on a surface and the height, h, of a monolayer can be determined. The atomic volume can be calculated without the use of a unit cell of crystallography. The atomic density, n atom/cm^3, is given by

$$n = \frac{N_A \rho}{A} \tag{3.46}$$

where N_A is Avogadro's number, ρ is the mass density in gram/cm^3, and A is the atomic mass (the number of protons and neutrons). For most common electronic materials, the atomic density varies between 4 and 9×10^{22} atom/cm^3. The semiconductors Si and GaAs have atomic densities of 5.0 and 4.4×10^{22} atom/cm^3, respectively; the metal Al has atomic density around 6×10^{22} while metals such as Co, Ni, and Cu have densities approximately 9×10^{22} atom/cm^3. The atomic volume Ω is given by

$$\Omega = l/n \tag{3.47}$$

with a typical value of 20×10^{-24} $cm^3 = 0.02$ nm^3; see Section 2.3.

A crystal is composed of atoms arranged in a pattern periodic in space and is defined by a set of lattice points. The space containing this set of points can be divided by three sets of planes into a set of cells each identical in size, shape, and orientation; such a cell is called a unit cell. This cell can be described by three unit vectors $\mathbf{a}$, $\mathbf{b}$, and $\mathbf{c}$, called crystallographic axes, which are related to each other in terms of their lengths a, b, and c, and the angles α, β, γ (see Fig. 3.9). Any direction in the cell can be described as a linear combination of the three axes:

$$r = n_1 \mathbf{a} + n_2 \mathbf{b} + n_3 \mathbf{c}. \tag{3.48}$$

where n_1, n_2, n_3 are integers.

Seven different cells are necessary to describe all possible point lattices. These define the seven crystal systems, shown in Fig. 3.9. Each corner of the unit cell of these seven systems has a lattice point. It is possible to place more lattice points either in the center of the unit cell or on the cell faces without violating the general definition of a lattice point. Based on this arrangement of points, a total of 14 Bravais lattices can be produced for

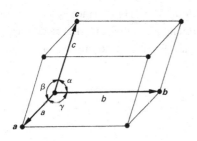

System	Axial Lengths and Angles	Bravais Lattice
Cubic	Three equal axes at right angles, $a = b = c$, $\alpha = \beta = \gamma = 90°$	Simple Body-centered Face-centered
Tetragonal	Three axes at right angles, two equal, $a = b \neq c$, $\alpha = \beta = \gamma = 90°$	Simple Body-centered
Orthorhombic	Three unequal axes at right angles, $a \neq b \neq c$, $\alpha = \beta = \gamma = 90°$	Simple Body-centered Base-centered Face-centered
Rhombohedral*	Three equal axes, equally inclined, $a = b = c$, $\alpha = \beta = \gamma \neq 90°$	Simple
Hexagonal	Two equal coplanar axes at 120°, third axis at right angles $a = b \neq c$, $\alpha = \beta = 90°$, $\gamma = 120°$	Simple
Monoclinic	Three unequal axes, one pair not at right angles, $a \neq b \neq c$, $\alpha = \gamma = 90° \neq \beta$	Simple Base-centered
Triclinic	Three unequal axes, unequally inclined and none at right angles, $a \neq b \neq c$, $\alpha \neq \beta \neq \gamma \neq 90°$	Simple

*Also called trigonal

Fig. 3.9   Unit cell, seven crystal systems, and 14 Bravais lattices.

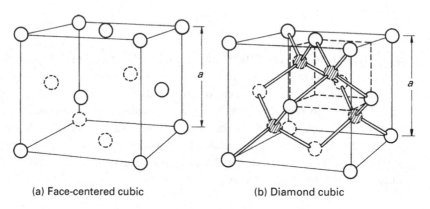

(a) Face-centered cubic   (b) Diamond cubic

Fig. 3.10   The fcc lattice and the diamond lattice of silicon with lattice parameter a. The shading of the atoms denotes Ga atoms in the zinc-blended structure of GaAs.

the seven crystal systems. Fig. 3.10 shows the fcc and the associated diamond structure of silicon and germanium. The number of lattice points in a unit cell, n_u, is given by

$$n_u = n_i + n_f/2 + n_c/8 \tag{3.49}$$

where n_i is the number of points in the interior, n_f is the number of points on faces (each n_f is shared by two cells), and n_c is the number of points on corners (each n_c point is shared by eight cells).

In considering the Bravais lattice of the fcc lattice shown in Fig. 3.10, it is customary to use the conventional unit cell of the fcc cell rather than the primitive cell. Since it is a cubic system ($a = b = c, \alpha = \beta = \gamma = 90°$), the length a is called the lattice parameter. The number of atoms per unit cell with $n_i = 0, n_f = 6$, and $n_c = 8$ is

$$n_u = 4 \text{ atom/unit cell}$$

We can calculate the atomic volume, Ω, by dividing the unit cell volume by the number of atoms in the unit cell as discussed in Section 2.3,

$$\Omega = \frac{a^3}{n_u} \tag{3.50}$$

A number of commonly used metals in thin-film technology belong to the fcc crystal structure, such as Al ($a = 0.405$ nm) and Cu ($a = 0.365$ nm). The height of a monolayer on the top surface of the fcc structure in Fig. 3.10 is $a/2$.

Many semiconductors have a diamond cubic structure, which is not one of the Bravais lattices. The diamond structure can be considered to be two interpenetrating fcc lattices with two atoms associated with one lattice point. The crystal structure of Si is diamond cubic with a lattice parameter $a = 0.543$ nm at room temperature. The number of atom/unit cell of the diamond lattice is given by $n_i = 4, n_f = 6$, and $n_c = 8$, so that

$$n_u = 4 + 6/2 + 8/8 = 8 \text{ atom/unit cell} \tag{3.51}$$

The height h of a monolayer of atoms on the Si surface of Fig. 3.10 is $a/4$. The number of atom/cm^3, n, is given by

$$n = \frac{n_u}{a^3} \frac{\text{atom}}{\text{unit cell}} \tag{3.52}$$

For Si, $n = 8/(0.543 \times 10^{-7})^3 = 5 \times 10^{22}$ atom/cm^3: Eq. (3.52) and Eq. (3.46) give the same result. To determine the number N_s of atom/cm^2 on a crystal surface, the number of surface atom/unit cell for a given orientation is divided by the surface cell area. For example, the surface of the Si lattice, Fig. 3.10, has an area a^2 and 2 atom/unit cell (1 center atom and 4 corner atoms shared with four adjacent cells so that each contributes 1/4). Then, $N_s = 2/a^2$. The values of N_s and h depend on the crystal structure and the orientation of the surface plane.

The III–V compounds such as GaAs and AlSb and the II–VI compounds such as ZnS have a crystal structure very similar to the diamond lattice, the zinc blende structure. In the zinc blende structure, the interior atoms of the cubic unit cell are different from those at the corners; other than that, the zinc blende structure is identical in atom position and stacking sequence to the diamond lattice. In Fig. 3.10, one of the fcc sublattices of zinc blende consists of shaded atoms of one element (e.g. Ga) and the other fcc sublattice consists of atoms of a different element (e.g. As).

Table 3.5. Conventions used to indicate directions and planes in crystallographic systems

A. **Directions: line from origin to point at *u, v, w***
    1. Specific directions are given in brackets [*uvw*].
    2. Indices *uvw* are the set of smallest integers. [$\frac{1}{2}$ $\frac{1}{2}$ 1] goes to [112].
    3. Negative indices are written with a bar.
    4. Directions related by symmetry are given by <*uvw*>.

B. **Planes: plane that intercepts axes at 1/*h*, 1/*k*, 1/*l***
    1. Orientation is given by parentheses (*hkl*).
    2. *hkl* are Miller indices.
    3. Negative indices are written with a bar.
    4. Planes related by symmetry are given by {hkl}.

C. **In cubic systems: bcc, fcc, diamond**
    1. Direction [*hkl*] is perpendicular to plane (*hkl*).
    2. Interplanar spacing: $d_{hkl} = a/\sqrt{h^2 + k^2 + l^2}$

3.10.2 Directions and planes

The direction of any line in a lattice may be described by drawing a line through the origin parallel to the given line and then assigning the coordinates of any point on the line. If the line goes through the origin and the point with coordinates u, v, w, where these numbers are not necessarily integers, the directions [*uvw*], written in brackets, are the indices of the direction of the line. Since this line also goes through $2u, 2v, 2w$, and $3u, 3v, 3w$, and so on, it is customary to convert u, v, w, to a set of smallest integers.

The orientation of planes in a lattice can also be defined by a set of numbers called the Miller indices. We can define the Miller indices of a plane as the reciprocals of the fractional intercepts that the plane makes with the crystallographic axes. Table 3.5 gives conventions used to indicate directions and planes used in crystallographic systems. In these conventions, if the plane is parallel to the axis, the intercept is taken to be at infinity, the reciprocal of ∞ is 0 and is the number used to designate the plane.

Fig. 3.11 shows the Miller indices of the three lattice planes most referred to in electronic materials technology. The planes are conventionally written as (*hkl*). The plane (*hkl*) is parallel to the ($\bar{h}\,\bar{k}\,\bar{l}$) plane, which is on the opposite side of the origin.

3.10.3 Surface reconstruction

The structure of the surface layer of atoms generally differs from that of the crystal substrate. The surface can reconstruct (Fig. 3.12) so that the surface atoms can share bonds. This reconstruction results in a two-dimensional symmetry with a periodicity that differs from that of the underlying atoms of the crystal. The surface atoms can also seek new equilibrium positions (called relaxation) that change the interlayer distance between the first and second layer of atoms. Relaxation changes the bond angles but not the number of nearest neighbors, nor the rotational symmetry of the surface atoms as

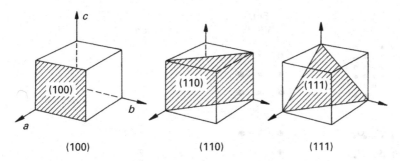

Fig. 3.11   Miller indices of the three principal planes in the cubic structure.

Fig. 3.12   Surface reconstruction on the Si (100) surface.

compared to the underlying atoms. Surface structure can be revealed directly by scanning tunneling microscopy (STM).

However, the notation used to describe surface structure is based on low-energy electron diffraction (LEED). In LEED, electrons of well-defined energy in the range of 10 to 500 eV and directions of propagation diffract off the crystal surface. The low-energy electrons are scattered mainly by individual atoms on the surface and produce a pattern of spots on the fluorescent observation screen because of wave interference. The spots in the pattern correspond to the points of the two-dimensional reciprocal lattice of the repetitive surface structure. As in any diffraction experiment, the LEED pattern is a reciprocal map of the surface periodicity as determined by the size and orientation of the surface unit cell. Although the assignment of atom positions on the basis of LEED patterns is not unique, it is possible to predict the symmetry of a LEED pattern from the real space configuration of atoms. Fig. 3.13 gives examples of overlayers on the (100) surface of a cubic crystal. The letter p in the figure indicates that the unit cell is primitive and the LEED pattern for $p(2 \times 2)$ has extra, half-order spots. The letter c indicates that the unit has an additional scatter in the center which gives rise to 1/2, 1/2 spots in the diffraction pattern.

In general, changes in the periodicity of the surface will result in changes in the diffraction pattern that are easily observable and interpretable in terms of the new two-dimensional symmetry. Such changes are often observed, for example, when gases are adsorbed on crystal surfaces. Electron diffraction is routinely used to evaluate surface cleanliness in ultrahigh vacuum chambers where contamination-free substrates are a prerequisite for subsequent crystal growth.

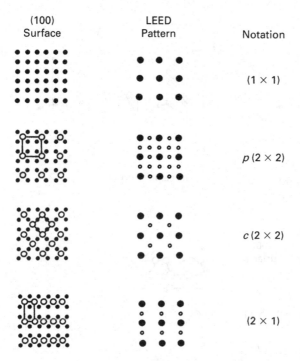

(100) Surface	LEED Pattern	Notation
		(1×1)
		$p\,(2 \times 2)$
		$c\,(2 \times 2)$
		(2×1)

Fig. 3.13   (100) surface of a cubic crystal with different atom configurations and the associated LEED patterns in reciprocal space. The notation for the patterns is indicated.

References

[1] B. H. Flowers and E. Mendoza, *Properties of Matter* (Wiley, New York, 1970).

[2] D. L. Goodstein, *States of Matter* (Prentice-Hall, Englewood Cliffs, NJ, 1975).

[3] P. Haasen, *Physical Metallurgy* (Cambridge University Press, Cambridge, 1978).

[4] C. Kittel, *Introduction to Solid State Physics*, 6th edn (Wiley, New York, 1986).

[5] C. Kittel and H. Kroemer, *Thermal Physics*, 2nd edn (W. H. Freeman, New York, 1980).

[6] A. Guinier and R. Jullien, *The Solid State* (Oxford University Press, Oxford, 1989).

[7] A. B. Pippard, *The Elements of Classical Thermodynamics* (Cambridge University Press, Cambridge, 1966).

[8] R. E. Hummel, *Electronic Properties of Materials*, 3rd edn (Springer, New York, 2001).

[9] W. D. Callister, Jr., *Materials Science and Engineering an Introduction*, 5th edn (Wiley, New York, 2000).

[10] H. Udin, "Measurement of solid/gas and solid/liquid interfacial energies", in *Metal Interfaces* (American Society for Metals, Cleveland, OH, 1952).

[11] L. E. Murr, *Interfacial Phenomena in Metals and Alloys* (Addison-Wesley, Reading, MA, 1975).

[12] G. A. Somorjai, *Chemistry in Two Dimensions: Surfaces* (Cornell University Press, Ithaca, NY, 1981).

[13] A. W. Adamson, *Physical Chemistry of Surfaces*, 4th edn (Wiley, Newark, NJ, 1982).

[14] A. Zangwill, *Physics at Surfaces* (Cambridge University Press, Cambridge, 1988).

[15] A. P. Sutton and R. W. Balluffi, *Interfaces in Crystalline Materials* (Oxford University Press, Oxford, 1995).

[16] J. A. Venables, *Introduction to Surface and Thin Film Processes* (Cambridge University Press, Cambridge, 2000).

Problems

3.1   What is the relationship between thermal expansion and anharmonicity?

3.2   Germanium is a diamond lattice with atomic density of $4.4.2 \times 10^{22}$ atom/cm^2, while aluminum is a fcc lattice with atomic density of 6.02×10^{22} atom/cm^3.
(a) Calculate the lattice parameter a for both.
(b) Determine the number of atom/cm^2 on the (100) surface for both.
(c) What is the height of a monolayer in terms of parameter a for both?
(d) What is the atomic volume for both in terms of the lattice parameter a?

3.3   What is the surface density of atoms, N_s, in terms of the lattice parameter a for (111) planes in the diamond lattice?

3.4   Si has a lattice parameter of $a = 0.543$ nm. On the (100) surface of Si, what is the surface density of atoms and what is the miscut angle $\Delta\theta$ to produce a monolayer-high step h every 50 nm? What would the angle be in units of lattice parameter a if the structure were a fcc or a simple cubic lattice? Use the relationship of step length $L_0 = h/\tan\theta$.

3.5   Nickel is a fcc metal with atomic density of 9.14×10^{22} atom/cm^3, atomic weight $= 58.73$ g, and density $= 8.91$ g/cm^3.
(a) What is the lattice parameter a, height h of a monolayer, and the atomic volume Ω?
(b) What is the number of atom/cm^2 on the (110) plane in terms of the lattice parameter a?

3.6   (a) What is the interplanar spacing d for the (100), (110), and (111) planes of Al which has fcc lattice with a lattice parameter of $a = 0.405$ nm?
(b) What are the Miller indices of a planes that intercepts the x-axis at a, the y-axis at $2a$ and the z-axis at $2a$?

3.7   (a) Calculate the surface tension of γ of Ni ($T_m = 1453\,°$C) at 300 K.
(b) Would Ni deposited on Si or Si deposited on Ni form clusters? In both cases a uniform adherent layer is formed. Suggest an explanation.
(c) Ag ($T_m = 962°$C) wets Cu. Compared to the SnBi solder discussed in the text, how far up a Cu hole ($d = 0.5$ mm) would an Ag solder be pulled (contact angle $\theta = 40°$ and density of Ag $= 10.5$ g/cm^3)?

3.8   Plot the Lennard-Jones potential from $r = 0.8a_0$ to $r = 4a_0$ for Au where $\varepsilon_b = 0.47$ eV/atom.

3.9   For an Au fcc metal with 6.0×10^{22} atom/cm^3 and a (100) surface energy of 0.5 eV/atom, calculate
(a) the latent heat of sublimation ΔE_s, and
(b) the interatomic potential energy ε_b.

3.10   What is the surface tension of benzene in the liquid state if in a liquid column measurement, the column height $h = 1.2 \times 10^{-2}$ m, the contact angle $\theta = 0°$, $r = 5 \times 10^{-14}$ m and the benzene density $= 800$ kg/m^3 (neglect the influence of air and its density). Compare your answer with that of Table 3.2.

3.11   For an fcc metal, determine the ratio $\gamma_{111}/\gamma_{100}$ of the surface energies on the (111) and (100) surfaces by considering the first (ϕ_1) and second (ϕ_2) nearest-neighbor bond energies and assuming that $\phi_2 = \frac{1}{2}\phi_1$.

3.12   Two elastically isotropic fcc materials A and B have the properties shown in the table below.

	a (nm)	Y (10^{12} dyne/cm^2)
A	0.566	1.03
B	0.543	1.30

(a) Calculate the surface energies γ, assuming that $R_F = a$.
(b) Which material wets the other?

3.13   Using the values of γ in Table 3.2, find the following for a drop of liquid mercury (Hg) on solid sodium chloride (NaCl):
(a) the surface tension γ given a contact angle θ of 80°;
(b) the maximum allowable surface tension and minimum allowable contact angle.
     Compare your answers to (b) with those to (a).

3.14   If you have a liquid droplet on a substrate surface, what will be the range of contact angles in the following?
(a) $\gamma_{LV} > \gamma_{SV} > \gamma_{SL}$
(b) $\gamma_{LV} > \gamma_{SV} = \gamma_{SL}$
(c) $\gamma_{LV} = \gamma_{SV} < \gamma_{SL}$

3.15   For Cu, a fcc structure with density of 8.93 g/cm^3 and atomic mass of 63.55, determine the atomic density n and lattice parameter a. Using the values in Table 2.1, determine the solid–vapor surface tension and heat of sublimation in terms of eV/atom for the (100) surface.

3.16   For zero-creep measurements, what is the length of a 0.01 cm radius Au wire, density 19.3 g/cm^3? Use the value of surface energy in Table 2.1 and ignore the influence of grain boundaries.

3.17   Consider the "spherical cap" droplet with surface tensions at the edge of the cap as shown in Fig. 3.4.
(a) Show that the total surface energy of the system can be expressed as

$$E_T = (\gamma_{SL} - \gamma_{SV})\left[\frac{6V - \pi h^3}{3h}\right] + \gamma_{LV}\left[\frac{2V}{h} + \frac{2\pi h^2}{3}\right] + \gamma_{SV} A$$

In this expression the volume of the droplet, $V = \pi/6(h^3 + 3ha^2)$ or $V = (\pi h^2/3)(3R - h)$ and A is the total area of the slab. The surface area of such a spherical cap is $S = 2\pi Rh$ and $R = (a^2 + h^2)/2h$.

(b) Show that the surface energy minimization condition, $dE_T/dh = 0$ (holding V and A constant), results in

$$h^3 = \frac{3V}{\pi} \frac{(\gamma_{SL} + \gamma_{LV} - \gamma_{SV})}{(-\gamma_{SL} + 2\gamma_{LV} + \gamma_{SV})}$$

4 Atomic diffusion in solids

4.1 Introduction

Atomic diffusion or atomic rearrangement in thin films is the basic kinetic process in microelectronic device manufacturing and reliability. Pure Si is not useful until we can diffuse electrically active n-type and p-type dopants into it. In fact, the fundamental behavior of a transistor, i.e. the p–n junction in silicon, is obtained by a non-uniform distribution of both n- and p-type dopants in Si in order to achieve the built-in potential which guides the motion of electrons and holes in the transistor. Thus, the diffusion of dopants in Si has been a very important subject in microelectronics, both in device characteristics and in device manufacturing. Indeed, there are some very sophisticated programs to simulate and to analyze the dopant diffusion profile in junction formation in Si devices.

In classical metallurgy, a blacksmith inserts a bar of iron into a charcoal furnace to allow the gas phase of carbon to diffuse into the iron. The diffusion time is typically short, just several minutes' heating in the furnace, so the blacksmith has to take out the red-hot bar and hammer it in order to homogenize the carbon in the bar. This process of "heat and beat" is to diffuse and to redistribute carbon in iron to make the Fe–C alloy.

In this chapter, we shall connect microscopic atomic jumps in a crystalline lattice to macroscopic diffusion behavior as described by the Fick's first and second laws. To analyze atomic diffusion in solids, we shall consider the vacancy mechanism of diffusion in a fcc metal [1–11]. We make the following assumptions in order to develop the analytical model.

(1) It is a thermally activated unimolecular process. By unimolecular, we mean that we consider the jumping of a single atom in the diffusion process, unlike both the gas phase diffusion in which it occurs by the collision of two molecules, and also the chemical reactions in liquid solutions, which are bimolecular processes, such as rock-salt formation, in which the collision of two atoms of Na and Cl are involved.

(2) It is a defect-mediated process. Here, the defect is a vacancy or a vacant lattice site. Note that vacancy is a thermal equilibrium defect, thus there is always an equilibrium vacancy concentration in the solid to facilitate atomic diffusion. Atomic diffusion occurs by the exchange of lattice site positions between a vacancy and one of its nearest-neighbor atoms. Vacancy concentration is given in Appendix B.

(3) The thermally activated process requires activation energy to pass through an activated state. The distribution of the activated state is assumed to obey Boltzmann's equilibrium distribution function on the basis of transition state theory. Hence, the Boltzmann distribution function is used to represent the activated states.

(4) It is assumed that the probability of reverse jumps is large, due to a small driving force in solid-state diffusion, so we have to consider reverse processes. After an atom and a vacancy have exchanged positions, the probability of them exchanging back is high. In other words, the process is near-equilibrium or not far from the equilibrium state.

(5) Statistically, atomic diffusion obeys the principle of random walk.

(6) A long-range directional diffusion of atoms requires a driving force.

4.2 Jump frequency and diffusional flux

Fig. 4.1 is a sketch of a two-dimensional arrangement of a square lattice. Several diffusion mechanisms are depicted; (a) an atom diffuses by jumping into a neighboring vacant lattice site, (b) an interstitial atom goes to a neighboring interstitial site, (c) an interstitial atom pushes an atom from its lattice site to an interstitial site, (d) two neighboring atoms swap position directly, and (e) there is a ring rotation of four atoms. The last two mechanisms, (d) and (e), are energetically too high to occur and we will not consider them. In the following, we shall consider the vacancy mechanism in (a), since it is the most common one in metals diffusion and most experimental data have been obtained for theoretical analysis and understanding.

To quantify the diffusion picture, we consider a simple one-dimensional case and represent the minimum potential energy of a row of atoms by repeating the pair potential curves between every two atoms as shown in Fig. 4.2(a). Recall that in Chapter 2

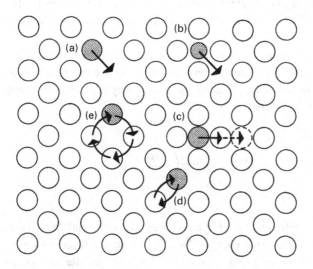

Fig. 4.1   Sketch of a two-dimensional arrangement of a square lattice.

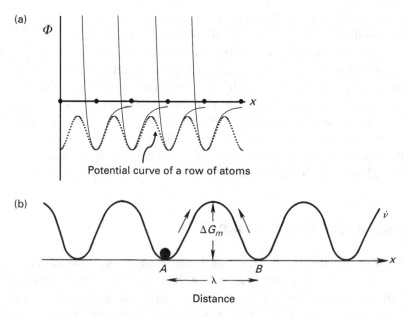

(a) Φ

Potential curve of a row of atoms

(b)

ΔG_m

A   B

λ

Distance

Fig. 4.2   (a) Schematic diagram of a simple one-dimensional case representing the pair potential energy of a row of atoms by repeating the pair potential curves between every two atoms as shown in Fig. 3.1(a). (b) For the row of atoms, we obtain the net periodic potential energy curve as shown by the dotted curve in (a).

we have discussed the pair potential energy curve between two atoms. For a row of atoms, if we draw the pair potential energy curve between every two atoms, we shall obtain the net periodic potential energy curve as shown schematically in Fig. 4.2(b). We assume at positions A and B, there is an atom and a vacancy, respectively, for diffusion consideration. The relaxation or the stress-free strain of atoms around a vacancy is ignored.

At equilibrium, the atom at A is attempting to jump over the potential energy barrier with the attempt frequency, v_0, to exchange position with the neighboring vacancy at B. The successful exchange or the exchange jump frequency is given below on the basis of Boltzmann's distribution function:

$$v = v_0 \exp\left(\frac{-\Delta G_m}{kT}\right) \tag{4.1}$$

where v_0 is the attempt frequency, v is the exchange frequency, and ΔG_m is the activation energy of motion (saddle point energy). Note that there is a reverse jump at the same frequency attempt. Such diffusion in an equilibrium state will lead to random walk of the vacancy in the lattice.

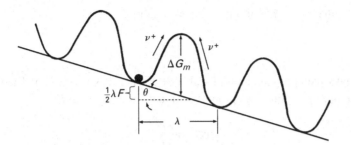

Fig. 4.3 To have a directional diffusion, we must introduce a driving force to drive the diffusion in a given direction. This can be represented by tilting the baseline of the potential energy. The tilting introduces a gradient of the potential energy, which is the driving force of diffusion.

To have a directional diffusion, we must introduce a driving force to drive the diffusion in a given direction. This can be represented by tilting the baseline of the potential energy as shown in Fig. 4.3. The tilting introduces a gradient of the potential energy, so the atom will have a higher tendency to jump from left to right. Recall that force is given as a potential gradient or the slope of the potential energy curve,

$$F = -\frac{\Delta\mu}{\Delta x} = \tan\theta$$

where $\tan\theta$ is the slope of the baseline and is equal to $\Delta\mu/\Delta x$ as shown in Fig. 4.3. If we assume that the interatomic distance or the jump distance is λ, the tilting of an angle of θ has raised the potential energy by the amount of $\Delta\mu$ to help the atom to jump into the vacancy. By taking $\Delta x = \lambda/2$, we have

$$\Delta\mu = \Delta x(\tan\theta) = \frac{\lambda}{2}\tan\theta = \frac{\lambda F}{2}$$

In other words, under a driving force F (the physical meaning of F will be discussed) by tilting the baseline as shown, we can define the forward jump which is increased by

$$v^+ = v_0\exp\left(\frac{-\Delta G_m + \Delta\mu}{kT}\right) = v\exp\left(+\frac{\lambda F}{2kT}\right) \qquad (4.2)$$

The reverse jump is decreased by

$$v^- = v_0\exp\left(\frac{-\Delta G_m - \Delta\mu}{kT}\right) = v\exp\left(-\frac{\lambda F}{2kT}\right) \qquad (4.3)$$

and the net frequency is

$$v_n = v^+ - v^- = 2v\sinh\left(\frac{\lambda F}{2kT}\right) \qquad (4.4)$$

Now, we take the "condition of linearization", that is,

$$\frac{\lambda F}{kT} \ll 1$$

So we can take the approximation of sinh $x = x$ when x is very small. Then, the net frequency jump, v_n, is "linearly" proportional to the driving force F:

$$v_n = v\frac{\lambda F}{kT} \tag{4.5}$$

We now define a drift velocity, which is equal to jump frequency times jump distance. The physical implication of drift velocity is that under the driving force, all the equilibrium vacancies in the solid will exchange position with a neighboring atom at the net frequency in the direction of the force, resulting in a flux of atoms moving with velocity v in the direction of the force.

$$v = \lambda v_n = \frac{v\lambda^2}{kT}F \tag{4.6}$$

The atomic flux, J, is given as

$$J = Cv = \frac{Cv\lambda^2}{kT}F = CMF \tag{4.7}$$

The atomic flux J is "linearly" proportional to the driving force, F. Since $v = MF$, where $M = v\lambda^2/kT$, M is defined as atomic mobility. The unit of mobility is cm^2/J s, or cm^2/eV s, where e is electric charge and V is volt, and eV is electric energy. Since kT is energy, we can give kT the unit of joule or eV.

For comparison, in electric charge conduction, the unit of mobility of a charge is given as cm^2/V s, not cm^2/eV s. This is because in electrical conduction the charge carrier flux or current density $j = -\sigma(d\phi/dx)$, where σ is electrical conductivity and ϕ (or V) is electric potential. So the electric force is defined as $d\phi/dx$, rather than deϕ/dx. Thus in the mobility, the charge e is cancelled, so we have the unit of electric mobility as cm^2/V s.

4.3   Fick's first law (flux equation)

Recall that in a field the physical meaning of driving force is generally defined as a potential gradient,

$$F = -\frac{\partial \mu}{\partial x}$$

In atomic diffusion, μ is the chemical potential of an atom in the diffusion field and is defined at constant temperature and constant pressure to be

$$\mu = \left(\frac{\partial G}{\partial C}\right)_{T,p}$$

where G is Gibbs free energy and C is concentration. For an ideal dilute solid solution, we have (we shall derive this equation a little later)

$$\mu = kT \ln C$$

$$F = -\frac{\partial \mu}{\partial C}\frac{\partial C}{\partial x} = -\frac{kT}{C}\frac{\partial C}{\partial x}$$

$$J = \frac{Cv\lambda^2}{kT}F = \frac{Cv\lambda^2}{kT}\left(-\frac{kT}{C}\frac{\partial C}{\partial x}\right) = -v\lambda^2\left(\frac{\partial C}{\partial x}\right) = -D\left(\frac{\partial C}{\partial x}\right) \tag{4.8}$$

Hence, we have obtained Fick's first law of diffusion

$$\frac{J}{-(\partial C/\partial x)} = D = v\lambda^2$$

where D is the diffusion coefficient (or diffusivity) in units of cm^2/s. Then, $M = D/kT$. In the above, we connect the atomic jump to Fick's first law of diffusion and link the two flux equations of $J = Cv$, which has been discussed in Chapter 2, and $J = -D(dC/dx)$, Fick's first law.

Before we move on, we show below how the potential of an ideal dilute solution, $\mu = kT \ln C$, is obtained. In thermodynamics, an ideal solution is defined by mixing two elements in which the enthalpy of mixing is zero and the entropy of mixing is ideal, given as

$$\Delta H = 0 \text{ and } \Delta S = -k[C \ln C + (1 - C) \ln(1 - C)]$$

We have $\Delta G = \Delta H - T\Delta S$, so the potential is obtained when we take the differentiation of ΔG and take $C \ll 1$ because of the dilute solution,

$$\mu = \left(\frac{\partial \Delta G}{\partial C}\right)_{T,p} = kT \ln C \tag{4.9}$$

4.4   Diffusivity

In the above derivation, as depicted in Fig. 4.1, we have assumed that the diffusing atom has a neighboring vacancy. For the majority of atoms in the lattice, this is not true, because they do not have a vacancy as a nearest neighbor, so we must define the probability of an atom having a neighboring vacancy in the solid as

$$\frac{n_v}{n} = \exp\left(-\frac{\Delta G_f}{kT}\right) \tag{4.10}$$

where n_v is the total number of vacancies in the solid, n is the total number of lattice sites in the solid, and ΔG_f is the Gibbs free energy of formation of a vacancy. Since in a fcc metal, a lattice atom has 12 nearest neighbors, the probability for a particular atom to have a vacancy as a neighbor is

$$n_c \frac{n_v}{n} = n_c \exp\left(-\frac{\Delta G_f}{kT}\right)$$

where $n_c = 12$ is the number of nearest neighbors.

Next, we have to consider the correlation factor in the fcc lattice. The physical meaning of the correlation factor is the probability of reverse jump; after the atom has exchanged positions with a vacancy, it has a high probability of returning to its original position before the activated configuration is relaxed. The factor, f, has a range between zero and unity. When $f = 0$, it means that the probability of reverse jump is 100%, so the atom and the vacancy are exchanging positions back and forth, which will not lead to any random walk; it is a correlated walk. When $f = 1$, it means that after the jump, the atom will not return back to its original position, and it is a random walk because the next jump will depend on the random probability of a vacancy coming to the neighborhood of this atom, or the vacancy will exchange position randomly with one of its 12 nearest-neighbor atoms. In fcc metal, $f = 0.78$, so ~80% of jumps are random walk, and ~20% are correlated walk. Finally, the diffusivity is given as

$$D = f n_c \lambda^2 v_0 \exp\left(-\frac{\Delta G_m + \Delta G_f}{kT}\right) \tag{4.11}$$

$$D = f n_c \lambda^2 v_0 \exp\left(\frac{\Delta S_m + \Delta S_f}{k}\right) \exp\left(-\frac{\Delta H_m + \Delta H_f}{kT}\right) = D_0 \exp\left[-\frac{\Delta H}{kT}\right]$$

Thus, we have

$$D_0 = f n_c \lambda^2 v_0 \exp\left(\frac{\Delta S_m + \Delta S_f}{k}\right) \tag{4.12}$$

and

$$\Delta H = \Delta H_m + \Delta H_f$$

4.5   Fick's second law (continuity equation)

We derived Fick's first law or the flux equation, Eq. (4.8), under a constant driving force. That equation can describe diffusion phenomena under a constant chemical potential, or constant concentration gradient, and it has a constant flux, independent of time. However, in most diffusion problems, the atomic flux changes with position and time. Fick's first law cannot be used to describe diffusion where the flux varies with position and time, or where the flux has a changing driving force. A simple example is a drop of ink in water. It spreads out and eventually reaches homogenization in the water; the concentration and concentration gradient of ink changes with time and position. To handle such a non-steady-state problem, we need to derive the continuity equation from the "principle of

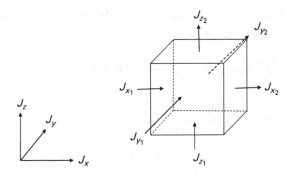

Fig. 4.4   Simple cube of six square surfaces; we consider six atomic fluxes passing through the six surfaces of the cube.

conservation of mass" and we need to use the Gauss theorem of flux divergence. We shall review the idea of flux divergence below.

We recall that in Chapter 2 we defined atomic flux as a number of atoms per unit area and unit time passing through a surface or deposited on a surface. Now we consider a simple cube of six square surfaces and we consider six atomic fluxes passing through the six surfaces of the cube, shown in Fig. 4.4. In a period Δt, the number of atoms entering (or leaving) one of the square surfaces, A_1, will be

$$\Delta N_1 = J_1 A_1 \Delta t$$

If we sum all the atomic fluxes, J_i, passing through each surface of area, A_i, for a given period of time, Δt, or we sum all the atoms in and out of the cubic volume, we have

$$\sum_{i=1}^{6} J_i A_i \Delta t = \sum_{i=1}^{6} \Delta N_i = \Delta N$$

where ΔN is the net change of total number of atoms inside the cubic box. If the volume of the cubic box is V and C is the concentration in the box, the change in concentration is

$$V \Delta C = \Delta N$$

From the last two equations, we obtain

$$\frac{dN}{dt} = V \frac{dC}{dt} = \sum_{i=1}^{6} J_i A_i$$

According to the Gauss theorem, the last term should equal the divergence inside the cubic volume, so we have

$$\sum_{i=1}^{6} J_i A_i = (\nabla \cdot J) V \qquad (4.13)$$

where V is the volume of the cubic. In general, the Gauss theorem applies to an arbitrary shape of a volume of V enclosed by an area of A,

$$\sum_i J_i A_i = \oint_A \vec{J} \cdot \vec{n} \, dA = (\nabla \cdot J)V \tag{4.14}$$

Combining the last three equations, we have

$$V\frac{dC}{dt} = \oint_a \vec{J} \cdot \vec{n} \, dA = (\nabla \cdot J)V \tag{4.15}$$

and so the well-known continuity equation is obtained,

$$\frac{dC}{dt} = \nabla \cdot J$$

Typically, there is a negative sign in the last equation. Whether we have a positive or negative sign depends on whether the net change of atoms inside the cubic volume is positive or negative. Conventionally, we assume that the in-flux is less than the out-flux, so the concentration inside the cubic decreases with time and we have a negative sign. Since the Gauss theorem assumes that the derivatives are continuous functions, the equation is called a continuity equation. The formal derivation of the continuity equation is given in Section 4.5.1.

There are actually three equations in Eq. (4.15). The first is the Gauss equation relating the second and the third terms. The second is the continuity equation relating the first and the third terms. The third is the growth equation relating the first and the second terms and it assumes conservation of mass, or the conservation of the number of atoms ($\Delta N = JA\Delta t$), or the conservation of the number of vacancies when we consider the growth of a void. We illustrate in Fig. 4.5 the interrelationship of the three equations.

Often, we encounter the growth equation in solid-state reactions such as the growth of a precipitate. This is because we used the small cube in Fig. 4.4 to consider flux divergence, which is best for gas states since the concentration can change within the cube. On the other hand, if the cube is a solid, e.g. a pure metal, the concentration of the cube is $C = 1/\Omega$ which is constant and cannot change, where Ω is atomic volume. If the concentration cannot change (except by alloying), the cube has to grow when a flux of atoms comes to it. So we need the growth equation, which will be discussed further in Chapter 5, Section 5.4.

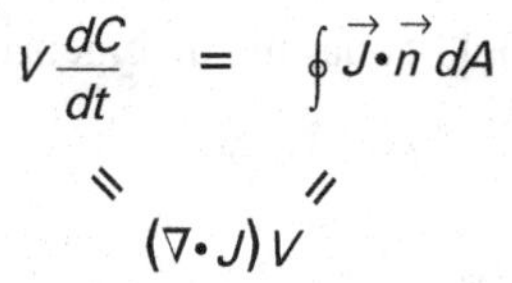

$$V\frac{dC}{dt} \quad = \quad \oint \vec{J} \cdot \vec{n} \, dA$$
$$\diagdown \qquad \diagup$$
$$(\nabla \cdot J)V$$

Fig. 4.5   Three equations that relate to each other in obtaining Fick's second law.

4.5.1 Derivation of the continuity equation

We use Cartesian coordinates to consider the fluxes going in and out of a cubic element, as shown in Fig. 4.6. If we use a vector to represent the flux, J, we can decompose the vector into three components, J_x, J_y, and J_z. The amount of material flowing into the cube through the surface x_1 of area of $dydz$ per unit time is equal to

$$\vec{J_{x1}} \cdot \vec{x_1} = J_{x1}x_1 \cos 180° = -J_{x1}dydz$$

Similarly, the amount of material flowing out of the cubic box through the surface x_2 (opposite to x_1) is equal to

$$\vec{J_{x2}} \cdot \vec{x_2} = \left(J_{x1} + \frac{\partial J_x}{\partial x}dx\right) x_2 \cos 0° = J_{x1}dydz + \frac{\partial J_x}{\partial x} dxdydz$$

If we add these together, the net flux out of the box in the x-direction becomes

$$(J_{x2} - J_{x1})dydz = \left(\frac{\partial J_x}{\partial x}\right) dxdydz$$

Note that we have used a continuous and differentiable function between x_1 and x_2, so that

$$J_{x2} = J_{x1} + \frac{\partial J_x}{\partial x}dx$$

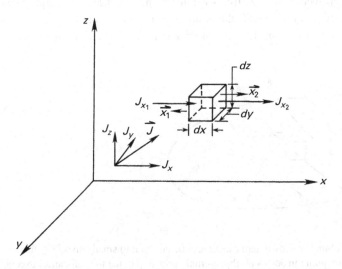

Fig. 4.6   Cartesian coordinates for considering the fluxes going in and out of a cubic element.

Furthermore, J_{x2} is larger than J_{x1}. As a consequence, more materials are flowing out of the cubic box. If we follow this approach in the y- and z-directions, we have

$$(J_{y2} - J_{y1})dxdz = \left(\frac{\partial J_y}{\partial x}\right)dxdydz$$

$$(J_{z2} - J_{z1})dxdy = \left(\frac{\partial J_z}{\partial x}\right)dxdydz$$

If we sum all of them together, we have

$$\sum_{i=1}^{6} J_i A_i = \left(\frac{\partial J_x}{\partial x} + \frac{\partial J_y}{\partial y} + \frac{\partial J_z}{\partial z}\right)dV \tag{4.16}$$

where $dV = dxdydz$. Now, instead of a cube, consider an arbitrary volume. An arbitrary volume can always be cut up into small cubes, as shown in Fig. 4.7, and the flux going out is equal to the flux going in across all the internal surfaces, so that they all cancel. The only fluxes we have to consider are those on the outer surface. For an arbitrary volume V bounded by area A, the summation can be expressed as an integral,

$$\int_A J \cdot n dA = (\nabla \cdot J) V \tag{4.17}$$

This is the well-known Gauss theorem, and the right-hand side is defined as the divergence of the flux, where

$$\nabla \cdot J = \frac{\partial J_x}{\partial x} + \frac{\partial J_y}{\partial y} + \frac{\partial J_z}{\partial z}$$

This quantity (divergence of J) times V must by mass conservation equal the change in the total number of atoms inside the volume. It needs a negative sign, since we derived this quantity when the out-flux was greater than the in-flux and there was no source

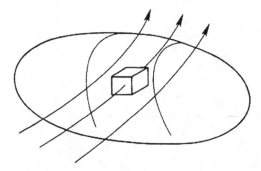

Fig. 4.7   An arbitrary volume as shown here can always be cut up into small cubes. The flux going out is equal to the flux going in across all the internal surfaces, so that they all cancel except those in and out of the surface of the volume.

within the volume; we lose material from this volume. Therefore, it must equal the time rate of decrease of concentration since $C = N/V$. This is the continuity equation in differential form,

$$-(\nabla \cdot J) = \frac{\partial C}{\partial t} \tag{4.18}$$

It describes a non-steady-state flux flow. It is a very well-known equation in fluid mechanics, as well as here in diffusion. Since we have already derived the flux equation in the one-dimensional case as

$$J = -D\frac{\partial C}{\partial x}$$

then

$$\frac{\partial C}{\partial t} = \frac{\partial}{\partial x}D\left(\frac{\partial C}{\partial x}\right)$$

and if D is independent of position, we have

$$\frac{\partial C}{\partial t} = D\left(\frac{\partial^2 C}{\partial x^2}\right) \tag{4.19}$$

This is Fick's second law of diffusion in one dimension.

4.6   A solution of the diffusion equation

As mentioned in Section 4.5, an example of three-dimensional diffusion is the case of a drop of ink spreading out in water. A two-dimensional example would be a drop of gasoline that spreads out on the surface of water. Here we consider a one-dimensional problem using Eq. (4.19). We take a long rod of a pure metal and place a small amount of its isotope, a tracer, on the end surface, as shown in Fig. 4.8(a). We shall determine how the isotope diffuses into the rod as a function of time and temperature. The standard method of setting up the problem is to put a reflecting mirror at the end surface, and thus change the problem into a symmetrical one as shown in Fig. 4.8(b). The solution is then a standard one,

$$C(x,t) = \frac{Q}{(\pi Dt)^{1/2}} \exp\left(\frac{-x^2}{4Dt}\right) \tag{4.20}$$

The constant Q satisfies the following boundary condition for a fixed amount of the tracer material. The initial conditions of the problem are that

$$\text{at } x = 0, C \rightarrow Q \text{ as } t \rightarrow 0$$
$$\text{for } |x| > 0, C \rightarrow 0 \text{ as } t \rightarrow 0$$

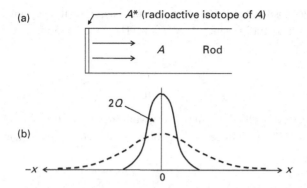

Fig. 4.8   (a) Schematic diagram of a long rod of a pure metal with a small amount of its isotope, a tracer, on the end surface. We shall determine how the isotope diffuses into the rod as a function of time and temperature. (b) The standard method of setting up the problem is to put a reflecting mirror at the end surface, and thus change the problem into a symmetrical one.

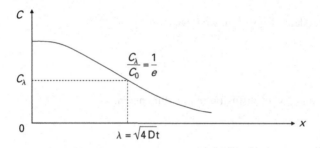

Fig. 4.9   Concentration profile of diffusion.

Two important concentration values are

$$C(0, t) = C_0 = \frac{Q}{(\pi Dt)^{1/2}} \text{ at } x = 0 \tag{4.21}$$

$$C(\lambda_D, t) = C_\lambda = C_0/e \text{ at } x = \lambda_D = (4Dt)^{1/2} \tag{4.22}$$

The last equation shows that the position where the ratio of local concentration at $x = \lambda$ to concentration at the source of $x = 0$ is always $1/e$, which occurs at $x^2 = 4Dt$. Here we present one solution which shows such a unique relation of diffusion that x^2 is proportional to Dt, and many other solutions of the diffusion equation using different boundary conditions always involve the same relation. This proportionality is one of the most important relationships of diffusion. If a kinetic process is controlled by diffusion, it must obey this relationship, $x^2 \sim Dt$.

On the basis of the solution in Eq. (4.20), we can measure diffusivity, for example, from the concentration profile shown in Fig. 4.9. Since we know the time of diffusion, we plot $\ln C$ versus x^2, and the slope equals $1/(4Dt)$ according to Eq. (4.20), as shown in Fig. 4.10, and we obtain D at the temperature of the experiment. When we measure D

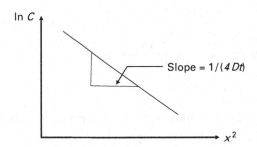

Fig. 4.10   Since we know the time of diffusion, we plot $\ln C$ versus x^2, and the slope equals $1/(4Dt)$ according to Eq. (3.27) and we obtain D at the temperature of the experiment.

as a function of temperature, at four or five temperatures, we can plot $\ln D$ versus $1/kT$ and the slope will give us the activation energy of the diffusion, ΔH. We can express the diffusion coefficient as

$$D = D_0 \exp\left(\frac{-\Delta H}{kT}\right)$$

where D_0 is the pre-exponential factor.

4.7   Diffusion coefficient

In general, it is found that the measured diffusion coefficient can be expressed as

$$D = D_0 \exp\left(\frac{-\Delta H}{kT}\right) \tag{4.23}$$

This is a Boltzmann distribution function, and the coefficient consists of two parts: the pre-exponential factor D_0 and the activation energy ΔH, where ΔH does not depend on temperature and D_0 may depend on temperature only very slightly.

Previously, we derived a simple flux equation, Eq. (4.8), where we have

$$D = \lambda^2 v$$

which gives diffusivity the dimension of cm^2/s. We have the same dimension in the relationship of

$$x^2 = 4Dt \tag{4.24}$$

In Section 4.4 on diffusivity, we expressed the diffusion coefficient as

$$D = f n_c \lambda^2 v_0 \exp\left(\frac{\Delta S_m + \Delta S_f}{k}\right) \exp\left(-\frac{\Delta H_m + \Delta H_f}{kT}\right) \tag{4.25}$$

Thus, we have

$$D_0 = fn_c\lambda^2 v_0 \exp\left(\frac{\Delta S_m + \Delta S_f}{k}\right)$$

$$\Delta H = \Delta H_m + \Delta H_f$$

In the above, we have connected the continuum mechanics approach of diffusion, Eq. (4.23), to the atomistic mechanism of diffusion, Eq. (4.11).

4.8   Calculation of the diffusion coefficient

Table 4.1 gives the self-diffusion coefficient D_0 (in cm^2/s) for some elements. The activation energies ΔH, ΔH_m, and ΔH_f are given in units of eV/atom. Below, we calculate the self-diffusivity of Al($D_0 = 0.047$ cm^2/s, $\Delta H = 1.28$ eV/atom) at room temperature:

$$D = D_0 \exp\left(-\frac{\Delta H}{kT}\right) = 0.047 e^{-\frac{1.28 \times 23000}{2 \times 300}}$$

$$= 0.047 \times 10^{-\frac{1.28 \times 23000}{2.3 \times 2 \times 300}} = 0.047 \times 10^{-22} \text{cm}^2/\text{s}$$

With this diffusivity, if we take time to be a day or 10^5 s, we will have $(Dt)^{1/2}$ less than 10^{-8} cm. In other words, it is less than one atomic jump. Thus, Al metal is very stable at room temperature. Most metals and semiconductors have a diffusivity smaller than 10^{-22} cm^2/s at room temperature, except the low-melting-point metals such as Pb and Sn.

When Al is used as an electrical conductor on Si devices, joule heating during device operation raises the temperature, and 100 °C has often been chosen as the upper limit, for reliability reasons. The diffusivity at this temperature is given as

$$D = D_0 e^{-\Delta H/kT} \cong 0.047 \times 10^{-\frac{1.28 \times 5000}{100+273}} \cong 0.33 \times 10^{-18} \text{cm}^2/\text{s}$$

which means that by lattice diffusion, Al atoms can diffuse a distance of about 2 nm at 100 °C in a day, which is not negligible.

We shall make a rough estimate of the upper and lower bounds of diffusivity in solids. From Table 4.1, we estimate the diffusivity of the fcc metals at their melting points to be around 10^{-8} cm^2/s; it is the upper bound. This value is smaller than the diffusivity found in liquid or molten metals which is about 10^{-5} cm^2/s. As a lower bound if we take the jump distance to be an interatomic distance of ~0.1 nm and the time to be 10 days, we have $x^2/t = 10^{-22}$ cm^2/s. Any diffusivity of this order of magnitude is not of practical interest and is difficult to measure. Using superlattice structures or layer removal techniques for concentration profiling, we can measure diffusivity around 10^{-19} to 10^{-21} cm^2/s. In an intermediate range of diffusivities, we find interdiffusion distances of 10^{-6} to 10^{-5} cm in thin-film reactions in times of 1000 s, giving values of

Table 4.1. Lattice self-diffusion in some important elements*

Element	D_0 (cm^2/s)	ΔH (eV)	ΔH_f (eV)	ΔH_m (eV)	$\Delta S/k$
FCC					
Al	0.047	1.28	0.67	0.62	2.2
Ag	0.04	1.76	1.13	0.66	—
Au	0.04	1.76	0.95	0.83	1.0
Cu	0.16	2.07	1.28	0.71	1.5
Ni	0.92	2.88	1.58	1.27	—
Pb	1.37	1.13	0.54	0.54	1.6
Pd	0.21	2.76	—	—	—
Pt	0.33	2.96	—	1.45	—
BCC					
Cr	970	4.51	—	—	—
α-Fe	0.49	2.95	—	0.68	—
Na	0.004	0.365	0.39/0.42	—	—
β-Ti	0.0036	1.35	—	—	—
V	0.014	2.93	—	—	—
W	1.88	6.08	3.6	1.8	—
β-Zr	0.000085	1.2	—	—	—
HCP					
Co	0.83	2.94	—	—	—
α-Hf	0.86/0.28	3.84/3.62	—	—	—
Mg	1.0/1.5	1.4/1.41	0.79/0.89	—	—
α-Ti	0.000066	1.75	—	—	—
Diamond Lattice					
Ge	32	3.1	2.4	0.2	10
Si	1460	5.02	$\sim$3.9	$\sim$0.4	

*From Chapter I of *Diffusion Phenomena in Thin Film and Micro-electronic Materials* (Gupta and Ho, 1988) (Courtesy of D. Gupta) [7]

$x^2/t = 10^{-15}$ to 10^{-13} cm^2/s, which is the value of diffusivity to be expected. Knowing D at a given temperature, we can easily estimate the diffusion distance for a given time by using the relation $x^2 = 4Dt$.

Substitutional dopants in Si have diffusivities close to that of self-diffusion in Si. In Fig. 4.11, we show a plot of diffusivity versus temperature for dopants in Si. To produce a p–n junction in Si by dopant diffusion, the diffusion temperature must be close to 1000 °C. There is a question about the mechanism of such diffusion. The measured diffusivity could be a combination of two mechanisms (e.g. two kinds of defect such as vacancies and divacancies may coexist in the sample). Whether self-diffusion in Si is by vacancies exclusively or may occur by interstitials or direct exchanges is also unclear. However, Fig. 4.11 shows that noble and near-noble metal elements such as Cu and Li have diffusivities several orders of magnitude higher than that of Si self-diffusion. They diffuse interstitially in Si. For bcc metals, the activation energy of diffusion shows a small dependence on temperature. It is a subject not to be discussed here.

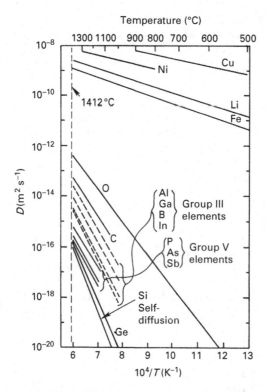

Fig. 4.11   Plot of diffusivity versus temperature for dopants in Si.

In the Simmons–Balluffi experiment (see Section 4.9.2) the lattice parameter expansion and the change of sample dimension due to the defect generation at high temperatures were determined simultaneously. It was concluded that in fcc metals such as Al and Au vacancies are the dominant point defects which mediate diffusion, and that the concentration of vacancies near the melting point is about 10^{-4}. Next, we shall consider the parameters in the diffusion coefficient.

4.9   Parameters in the diffusion coefficient

4.9.1   Atomic vibrational frequency

We have derived an expression for the diffusion coefficient,

$$D = D_0 \exp\left(\frac{-\Delta H}{kT}\right)$$

with the pre-exponential factor,

$$D_0 = f n_c \lambda^2 \nu_0 \exp\left(\frac{\Delta S_m + \Delta S_f}{k}\right)$$

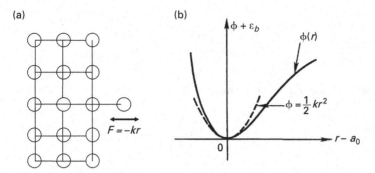

Fig. 4.12   (a) Sketch of a single-surface atom which is bonded to a solid by a pair potential ϕ. (b) We approximate the bottom portion of ϕ (i.e. around $\phi(a_0)$), by a parabolic potential and move the point $(-\varepsilon_b, a_0)$ to the origin of the coordinates.

and the activation enthalpy

$$\Delta H = \Delta H_m + \Delta H_f$$

Among all the parameters in the pre-exponential factor and the activation enthalpy, the atomic vibrational frequency v_0 is a fundamental parameter of the solid. We present here an order of magnitude calculation of v_0. It serves to provide a simple physical picture of atomic vibration in the solid and an estimate of the magnitude of the diffusion coefficient.

In Fig. 4.12(a), we sketch a single surface atom which is bonded to a solid by a pair potential ϕ. The potential function is given by, for example, that shown in Fig. 3.l(a). We assume that the atom undergoes harmonic motion. It is a simplified model, yet it gives a vibrational frequency of the same order of magnitude as that of atoms in the solid.

In Chapter 3, Section 3.2, a simple harmonic vibration of atoms was presented, and the frequency of vibration was given as

$$v_0 = \frac{1}{t} = \frac{\omega}{2\pi} = \frac{1}{2\pi}\sqrt{\frac{k}{m}} \tag{4.26}$$

We can calculate v_0 if we know k and m, where k is force constant and m is atomic weight. The latter is given by knowing the atomic weight of the solid and Avogadro's number. To determine k, we take the second derivative of ϕ in Eq. (3.2) and we have

$$\frac{d^2\phi}{dr^2} = +k \tag{4.27}$$

In Chapter 3, we have given Eq. (3.7) for a solid that obeys the Lennard-Jones potential, from which we obtain

$$\frac{d^2\phi}{dr^2} = \frac{12\varepsilon_b}{a_0^2}\left(\frac{a_0}{r}\right)^8\left[13\left(\frac{a_0}{r}\right)^6 - 7^6\right]$$

Then

$$k = \frac{72\varepsilon_b}{a_0^2} \text{ at } r = a_0 \tag{4.28}$$

Substituting k into Eq. (3.5), we have

$$v_0 = \frac{3}{\pi} \sqrt{\frac{2\varepsilon_b}{ma_0^2}} \tag{4.29}$$

Now, to calculate v_0, we consider the atoms shown in Fig. 4.12(a) to be gold atoms, then

$$m = \frac{197 \text{g/mole}}{6.02 \times 10^{23} \text{atoms/mole}} = 32.8 \times 10^{-23} \text{g/atom}$$

The interatomic distance a_0 in fcc Au is given by $a_0/\sqrt{2} = 0.288$ nm, so that $a_0^2 = 8.3 \times 10^{-16}$ cm^2 and

$$ma_0^2 = 2.72 \times 10^{-37} \text{ gm-cm}^2/\text{atom}$$

$$= 2.72 \times 10^{-37} \text{ dyne-cm-s}^2/\text{atom}$$

$$= 2.72 \times 10^{-37} \text{ erg-s}^2/\text{atom}$$

where we have used the conversion 1 dyne = 1 g-cm/s^2. From Table 3.1, the interatomic potential energy of Au is

$$\varepsilon_b = 0.47 \text{ eV/atom} = 0.47 \times 1.6 \times 10^{-12} \text{ erg/atom}$$

We obtain

$$v_0 = \frac{3}{\pi} \sqrt{\frac{2\varepsilon_b}{ma_0^2}} = 2.24 \times 10^{12} \text{ s}^{-1} \tag{4.30}$$

If we extend the above simple calculation to an atom within the fcc lattice where it has 12 nearest neighbors, the force constant has to be multiplied by a factor of 6 for vibration along a close-packed direction. The factor of 6 comes in because we use the principle of superposition and sum the projections ($\cos \theta$) of the interatomic force of all 12 atoms. In turn, we have to multiply v_0 by $\sqrt{6}$, giving

$$v_0 = \frac{6}{\pi} \sqrt{\frac{3\varepsilon_b}{ma_0^2}} \tag{4.31}$$

This is called the Einstein frequency. For an Au atom within its lattice, we have $v_0 = 5.5 \times 10^{12}$ cycle/s.

The formal treatment of elastic vibrations (phonons) in a finite piece of solid has been given by Debye, and the subject is covered in textbooks about solid-state physics. The Debye frequency ν_D is defined by

$$h\nu_D = kT_\theta \tag{4.32}$$

where h is Planck's constant ($h = 6.626 \times 10^{-27}$ erg s) and T_θ is the Debye temperature at which all the $3N$ modes of elastic waves are operative. For metal Au, $T_\theta = 165$ K which is given, for example, in Table 4.1 of *Thermal Physics* by Kittel and Kroemer [8]. Hence

$$\nu_D = \frac{kT_\theta}{h} = \frac{1.38 \times 10^{-16} \text{ erg K}^{-1} \times 165 \text{ K}}{6.626 \times 10^{-27} \text{ erg-s}} = 3.42 \times 10^{12} \text{ s}^{-1}$$

which is not far from the frequencies we have calculated. Since the Debye temperatures of common metals vary only by a factor of 2 to 3 (e.g. $T_\theta = 428$ K for Au), the atomic vibrational frequency of metals from the viewpoint of diffusion is typically taken to be 10^{13} cycle/s or 10^{13} Hz.

4.9.2 Activation enthalpy

In Eq. (3.37), the activation enthalpy of vacancy diffusion consists of two components,

$$\Delta H = \Delta H_f + \Delta H_m$$

Since knowing the activation energies is important in understanding the mechanism of diffusion and in identifying the types of defect which mediate the diffusion, their measurement has been a key activity in studying diffusion. The values ΔH can be determined by measuring D at several temperatures and by plotting $\ln D$ versus $1/kT$; from the slope of the straight-line plot we obtain ΔH. Experimental techniques of thermal expansion, quenching plus resistivity measurement, and positron annihilation have been used to determine ΔH_f. A quenching technique has also been used to measure ΔH_m. These techniques are well covered in textbooks and reference books about diffusion. Here we shall only discuss briefly the measurement of ΔH_f by thermal expansion.

The concentration of vacancies in a solid is an equilibrium quantity. The concentration n_v/n increases with temperature as given by Eq. (3.33),

$$\frac{n_v}{n} = \exp\left(\frac{-\Delta G_f}{kT}\right) = \exp\left(\frac{\Delta S_f}{k}\right) = \exp\left(\frac{-\Delta H_f}{kT}\right)$$

where n_v and n are, respectively, the number of vacancies and atoms in the solid. As temperature increases, more vacancies are formed by removing atoms from the interior to the surface of the solid. Consequently, the volume of the solid increases. Provided

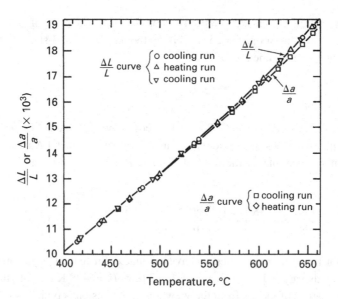

Fig. 4.13   Plot of $\Delta L/L$ and $\Delta a/a$ of Al wire as a function of temperature up to the melting point.

that we can decouple this volume increase from that due to thermal expansion, we can measure the vacancy concentration. Thermal expansion can be determined by measuring the lattice parameter change Δa as a function of temperature using X-ray diffraction, and we can express the fractional change by $\Delta a/a$. Similarly, for the volume change, we can use a wire of length L and measure the length change ΔL as a function of temperature. We then have

$$\frac{\Delta n_v}{n} = 3\left(\frac{\Delta L}{L} - \frac{\Delta a}{a}\right) \tag{4.33}$$

The factor of 3 comes in because both ΔL and Δa are linear changes. Fig. 4.13 shows $\Delta L/L$ and $\Delta a/a$ of Al wire as a function of temperature up to the melting point, as measured by Simmons and Balluffi. The results show that near the melting point, the vacancy concentration $n_v/n = 10^{-4}$. Since $(\Delta L/L - \Delta a/a)$ is positive, the defect is predominantly vacancies. For interstitials, the difference is expected to be negative. In the derivation, we have ignored the effect of compensation between a vacancy and an interstitial, and also the effect of divacancies. From the two curves shown in Fig. 4.13,

$$\frac{\Delta n_v}{n} = \exp(2.4)\exp(-0.76\,\text{eV}/kT) \tag{4.34}$$

so we have $\Delta H_f = 0.76\,\text{eV}$ and $\Delta S_f/k = 2.4$ for Al. These values are in good agreement with those listed in Table 4.1.

4.9.3 The pre-exponential factor

Using $v_0 = 10^{13}$ Hz, we can estimate the entropy factor by measuring the pre-exponential factor of diffusion D_0, provided that we accept the theoretical value of the correlation factor f $(= 0.78$ for fcc metals). This is shown by the form of Eq. (4.12) presented near the end of Section 4.4:

$$\exp\left(\frac{\Delta S}{k}\right) = \exp\left(\frac{\Delta S_m + \Delta S_f}{k}\right) = \frac{D_0}{f v_0 \lambda^2 n_c}$$

Since ΔS has to be positive, we have $D_0 > f v_0 \lambda^2 n_c$. For Au,

$$D_0 > 0.78 \times 3.42 \times 10^{12} \times (2.88 \times 10^{-8})^2 \times 12 = 0.027 \text{ cm}^2/\text{s}.$$

In general, D_0 for self-diffusion in metals is of the order of 0.1 to 1 cm^2/s, so the entropy change per atom is of the order of unity times k. In textbooks on diffusion, ΔS is sometimes given in terms of R, the gas constant (instead of Boltzmann's constant k), when the activation enthalpy is given in kcal/mole or kJ/mole rather than eV/atom.

We consider below the theoretical calculation of ΔS for interstitial diffusion such as carbon in iron. In such a case, diffusion requires only the activation energy of motion because in Fe the interstitial sites neighboring an interstitial C solute atom are always available. To calculate the entropy change in this case, for a constant pressure process, we have

$$\Delta S_m = -\frac{\partial \Delta G_m}{\partial T} \tag{4.35}$$

Zener reasoned that the Gibbs free energy change during the diffusion of a carbon atom is essentially the strain energy needed to push out the Fe atoms in order to open up a passage wide enough for the interstitial carbon atom to pass to a neighboring interstitial site. We shall consider the diffusion process shown schematically in Fig. 4.14(a), where two unit cells of bcc Fe contain an interstitial carbon atom at the center position. The carbon atom is surrounded by six Fe atoms, labeled $a, b, c, d, e,$ and f, where the pair e-f is along the vertical direction, and there is a strain in this direction because of the carbon atom.

The lattice parameter of the bcc unit cell of Fe is $a = 0.2866$ nm. The distance of closest approach between two Fe atoms is 0.2481 nm; we regard this distance as the diameter of an iron atom. The distance of closest approach between two carbon atoms in the basal plane of graphite is 0.142 nm, and we shall take this value to be the diameter of a carbon atom. On the bcc structure shown in Fig. 4.14(a), the interstitial carbon atom fits comfortably with the four neighboring Fe atoms labeled $a, b, c,$ and d. This is because the distance between a and c, or between b and d, is $0.2866 \times \sqrt{2} = 0.4052$ nm, which is slightly greater than the sum of the diameters of an iron atom and a carbon atom (which is 0.3901 nm). On the other hand, the other two neighboring Fe atoms e and f must be stretched out (as indicated by the pair of short arrows) by an amount of about $0.1035/0.2866 = 36\%$. It is reasonable to expect that the

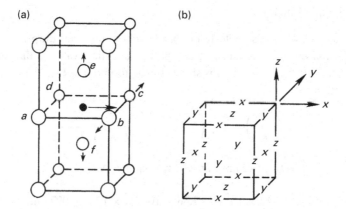

Fig. 4.14   (a) Two bcc unit cells of Fe containing an interstitial carbon atom in the center of four Fe atoms labeled a, b, c, and d. The carbon atom has caused a displacement of the other two neighboring Fe atoms labeled e and f. When the carbon atom jumps to the interstitial site between b and c, they move apart as shown by the pair of arrows shown. (b) The equivalent interstitial sites on the bcc lattice are shown by the labels of x, y, and z.

distance between the carbon atom and the two iron atoms can be compressed a bit closer because of hybridization, but note that in iron carbide, Fe$_3$C, the closest interatomic distance between Fe and C is 0.39 nm. Therefore the strain in the e–f pair is of the order of 36%.

To consider the diffusion of the interstitial carbon atom to another interstitial site, we first note that the equivalent interstitial sites in the bcc structure are located in the face-centered and edge-centered positions in the lattice; they are indicated in Fig. 4.14(b). They are equivalent because, except for a rotation of 90°, they have similar surrounding Fe atoms, that is, the interstitial site between b and c is the same as the center one, except that when the centered carbon atom jumps to the site between b and c, the stretch (as indicated by the pair of short arrows) is now along the b–c direction (y-direction) rather than the e–f direction (z-direction). For such a diffusion jump, we see that before and after the jump, the e–f and b–c pairs are strained, respectively. Yet, during the diffusion, both must be strained; this is because without stretching the b–c pair, the carbon atom cannot jump into the interstitial positions between b and c, so an activation energy of diffusion is needed.

If we assume that the process is elastic, we can regard the activation energy as the strain energy involved,

$$\Delta G_m = -\frac{1}{2} K \varepsilon^2 \tag{4.36}$$

where K is the bulk modulus and ε is the volume strain. Since the strain can be regarded as independent of temperature, we have

$$\Delta S_m = \frac{1}{2} \varepsilon^2 \frac{\partial K}{\partial T} \tag{4.37}$$

In Chapter 3, we have shown that surface energy and Young's modulus are closely related as illustrated by Table 3.4. Both are proportional to the interatomic potential energy. This relation is also true for the binding energy and the bulk modulus K. Specifically, if a solid obeys a Lennard-Jones potential, we have shown that

$$K = \frac{8\Delta E_s}{V}$$

where ΔE_s is the latent heat of sublimation and $V(= N_A\Omega$, where Ω is the atomic volume and N_A is Avogadro's number) is the molar volume. Then,

$$\Delta S_m = \frac{4\varepsilon^2}{V}\frac{\partial \Delta E_s}{\partial T} = \frac{4\varepsilon^2}{V}c_v = \frac{4\varepsilon^2(3N_Ak)}{N_A\Omega} = 12\varepsilon^2k/\Omega \qquad (4.38)$$

where c_v is the heat capacity at constant volume and we take $c_v = 3N_Ak$. For interstitial carbon in iron, the strain occurs along one direction, so the volume strain can be approximated by the linear strain without a factor of 3. Then, if we take $\varepsilon = 0.36$, we have

$$\Delta S_m = 1.6k/\text{atom}$$

which is of the right order of magnitude even though $\varepsilon = 0.36$ is unreasonably large. Empirically, we can approximate Eq. (3.50) by

$$\Delta S_m = \beta \frac{\Delta H}{T_m} \qquad (4.39)$$

where ΔH is the activation energy of diffusion, T_m is the melting point, and β is a proportionality constant. This relationship holds well for interstitial diffusion of carbon, nitrogen, and oxygen in bcc transition metals.

References

[1] P. G. Shewmon, *Diffusion in Solids*, 2nd edn (The Minerals, Metals, and Materials Society, Warrendale, PA, 1989).

[2] R. J. Borg and G. J. Dienes, *An Introduction to Solid State Diffusion* (Academic Press, Boston, MA, 1988).

[3] H. S. Carslaw and J. C. Jaeger, *Conduction of Heat in Solids*, 2nd edn (Clarendon Press, Oxford, 1980).

[4] J. Crank, *Mathematics of Diffusion* (Oxford University Press, Fair Lawn, NJ, 1956).

[5] R. P. Feynman, R. B. Leighton, and M. Sands, *The Feynman Lectures on Physics*, vol. I (Addison-Wesley, Reading, MA, 1963).

[6] S. Glasstone, K. J. Laidler and H. Eyring, *The Theory of Rate Processes* (McGraw-Hill, New York, 1941).

[7] D. Gupta and P. S. Ho (eds), *Diffusion Phenomena in Thin Films and Micro-electronic Materials* (Noyes Publications, Park Ridge, NJ, 1988).

[8] C. Kittel and H. Kroemer, *Thermal Physics* (Wiley, New York, 1970).

[9] J. R. Manning, *Diffusion Kinetics for Atoms in Crystals* (Van Nostrand, Princeton, NJ, 1968).

[10] M. E. Glicksman, *Diffusion in Solids* (Wiley-Interscience, New York, 2000).

[11] R. W. Balluffi, S. M. Allen and W. C. Carter, *Kinetics of Materials* (Wiley-Interscience, New York, 2005).

Problems

4.1   What is the unit of atomic mobility in diffusion? Is it different from the unit of electron mobility?

4.2   Why are two laws, Fick's first and second laws, sufficient for solving most of the problems in diffusion-related phase transformations? Why don't we need a third law?

4.3   What are the attempt frequency and exchange frequency in atomic diffusion?

4.4   How do you measure the activation enthalpy of formation and motion in a vacancy diffusion mechanism?

4.5   Using the data in Table 4.1 for copper and aluminum,
(a) calculate D_{Cu} and D_{Al} at 600 K;
(b) at the melting temperature, T_m, calculate D_{Cu} ($T_m = 1083\,°C$) and D_{Al} ($T_m = 660\,°C$).

In comparing diffusion coefficients for fcc metals, which is the more appropriate scaling factor, t or T_m?

4.6   A sample is diffused at 1100 °C for 20 min with a total amount of radioactive tracer of 2×10^{15} atom/cm^2. The diffusion length λ_D is 10^{-4} cm.
(a) What is the diffusivity?
(b) Calculate the concentration at $0.3\lambda_D$ and $0.4\lambda_D$ and estimate the tracer flux J.

4.7   For a diffusion of 10^{15} radioactive tracer Cu atoms into Cu at 800 °C to a diffusion length λ_D of 10^{-5} cm, calculate using the data in Table 4.1,
(a) the diffusion coefficient D and time;
(b) at $0.5\lambda_D$, the driving force F, and the mobility M.

4.8   An isotope film of Au is applied to one side of a gold disc. The assembly is raised to 900 °C and the isotope begins to diffuse into the disc. After 1 h the assembly is quenched. The concentration of the isotope at a length of 10 μm microns into the disc is found to be 4.0×10^{-5} atom fraction. At an 80 μm depth, the isotope concentration is 2.3×10^{-6} atom fraction.
(a) What is the diffusivity?
(b) Given an activation energy of diffusion of 1.84 eV/atom, what is the pre-exponential factor D_0?

4.9 With the solution of the diffusion equation given by Eq. (4.20) and its boundary condition, evaluate

$$\bar{x} = \frac{\int_0^\infty xC(x)dx}{\int_0^\infty C(x)dx} = \frac{1}{\sqrt{\pi}}\sqrt{4dt}$$

4.10 Show that the equation below is a solution of the diffusion equation.

$$C(x,t) = \frac{Q}{\sqrt{\pi Dt}}\exp\left(-\frac{x^2}{4dt}\right)$$

5   Applications of the diffusion equation

5.1   Introduction

When the initial and boundary conditions are given, many solutions of the diffusion equation have been obtained in the literature and textbooks, and the solutions have been applied to many technical problems [1–6]. In this chapter, we shall select a few thin-film problems to illustrate the application of Fick's first and second laws.

Darken's analysis of the Kirkendall effect of interdiffusion in a bulk diffusion couple will be presented [3–5]. Why do we include the bulk diffusion behavior in a book on thin-film reliability? This is because interdiffusion is one of the most common behaviors in materials, whether it is in bulk or in thin films. More importantly, it will help us to understand the mechanism of failure in thin films. In Darken's analysis of interdiffsuion, specifically, no void formation and no stress are assumed, hence there is no failure in the microstructure. It is a constant volume kinetic process. While Darken's analysis explains the Kirkendall shift (or lattice shift) by assuming vacancy to be equilibrium everywhere in the sample, it does not allow Kirkendall (or Frenkel) void formation. Only when vacancies are supersaturated will voids nucleate and form. Thus, Kirkendall (or Frenkel) void formation will require a different condition from that in Kirkendall shift, so we assume that Kirkendall shift is missing in void formation. Then, this condition of void formation can help us to understand in failure analysis the important role of vacancy flux divergence and the lattice shift missing due to the lack of a combined action of sources and sinks for vacancies. For example, in electromigration-induced failure in interconnects, void formation occurs in the cathode and stress-induced hillock or whisker formation occurs in the anode. Darken's analysis, conceptually, will help us to understand in reliability failure why void formation and hillock growth can occur.

Then, how to use the diffusion equation to analyze the growth kinetics of a precipitate will be given. This is because in the growth of a solid precipitate of constant composition, there is no concentration change within the precipitate, so when a flux of atoms diffuses to the precipitate, it leads to growth of the precipitate in size, instead of changing the concentration of the precipitate, so there is flux divergence at the interface of the precipitate, but none within the precipitate. Two kinds of growth kinetics will be presented: one-dimensional planar growth and three-dimensional spherical growth. Many growth problems in thin films are planar such as silicide formation and can be treated as one-dimensional growth. The growth velocity of a planar interface can be given by combining the flux equations of $J = Cv$ and $J = -D(dC/dx)$, where v is the growth velocity of the interface. Spherical growth is essential to the understanding of the kinetics of ripening or the growth of a void.

The Kirkendall effect as well as inverse Kirkendall effect have been of interest in studying void formation in nanoparticles and nanowires, but are beyond the scope of this book and will not be covered.

5.2 Application of Fick's first law (flux equation)

Fick's first law assumes a constant driving force of diffusion or a constant concentration gradient. The flux is given by the product of the constant concentration gradient and diffusivity. A typical application is the permeation of a flux of gas through a diaphragm when the pressures on both sides of the diaphragm are given. In layered thin-film structures, the interdiffusion and reaction in most cases lead to the formation of intermetallic compounds (IMCs). Experimentally, the thickening of the layered IMCs obeys diffusion-controlled kinetics, or the simple relationship of $x^2 \sim Dt$. This is also true in many bulk diffusion couples. The kinetics of the diffusion-controlled growth of a layered compound can be solved more easily by using Fick's first rather than second law. We shall illustrate the approach below. In Chapter 8, the diffusion-controlled growth and interfacial-reaction-controlled growth of a planar interface will be given.

5.2.1 Zener's growth model of a planar precipitate

We consider the growth of the planar surface of a layer-type precipitate in a super-saturated matrix. Fig. 5.1(a) depicts the concentration profile of the precipitation. We have the precipitate phase β in the matrix of the α phase. The growth front of the β phase is planar. The concentration of the β phase is C_S. The average concentration in the α phase is C_m. The equilibrium concentration at the β/α interface is C_e. The precipitate grows when solute atoms diffuse to the β/α interface and are incorporated into the β phase. Typically, we should solve the diffusion equation and obtain the concentration profile in the α matrix as indicated by the curve of the concentration profile. Then we could obtain the flux of solute atoms arriving at the β/α interface, and we could calculate the growth velocity of the precipitate, $v = dx/dt$.

Here, we consider flux equations at the planar interface,

$$-J_{prep} = D\frac{dC}{dx} = (C_S - C_e)\frac{dx}{dt} \tag{5.1}$$

The equation above is obtained by equaling two flux equations; the middle term is Fick's first law, $J = -D(dC/dx)$, and the last term is based on the expression of $J = Cv$, and it is assumed that the concentration step across the interface times the interface velocity is equal to the flux arriving at the interface. To solve the above equation, we follow the analysis by Zener [3].

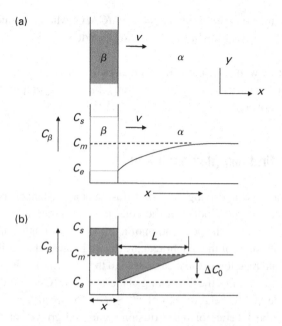

Fig. 5.1   (a) Schematic diagram of the concentration profile of precipitation. We have the precipitate phase β in the matrix of α phase. (b) We redraw the concentration profile in the matrix and we change the curved concentration profile between C_m and C_e to a straight line. The straight line is drawn so that the two shaded areas are the same.

In Fig. 5.1(b), we redraw the concentration profile in the matrix by changing the concentration profile between C_m and C_e from a curved line to a straight line. The straight line is drawn so that the two shaded areas are the same. Physically it means that all solute atoms in the precipitate have come from the reduction of solute concentration in the matrix. In other words, the length of L in the figure is defined by

$$(C_S - C_m)x = \frac{1}{2}(C_m - C_e)L \qquad (5.2)$$

Thus, we can take

$$\frac{dC}{dx} = \frac{C_m - C_e}{L}$$

From Eq. (5.1), we obtain

$$\frac{dx}{dt} = \frac{D}{(C_S - C_e)}\frac{(C_m - C_e)}{L} = \frac{D}{(C_S - C_e)}\frac{(C_m - C_e)^2}{2x(C_S - C_e)}$$

Hence,

$$-J_{prep} = (C_S - C_e)\frac{dx}{dt} = \frac{D}{2x}\frac{(C_m - C_e)^2}{(C_S - C_m)}$$

By assuming that C_S is much greater than C_m and C_e, we replace $(C_S - C_m)$ in the denominator in the above equation by $(C_S - C_e)$, and we have,

$$\frac{dx}{dt} = \frac{D}{2x} \frac{(C_m - C_e)^2}{(C_S - C_e)^2}$$

By integration and by taking $x = 0$ at $t = 0$, we obtain the solution,

$$x^2 = \frac{(C_m - C_e)^2}{(C_S - C_e)^2} Dt \tag{5.3}$$

It shows that the thickening rate of the planar precipitate is diffusion-limited, without using Fick's second law.

5.2.2 Kidson's analysis of planar growth in layered thin films

In thin-film reactions between two metallic thin films of Sn and Cu, the formation of Cu_6Sn_5 and Cu_3Sn is sequential, i.e. Cu_6Sn_5 forms first and alone at room temperature, and Cu_3Sn will form only at temperatures higher than 60 °C. If we examine the binary phase diagram of SnCu, we find that both intermetallic phases of Cu_6Sn_5 and Cu_3Sn exist in the temperature range from room temperature to 250 °C. On the basis of the phase diagram or thermodynamics, we cannot explain why at room temperature solid state reaction, only Cu_6Sn_5 forms alone first and Cu_3Sn will form sequentially at a higher temperature rather than simultaneously.

Sequential-phase formation in thin-film reactions has also been studied in the reaction between a Si wafer and metallic films to form IMC phases of silicide. Single-phase formation of a specific silicide on Si to serve as ohmic contacts and gates in FET devices has been a very important technological issue. There are millions or even billions of silicide contacts and gates on a Si chip of the size of a fingernail, having a very large-scale integration of circuits. These contacts and gates must have the same physical properties. In other words, we cannot have a contact that consists of a mixture of silicide phases. Therefore, the device application demands single-phase formation, which in principle is against thermodynamics. Thus, a kinetic rather than a thermodynamic reason has to be given. The kinetics of single-phase growth has been analyzed, assuming a layered model of competition of growth of coexisting phases by combining diffusion-controlled growth and interfacial-reaction-controlled growth.

Fig. 5.2 depicts the growth of a layered IMC phase between two pure elements, for example, the growth of Cu_6Sn_5 between Cu and Sn. We represent Cu, Cu_6Sn_5 and Sn by $A_\alpha B$, $A_\beta B$, and $A_\gamma B$, respectively. The thickness of Cu_6Sn_5 is x_β and the position of its interface with Cu and Sn is defined by $x_{\alpha\beta}$ and $x_{\beta\gamma}$, respectively. Across the interfaces, there is an abrupt change in concentration. In Fig. 5.3, the abrupt concentration change of Cu across the interfaces is shown. In a diffusion-controlled growth of x_β, the concentrations at its interface are assumed to have the equilibrium values, and the

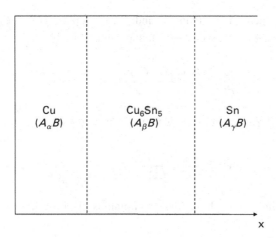

Fig. 5.2   Schematic diagram of the growth of a layered IMC phase between two pure elements, for example, the growth of Cu_6Sn_5 between Cu and Sn. We represent Cu, Cu_6Sn_5 and Sn by $A_\alpha B$, $A_\beta B$, and $A_\gamma B$, respectively.

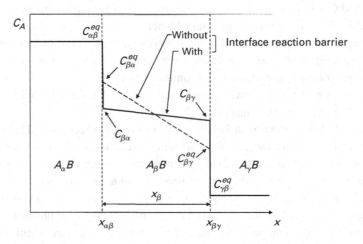

Fig. 5.3   Abrupt concentration change of Cu across the interfaces is shown. In a diffusion-controlled growth of x_β, the concentrations at its interface are assumed to have the equilibrium values, and the concentration across the layer is represented schematically by the broken curve in the x_β layer.

concentration across the layer is represented schematically by the broken curve in the x_β layer in Fig. 5.3.

    To consider a diffusion-controlled growth of a layered phase of x_β in Fig. 5.3, we shall use Fick's first law in one dimension. The corresponding fluxes within the layer are shown in Fig. 5.4 by

$$J = -D\frac{dC}{dx} \tag{5.4}$$

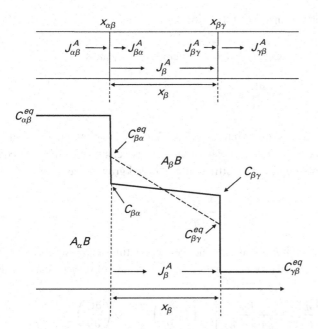

Fig. 5.4 Schematic diagram shows the corresponding fluxes within the layer.

and the flux equation across the interface is given by

$$J = \Delta C v \tag{5.5}$$

where J is atomic flux, having the unit of number of atom/cm^2 s; D is atomic diffusivity, cm^2/s; ΔC is concentration step across the interface, number of atom/cm^3; x is length, cm; and v is the velocity of the moving interface, cm/s. For example, $v = dx_{\alpha\beta}/dt$ of the interface $x_{\alpha\beta}$. For the growth of this interface, by considering the conservation of flux that enters and leaves the interface, we have on the basis of Eqs (5.5) and (5.4),

$$(C_{\alpha\beta} - C_{\beta\gamma})\frac{dx_{\alpha\beta}}{dt} = J_{\alpha\beta} - J_{\beta\gamma} = -D\frac{\partial C}{\partial x}\bigg|_{\alpha\beta} + D\frac{\partial C}{\partial x}\bigg|_{\beta\gamma} \tag{5.6}$$

Rearranging, we obtain the expression of velocity of the $x_{\alpha\beta}$ interface,

$$\frac{dx_{\alpha\beta}}{dt} = \frac{1}{C_{\alpha\beta} - C_{\beta\gamma}}\left[\left(-D\frac{\partial C}{\partial x}\right)_{\alpha\beta} - \left(-D\frac{\partial C}{\partial x}\right)_{\beta\gamma}\right] \tag{5.7}$$

To overcome the unknown of the concentration gradients in the square bracket in the above equation, we shall make a transformation by combining the two variables of x and

t into one, i.e. Boltzmann's transformation,

$$C(x,t) = C(\eta), \text{ where } \eta = \frac{x}{\sqrt{t}}, \text{and so}$$

$$\frac{\partial C}{\partial x} = \frac{1}{\sqrt{t}} \frac{dC}{d\eta} \tag{5.8}$$

Since the concentrations at the interface, i.e. $C_{\alpha\beta}$ and $C_{\beta\gamma}$, can be assumed to remain constant with respect to time and position, because we can take them as the equilibrium values under the assumption of a diffusion-controlled growth, we have

$$\frac{dC(\eta)}{d\eta} = f(\eta) \tag{5.9}$$

where $f(\eta)$ is constant if η is constant, independent of time and position, at the interfaces for a diffusion-controlled process. Therefore, the equation of velocity can be rewritten as

$$\frac{dx_{\alpha\beta}}{dt} = \frac{1}{C_{\alpha\beta} - C_{\beta\gamma}} \left[-\left(D\frac{\partial C}{\partial \eta} \right)_{\alpha\beta} + D\left(\frac{\partial C}{\partial \eta} \right)_{\beta\gamma} \right] \frac{1}{\sqrt{t}} \tag{5.10}$$

The quantity within the square bracket is independent of time, after we take the factor of time out of the square bracket. Integration of the above equation gives

$$x_{\alpha\beta} = A_{\alpha\beta}\sqrt{t} \tag{5.11}$$

where

$$A_{\alpha\beta} = 2 \left[\frac{(DK)_{\beta\alpha} - (DK)_{\alpha\beta}}{C_{\alpha\beta} - C_{\beta\alpha}} \right]$$

$$K_{ij} = \left(\frac{dC}{d\eta} \right)_{ij}$$

Following a similar approach, we can obtain at the other interface of $x_{\beta\gamma}$

$$x_{\beta\gamma} = A_{\beta\gamma}\sqrt{t} \tag{5.12}$$

By combining the two interfaces, we have the width of the β phase as

$$w_\beta = x_{\beta\gamma} - x_{\alpha\beta} = (A_{\beta\gamma} - A_{\alpha\beta})\sqrt{t} = B\sqrt{t} \tag{5.13}$$

which shows that the β phase has a parabolic growth rate or diffusion-controlled growth. Note that the above is a very simple derivation of a diffusion-controlled growth of a layered phase, or a relationship of $w^2 \propto t$ for a layered growth with abrupt change of composition at its interfaces by using only Fick's first law.

A fundamental nature of a diffusion-controlled growth layer is that it will not disappear or it cannot be consumed in growth competition in a multilayered structure, since its velocity of growth is inversely proportional to its thickness. As the thickness w approaches 0,

$$\lim \frac{dw}{dt} = \frac{B}{w} \to \infty \qquad (5.14)$$

The growth rate will approach infinity, or the chemical potential gradient to drive the growth will approach infinity.

Therefore, in a multilayered structure, for example, Cu/Cu$_3$Sn/Cu$_6$Sn$_5$/Sn, when both of Cu$_3$Sn and Cu$_6$Sn$_5$ exist and have diffusion-controlled growth, they will co-exist and grow together. For this reason, in a sequential growth of Cu$_6$Sn$_5$ followed by Cu$_3$Sn, we cannot assume that both of them can nucleate and grow by a diffusion-controlled process; otherwise they will coexist and we cannot have the growth of a single IMC phase. To overcome this difficulty, we introduce interfacial-reaction-controlled growth in Chapter 8.

5.3 Applications of Fick's second law (diffusion equation)

5.3.1 Effect of diffusion on composition homogenization

In Fick's second law, the concentration is a function of location and time, $C = C(x, t)$. Thus, the driving force dC/dx changes with location and time. However, an intrinsic or build-in property of the diffusion equation is that, with time, the diffusion leads to homogenization of concentration (outside the spinodal region). We demonstrate this nature by considering a simple concentration profile of $C = C_0 \sin x$. The first derivative is $C' = C_0 \cos x$ and the second derivative is $C'' = -C_0 \sin x$. We plot them in Fig. 5.5. We see from the diffusion equation below that, if we assume that D is positive, the change of dC/dt or the change of C as a function of time depends on the sign of the second derivative:

$$\frac{\partial C}{\partial t} = D \frac{\partial^2 C}{\partial x^2} = D \left(\pm \frac{\partial^2 C}{\partial x^2} \right)$$

In the plot of the concentration curve in Fig. 5.5, the first half of the curve is concave downward and its second derivative is negative, which means that the concentration will decrease with time. In the second half of the concentration curve, it is concave upward, and the second derivative is positive, which means that the concentration will increase with time. Therefore, the effect of diffusion time on concentration is to homogenize the concentration profile or to make the concentration flat. Hills and valleys will be evened out with time by diffusion. Therefore, if we have a periodic structure of concentration profile, the diffusion or interdiffusion will even out the profile.

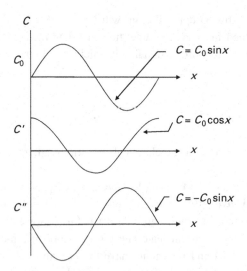

Fig. 5.5   Plot of a simple concentration profile of $C = C_0 \sin x$, and the first derivative of $C' = C_0 \cos x$ and the second derivative of $C'' = -C_0 \sin x$.

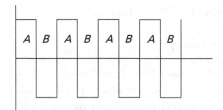

Fig. 5.6   A periodic structure of $A/B/A/B/A/B$.

It is worth mentioning that in spinodal decomposition, the free energy curve is concave downward, so G'' is negative. The opposite will then happen.

5.3.1.1   Homogenization in a periodic structure

A period of a periodic structure of $A/B/A/B/A/B$ is depicted in Fig. 5.6. At time $t = 0$, the concentration profile is presented by a sinusoidal curve, given as (see Fig. 7.9),

$$C = \overline{C} + \beta_0 \sin \frac{\pi x}{l} \tag{5.15}$$

where $\overline{C}$ is the mean composition, β_0 is the amplitude of the profile at $t = 0$, and l is the period of the periodic structure. Upon annealing, we expect homogenization, so that the region where the curvature is concave downward, the concentration will decrease, and the region where the curvature is concave upward, the concentration will increase.

If the diffusivity is independent of concentration, the solution to the diffusion equation which satisfies the initial condition at $t = 0$ is

$$C = \overline{C} + \beta_0 \sin \frac{\pi x}{l} \exp\left(-\frac{t}{\tau}\right) \qquad (5.16)$$

To check that the above equation is indeed a solution of the diffusion equation, we have

$$\frac{dC}{dt} = \left[\beta_0 \sin \frac{\pi x}{l} \exp\left(-\frac{t}{\tau}\right)\right]\left(-\frac{1}{\tau}\right)$$

$$\frac{dC}{dx} = \left[\beta_0 \cos \frac{\pi x}{l} \exp\left(-\frac{t}{\tau}\right)\right]\frac{\pi}{l}$$

$$\frac{d^2C}{dx^2} = \left[\beta_0 \sin \frac{\pi x}{l} \exp\left(-\frac{t}{\tau}\right)\right]\left(-\frac{\pi^2}{l^2}\right)$$

Now, if we take

$$-\frac{1}{\tau} = -D\frac{\pi^2}{l^2} \qquad (5.17)$$

then the diffusion equation below is satisfied:

$$\frac{\partial C}{\partial t} = D\frac{\partial^2 C}{\partial x^2}$$

We define $\tau = \frac{l^2}{D\pi^2}$ to be the relaxation time. Since the amplitude at $x = l/2$ is given as

$$C - \overline{C} = \beta = \beta_0 \exp\left(-\frac{t}{\tau}\right)$$

we find that after $t = \tau$, the amplitude of $\beta = \beta_0/e$. After $t = 2\tau$, the amplitude decreases to $\beta = \beta_0/e^2$. Since τ is proportional to l^2, it decays quickly with shorter and shorter period of l.

5.3.2   Interdiffusion in a bulk diffusion couple

If two pieces of bulk metal are joined and heated, they generally interdiffuse and lead to the formation of alloy or IMCs. We shall limit our analysis here to the formation of alloy between two very long rods of A and B. They are assumed to have the same crystal structure, such as Cu and Ni, so when they interdiffuse, they form a continuous solid solution that has the same crystal structure. We shall discuss those cases when the interdiffusion leads to IMC formation, such as Ni and Si reacting to form silicide compounds, in Chapter 10.

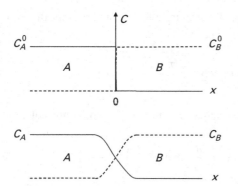

Fig. 5.7   Schematic diagram of an interdiffusion couple of A and B, and the concentration profile broadens as shown.

When A and B interdiffuse, the concentration profile broadens, as shown in Fig. 5.7. The mathematical solution of the profile is determined by considering the very general diffusion equation in the one-dimensional case:

$$\frac{\partial C_A}{\partial t} = D\frac{\partial^2 C_A}{\partial x^2}$$

The solution will be obtained by considering first the solution of a thin layer of A diffusing into a semi-infinite rod of B with the unit cross-sectional area as given in Chapter 4 when we discussed the diffusion of a tracer layer of A^* into a semi-infinite long rod of A, as shown below,

$$C_A = \frac{bC_A^0}{\sqrt{4\pi Dt}}\exp\left(-\frac{x^2}{4Dt}\right) \tag{5.18}$$

where C_A is the concentration of A atom/cm^3 and C_A^0 is the initial concentration of A atom/cm^3 in the thin layer of thickness b before interdiffusion.

Then, the thin-layer solution as given in the above is extended to a semi-infinite couple by dividing A into many vertical slices of thickness of $\Delta\alpha$ and integration will be taken to obtain the following solution:

$$C_A(x,t) = \frac{C_A^0}{\sqrt{4\pi Dt}}\int_{-\infty}^{0}\exp\left[-\frac{(x+\alpha)^2}{4Dt}\right]d\alpha \tag{5.19}$$

Let a new dimensionless variable η be

$$\eta = \frac{x+\alpha}{\sqrt{4dt}}, \quad d\eta = \frac{d\alpha}{\sqrt{4dt}}$$

then

$$\eta = \frac{x}{\sqrt{4dt}} \quad \text{when } \alpha = 0$$

$$\eta = -\infty \quad \text{when } \alpha = -\infty$$

The last equation can be rewritten as

$$C_A(x,t) = \frac{C_A^0}{\sqrt{\pi}} \int_{-\infty}^{x/\sqrt{4Dt}} \exp(-\eta^2) d\eta = \frac{C_A^0}{2}\left[1 + \text{erf}\left(\frac{x}{\sqrt{4Dt}}\right)\right] \qquad (5.20)$$

The error function is defined as

$$\text{erf}(z) = \frac{2}{\sqrt{\pi}} \int_{x}^{0} \exp(-\eta^2) d\eta \qquad (5.21)$$

and we have $\text{erf}(0) = 0$, $\text{erf}(\infty) = 1$, and $\text{erf}(-z) = -\text{erf}(z)$. Thus, at $x = 0$, we have

$$C_A = \frac{C_A^0}{2}$$

As a function of time and temperature, interdiffusion leads to a change of composition profile. On the basis of the mathematical solution in the above, the midpoint of composition is always at $x = 0$, i.e. at the original interface of the couple, as shown in Fig. 5.8.

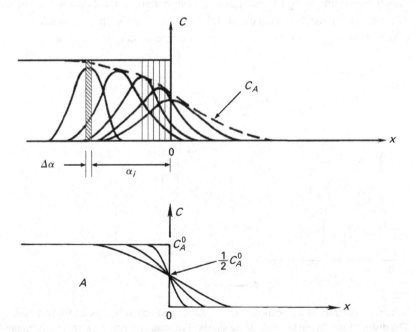

Fig. 5.8   On the basis of the mathematical solution, the midpoint of composition curve C_A is always at $x = 0$, i.e. at the original interface of the couple.

It means that when one A atom diffuses to B, one B atom will diffuse to A; in other words, the exchange between A and B atoms is equal. The original interface remains at the original position for the interdiffusion of A and B.

But the above implication is wrong because the mathematical solution does not consider the physical mechanism of diffusion. Experimentally it was found that the midpoint concentration will not locate at the original interface, and it move from the original interface, toward either A or B.

Kirkendall performed the classic interdiffusion experiment between Cu and CuZn (brass) with Mo markers placed at the original interface. Using a sandwiched structure of Cu/CuZn/Cu with markers in the two interfaces, he observed that Mo markers at the two interfaces moved closer to each other with annealing. It indicates that more Zn has diffused out than Cu has diffused in. By assuming that the diffusivity of Zn is faster than that of Cu in the CuZn alloy, Darken provided the analysis of interdiffusion given below.

5.3.2.1   Darken's analysis of Kirkendall shift and marker motion

If we assume more A atoms are diffusing into B, the lattice sites in the B must increase in order to accommodate the incoming A atoms. Similarly, the lattice sites in A will decrease because more A atoms have left. The increase and decrease of lattice sites can be achieved by the creation or elimination of vacant sites or vacancies, respectively. A very important assumption in Darken's analysis is that vacancy is at equilibrium everywhere in the couple. In other words, the vacancy sources in B are effective in creating vacant sites to accommodate the added A atoms and the vacancy sinks in A are effective in eliminating vacant sites to accommodate A's lost atoms. One of the atomic mechanisms of vacancy source and sink is dislocation climb as depicted in Fig. 5.9. The climb of an edge dislocation loop in the right-hand side by taking up atoms at the end

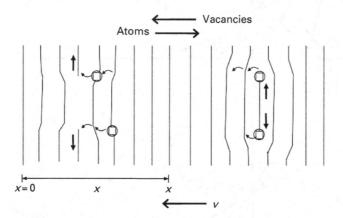

Fig. 5.9   Schematic diagram of dislocation climb as source and sink of vacancies interdiffusion. The dislocation loop on the right-hand side will expand to create an extra atomic plane. The dislocation loop on the left-hand side will shrink to remove an atomic plane. The consequence will lead to lattice shift from the right to the left by an atomic plane.

of the edge dislocation will lead to the addition of an atomic plane. The climb of an edge dislocation loop in the left-hand side by taking up vacancies at the end of the edge dislocation will lead to the removal of a lattice plane. Their combination will cause all the lattice planes in between them to shift to the left by the distance of one atomic plane thickness. This is defined as "lattice shift" or "Kirkendall shift" in interdiffuison. If we place markers in the lattice having a shifting velocity of v, the markers will move with a velocity v, moving towards A (or against the faster diffusing species). We shall define the marker velocity v later. It is the atomic mechanism of marker motion in Kirkendall shift.

Because of the lattice shift, the vacancy is at equilibrium everywhere. Thus, there is no vacancy supersaturation, so there will be no void formation. Also, because of the lattice shift, the added lattice site on the B side is balanced by the loss of a lattice site on the A side, so the net change in lattice sites is zero. If we further assume that the partial molar volume of A atoms and B atoms in the AB alloy is the same, it is a constant volume process. Then it also means that there is no stress in the interdiffusion, so in Darken's analysis, there is no void formation and no stress.

On the other hand, many experimental observations of interdiffusion in bulk diffusion couples have found void formation (Kirkendall or Frenkel). Also, many an interdiffusion between a thin film and a Si wafer has led to the bending of the wafer, which indicates that stress was produced in the interdiffusion. In these cases, we have to remove the assumption of lattice shift, so we do not have vacancy equilibrium. In Chapter 15, when we discuss reliability failure due to void or hillock formation, we assume that lattice shift does not occur and, in turn, the net lattice site change is not zero and it becomes a non-constant volume process.

Because of the motion of the lattice, Darken used two reference coordinates for the analysis of the atomic fluxes: the laboratory frame (the fixed frame) and the marker frame (the moving frame). In the laboratory frame, the atomic fluxes are examined from a distance and the origin of the coordinate is located at one end of the diffusion couple where no interdiffusion occurs. In the marker frame, the atomic fluxes are examined by sitting on a marker, so the frame moves with the marker. In other words, the origin of the coordinate is located on a moving marker, so it is a moving frame.

If we examine the atomic fluxes, expressed by J_A and J_B, from the laboratory frame (the fixed frame), the atomic fluxes consist of two terms as given below, the first due to diffusion and the second due to motion of the lattice with a velocity v:

$$J_A = -D_A \frac{\partial C_A}{\partial x} + C_A v \tag{5.22}$$

$$J_B = -D_B \frac{\partial C_B}{\partial x} + C_B v \tag{5.23}$$

Thus,

$$J_A + J_B = -D_A \frac{\partial C_A}{\partial x} - D_B \frac{\partial C_B}{\partial x} + C v \tag{5.24}$$

where D_A and D_B are intrinsic diffusivities, and C_A and C_B are concentrations of A and B in the AB alloy, respectively. Thus, $C_A + C_B = C$. In the equations, we use lower-case x as the x-axis of the diffusion coordinate. We shall use capital X as the atomic fraction of A and B as shown below. We define

$$X_A = \frac{C_A}{C_A + C_B} = \frac{C_A}{C} \tag{5.25}$$

$$X_B = \frac{C_B}{C_A + C_B} = \frac{C_B}{C} \tag{5.26}$$

where $X_A + X_B = 1$, and $CX_A + CX_B = C_A + C_B = C$. We can take C to be the total number of atoms per unit volume, which is a constant. Or, if we consider one mole of alloy, we can take $C = N_A$, where N_A is Avogadro's number and is equal to the number of atoms per mole. Since C is a constant, it implies that

$$\frac{\partial(C_A + C_B)}{\partial x} = 0$$

$$\frac{\partial C_A}{\partial x} = -\frac{\partial C_B}{\partial x}$$

$$\frac{\partial X_A}{\partial x} = -\frac{\partial X_B}{\partial x}$$

In the laboratory frame, the net flux is

$$J = J_A + J_B$$

and

$$\frac{\partial C}{\partial t} = -(\nabla \bullet J) = \frac{\partial}{\partial x}\left[D_A\frac{\partial C_A}{\partial x} + D_B\frac{\partial C_B}{\partial x} - Cv\right] \tag{5.27}$$

Since C is constant and is independent of time, $\partial C/\partial t = 0$, and we have

$$D_A\frac{\partial C_A}{\partial x} + D_B\frac{\partial C_B}{\partial x} - Cv = \text{Const.} = K \tag{5.28}$$

To determine the constant K, we consider the origin at the end of the sample, $x = 0$, where no interdiffusion occurs, so the concentration of C_A and C_B are constant, and the concentration gradient of $\partial C_A/\partial x$ and $\partial C_B/\partial x$ are zero. Thus, there is no marker motion, so $v = 0$. Therefore, the constant K in the last equation is zero. This means that the net flux in the laboratory frame is zero, so $J_A + J_B = 0$, and $J_A = -J_B$ Therefore, we see that in the laboratory frame, J_A and J_B are equal in magnitude but opposite in sign.

Now if we examine the atomic fluxes of A and B in the marker frame (or the moving frame), expressed respectively by j_A and j_B, we sit on a marker or sit on the origin of the frame, then we can take $v = 0$ and we have

$$j_A = -D_A \frac{\partial C_A}{\partial x} \tag{5.29}$$

$$j_B = -D_B \frac{\partial C_B}{\partial x} \tag{5.30}$$

So

$$j_A + j_B = -Cv \tag{5.31}$$

We see that in the marker frame, the net flux is not zero. To continue, we can rewrite the marker velocity as

$$v = \frac{1}{C} \left[D_A \frac{\partial C_A}{\partial x} + D_B \frac{\partial C_B}{\partial x} \right] = D_A \frac{\partial X_A}{\partial x} + D_B \frac{\partial X_B}{\partial x} = (D_B - D_A) \frac{\partial X_B}{\partial x} \tag{5.32}$$

In the following, we shall derive the atomic flux of B in the laboratory frame as

$$J_B = -\overline{D} \frac{\partial C_B}{\partial x}$$

$$\overline{D} = X_A D_B + X_B D_A \tag{5.33}$$

If we use the laboratory frame (the fixed frame) to examine the atomic flux, we have as given before

$$J_B = j_B + C_B v = -D_B \frac{\partial C_B}{\partial x} + C_B v = -D_B \frac{\partial C_B}{\partial x} + C_B (D_B - D_A) \frac{\partial X_B}{\partial x}$$

$$= -\frac{(C_A + C_B)}{C} D_B \frac{\partial C_B}{\partial x} + \frac{C_B}{C} (D_B - D_A) \frac{\partial C_B}{\partial x} = -\frac{C_A D_B}{C} \frac{\partial C_B}{\partial x} - \frac{C_B D_A}{C} \frac{\partial C_B}{\partial x}$$

$$= -\frac{1}{C} (C_A D_B + C_B D_A) \frac{\partial C_B}{\partial x} = -(X_A D_B + X_B D_A) \frac{\partial C_B}{\partial x} = -\overline{D} \frac{\partial C_B}{\partial x}$$

where $\overline{D} = X_A D_B + X_B D_A$ is defined as the interdiffusion coefficient. Similarly, we have in the laboratory frame

$$J_A = -\overline{D} \frac{\partial C_A}{\partial x} \tag{5.34}$$

Because $\partial C_A / \partial x = -\partial C_B / \partial x$, we have $J_A = -J_B$ as shown before. At this point, note the difference among the following three flux equations of B:

$$j_B = -D_B \frac{\partial C_B}{\partial x}$$

$$J_B = -D_B \frac{\partial C_B}{\partial x} + C_B v$$

$$J_B = -\overline{D} \frac{\partial C_B}{\partial x}$$

Similarly, the same kinds of equation can be written for the flux of A atoms too.

In the above analysis, we express the driving force of atomic flux in terms of concentration gradient. We can also express the driving force in terms of chemical potential gradient (see Appendix F), and we can obtain an expression of the interdiffusion coefficient as $\overline{D} = C_B M G''$, where M is mobility and G'' is the second derivative of Gibbs free energy against concentration, so the interdiffusion coefficient takes the sign of G''. Note that G'' is negative within the spinodal region, so $\overline{D}$ is negative, indicating that the diffusion is against the concentration gradient, i.e. an uphill diffusion. Outside the spinodal region, G'' is positive and the interdiffusion coefficient is positive as in the Darken's analysis. If we can measure $\overline{D}$ experimentally, we can solve D_A and D_B when we also measure the marker velocity experimentally. In the next section, we will show how to measure $\overline{D}$ by the Boltzmann and Matano analysis.

5.3.2.2 Boltzmann and Matano's analysis of interdiffusion

We can measure $\overline{D}$ experimentally by following the Boltzmann–Matano analysis. We begin by recalling Fick's second law,

$$\frac{\partial C_B}{\partial t} = \frac{\partial}{\partial x}\left(\overline{D}\frac{\partial C_B}{\partial x}\right) = \frac{\partial \overline{D}}{\partial x}\frac{\partial C_B}{\partial x} + \overline{D}\frac{\partial C_B}{\partial x}$$

To solve the above equation, we assume that

$$C(x,t) = C(\eta) \text{ and } \eta = \frac{x}{t^{1/2}}$$

By differentiation, we have

$$\frac{\partial C}{\partial t} = \frac{\partial C}{\partial \eta}\frac{\partial \eta}{\partial x} = -\frac{1}{2}\frac{x}{t^{3/2}}\frac{dC}{d\eta} = -\frac{\eta}{2t}\frac{dC}{d\eta}$$

$$\frac{\partial C}{\partial x} = \frac{dC}{d\eta}\frac{\partial \eta}{\partial x} = \frac{1}{t^{1/2}}\frac{dC}{d\eta}$$

$$\frac{\partial^2 C}{\partial x^2} = \frac{\partial}{\partial x}\left(\frac{\partial C}{\partial x}\right) = \frac{\partial}{\partial \eta}\frac{\partial \eta}{\partial x}\left(\frac{\partial C}{\partial x}\right) = \frac{1}{t}\frac{d^2 C}{dx^2}$$

$$\frac{\partial \overline{D}}{\partial x} = \frac{\partial \overline{D}}{\partial \eta}\frac{\partial \eta}{\partial x} = \frac{1}{t^{1/2}}\frac{d\overline{D}}{d\eta}$$

Substituting these terms into the diffusion equation, we have

$$-\frac{\eta}{2t}\frac{dC}{d\eta} = \frac{1}{t^{1/2}}\frac{d\overline{D}}{d\eta}\frac{1}{t^{1/2}}\frac{dC}{d\eta} + \frac{\overline{D}}{t}\frac{d^2 C}{d\eta^2}$$

We can cancel t in the above equation, and we have

$$-\frac{\eta}{2}\frac{dC}{d\eta} = \frac{d\overline{D}}{d\eta}\frac{dC}{d\eta} + \overline{D}\frac{d^2 C}{d\eta^2} = \frac{d}{d\eta}\left(\overline{D}\frac{dC}{d\eta}\right)$$

Because these are total differentials, we can drop $1/d\eta$ and integrate both sides to give

$$-\frac{1}{2}\int_0^{C'} \eta dC = \int_0^{C'} d\left(\overline{D}\frac{dC}{d\eta}\right) = \left[\overline{D}\frac{dC}{d\eta}\right]_0^{C'} \tag{5.35}$$

where C' is an arbitrary concentration, $0 < C' < C_0$, and C_0 is the concentration of A at $x = \infty$. Now we consider the physical picture of the interdiffusion. If we consider a given time (i.e. t is fixed, so $d\eta = t^{-1/2}dx$), we have at both ends of the diffusion couple, $dC/d\eta = 0$ at $C = 0$ and $C = C_0$. Therefore, in Eq. (5.35), if we integrate from $C = 0$ to $C = C_0$ along the "vertical" axis, it gives

$$-\frac{1}{2}\int_0^{C_0} \eta dC = \overline{D}\left.\frac{dC}{d\eta}\right|_{C=C_0} - \overline{D}\left.\frac{dC}{d\eta}\right|_{C=0} = 0 - 0 = 0$$

This means that

$$\int_0^{C_0} x dC = 0 \tag{5.36}$$

because the interdiffusion is considered at a fixed time, so that the variable η is the same as x when t is constant. This relationship, Eq. (5.36), defines the Matano interface, where the quantity of A atoms $(1 - C_A)$ that have been removed from the left of the Matano interface is equal to the quantity C_A that has been added to the right of the interface. Graphically, it means in the sketch in Fig. 5.10 that the shaded area in A (left of the interface) is equal to the shaded area in B (right of the interface). The reference interface is defined as the Matano interface. It is not at the same location as the original interface.

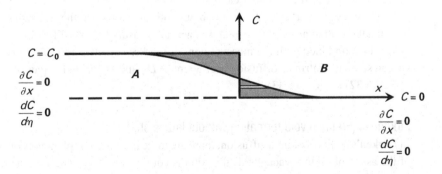

Fig. 5.10   Equation (5.36) defines the Matano interface. With respect to the Matano interface, the quantity of A atoms $(1 - C_A)$ that have been removed from the left of the Matano interface is equal to the quantity C_A that has been added to the right of the interface. Graphically, it means that the shaded area in A (left of the interface) is equal to the shaded area in B (right of the interface).

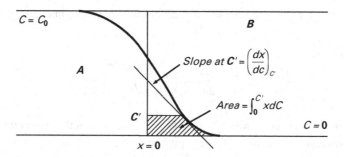

Fig. 5.11   The interdiffusion coefficient $\overline{D}$ at a concentration C' can be obtained by a graphical method (when t is given), using the slope and the shaded area as shown in the figure. On the measured profile, we can choose a concentration of C' and obtain both the slope and the shaded area.

Why is the Matano interface important? It defines the location of the origin of the x-axis, i.e. $x = 0$ for the integration in Eq. (5.35) and Eq. (5.36). Otherwise, the integral of

$$-\frac{1}{2} \int_0^{C'} \eta dC$$

is "undetermined" since it is integrated over C, i.e. over the vertical axis, and x is arbitrary until the origin of x is defined by the Matano interface.

In Eq. (5.35), if we convert η to x, and define $\overline{D}$ to be the interdiffusion coefficient, we have

$$\overline{D}(C') = -\frac{1}{2t}\left(\frac{dx}{dC}\right)_{C'} \int_0^{C'} xdC \qquad (5.36)$$

This equation indicates that the measurement of C as a function of x (i.e. the concentration profile of A) can give $\overline{D}$ at a concentration C' by a graphical method (when t is given), using the slope and the shaded area as shown in Fig. 5.11. On the measured profile, we can choose a concentration of C' and obtain both the slope and the shaded area. Thus, by using the Boltzmann–Matano analysis, we can measure the interdiffusion coefficient. When it is combined with the measurement of marker velocity in the diffusion couple, we can solve the intrinsic diffusion coefficient of D_A and D_B using the pair of equations of Eq. (5.32) and Eq. (5.33).

5.3.2.3   Kirkendall (Frenkel) void formation without lattice shift

In Darken's analysis of interdiffusion, there are three important implications on the basis of the assumption that vacancy distribution is equilibrium everywhere in the diffusion couple. First, lattice shift or Kirkendall shift occurs which can be measured by marker motion. Second, because of lattice shift, no stress is generated. Third, because equilibrium vacancy is assumed, there cannot be supersaturated vacancies, so no nucleation of void and no void formation can take place.

However, in actual diffusion couples, the vacancy sources and sinks are often partially effective, and voids have been found. The voids are often called Frenkel voids rather than Kirkendall voids. This is because, strictly speaking, under the assumption of lattice shift or Kirkendall shift, there should be no Kirkendall void. Void formation implies no lattice shift. But experimentally, Kirkendall shift and Kirkendall void can coexist. Furthermore, it is known that by using hydrostatic compression, void formation can be suppressed in interdiffusion. This implies stress. Indeed, diffusion couple bending occurs due to stress in particular in the interdiffusion between a metal thin film and a Si wafer, to be discussed in Chapter 8. Also, in electromigration to be discussed in Chapter 11, back stress is an issue.

Recall that in the marker frame the net atomic flux is not zero. If we assume the mechanism of atomic diffusion to be substitutional diffusion via a vacancy mechanism, we can add a flux of vacancy, j_V, to balance the fluxes.

$$j_A + j_B + j_V = 0 \tag{5.37}$$

Thus, we have $j_V = Cv$ from Eq. (5.31). It indicates that the marker velocity is controlled by the vacancy flux or the net flux between A and B:

$$j_V = (D_B - D_A)\frac{\partial C_B}{\partial x}$$

If we assume that the flux j_A is larger than j_B, we can rewrite $j_A = j_B + j_V$, which means that the flux of B and the flux of vacancy are moving against the flux of A. In Fig. 5.12, schematic curves are shown to depict: (a) the composition profile in interdiffusion of A and B, (b) the first derivative of composition for the fluxes of A, B, and vacancy as a function of x, (c) the second derivative of composition for the rate at which the vacancy concentration would increase in A and decrease in B in order to achieve vacancy equilibrium in the couple. Let C_V be the vacancy concentration; the rate can be expressed as

$$\frac{\partial C_V}{\partial t} = -\frac{\partial j_V}{\partial x}$$

The right-hand side of the above equation is a divergence of vacancies. If the distribution of vacancy or the vacancy concentration in the diffusion couple is equilibrium everywhere, there will be no void formation, although there is vacancy divergence. Thus, vacancy divergence alone may not lead to failure.

In order to have equilibrium vacancy everywhere in the sample, we must assume that the sources and sinks of vacancies are fully operative everywhere in the sample. Only if the sinks for vacancies in the A side cannot absorb all of them, do excess vacancies occur. It can lead to supersaturation of vacancies, so the nucleation of void can occur. The void grows by becoming a sink to the vacancies. However, the absorption of vacancies by the void, unlike that by dislocation, will not lead to lattice shift. The void growth takes lattice sites.

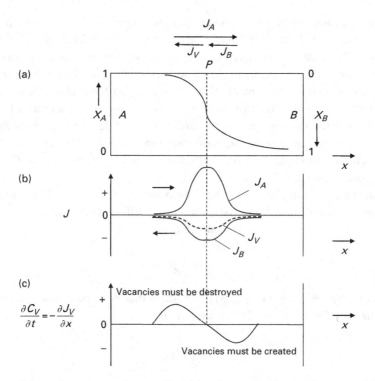

Fig. 5.12   Schematic curves are shown to depict: (a) the composition profile in interdiffusion of A and B, (b) the fluxes of A, B, and vacancy as a function of x, and (c) the rate at which the vacancy concentration would increase in A and decrease in B, if the excess vacancies were not destroyed in A and created in B due to ineffective sinks and sources, respectively.

On the B side, if it cannot generate all the vacancies needed for the incoming A atoms, the A atoms will take the lattice site of equilibrium vacancies, and it leads to the reduction of equilibrium vacancy concentration quickly because the number of equilibrium vacancies is quite small. Then a compressive stress state will be built up in the B side. We imagine that if we diffuse atoms into a fixed volume of V and if there is no lattice shift, each atom will add an atomic volume of Ω to the fixed volume, and a compressive stress will be introduced by

$$\sigma = -B_m \frac{\Omega}{V} \qquad (5.38)$$

where B_m is the bulk modulus and the negative sign is to indicate that it is compressive. Without lattice shift to generate vacancies in B, the only way to reduce the stress is to have a flux of atoms to diffuse out of the volume V to the free surface of B. A free surface has no normal stress, so there is a stress gradient to drive the diffusion and it leads to hillock or whisker growth in the B side. Similarly, in the A side we might expect a tensile stress to build up. But it is much more difficult to develop a hydrostatic tensile stress in real systems; instead, a void will form and this is the reason why we find void formation in bulk interdiffusion couples. The void surface is a free surface and has no

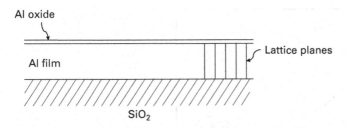

Fig. 5.13 Schematic diagram of the cross-section of a thin Al film having an oxide surface. When the oxide has a strong binding to the lattice planes, it will prevent the lattice plane motion; in turn there will be no lattice shift, so extra lattice sites will be created in the diffusion couple. Both void formation and stress occur.

normal stress. Thus, a stress gradient exists between the tensile region and the void, so the void can become the sink of vacancies in the tensile region.

Why is there no lattice shift? We have depicted in Fig. 5.9 the physical picture of dislocation climb to serve as the source and sink of vacancies in interdiffusion. The outcome of the interdiffusion is the successive motion of lattice planes from B to A.

In the above analysis, it is assumed that lattice planes can migrate freely. If not, there is no lattice shift. For example, as depicted in Fig. 5.13, if we assume that the two ends of a lattice plane are pinned by oxide on the sample surfaces, the plane will not be able to migrate freely. This happens in Al thin-film interconnect lines, where the thickness of the Al lines is very thin, less than half of a micron, so the near-surface effect due to its oxide is large. Since Al is known to have good adhesion on a glass surface and Al oxide is known to be protective, which means that the bonding between Al and SiO_2 and also the bonding between Al and Al_2O_3 are very strong, the oxides can prevent the lattice planes of Al from moving. Thus, when Al atoms are driven by electromigration from the cathode into the anode, a compressive stress is developed in the anode and a void is formed in the cathode because of the absence of lattice shift. In the process, lattice sites are created, so it is a non-constant volume process.

It is worthwhile mentioning that when the atomic sizes of A and B are different, $\Omega_A \neq \Omega_B$, there will be a volume change when we exchange A and B in the interdiffusion. For example, if we assume that the size of the A atom is smaller than that of B, and A diffuses faster into B, there could be void formation due to the molar volume change and we may mistake it for a Kirkendall or Frenkel void that is due to excess vacancies. But the molar volume change should occur in the middle of the diffusion couple where the exchange of A and B is large, or it should be more in the B side because more A have diffused into B. Yet the location of Kirkendall or Frenkel void formation follows the direction of diffusion; it should form in the region of the highest concentration of vacancy in the side of the faster diffusing species; it forms in the A side.

5.3.2.4 Interdiffusion coefficient

We have an interdiffusion coefficient given as $\overline{D} = X_A D_B + X_B D_A$. To understand the physical meaning of an interdiffusion coefficient, an intrinsic diffusion coefficient, and

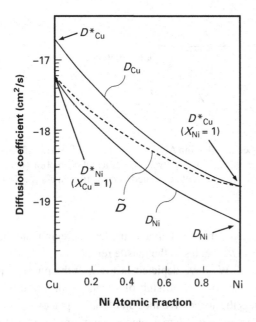

Fig. 5.14   Relations among interdiffusion coefficient, intrinsic diffusion coefficient, and tracer diffusion coefficient are shown by using a schematic plot of these diffusion coefficients of CuNi alloys.

a tracer diffusion coefficient, we show a schematic plot of these diffusion coefficients of CuNi alloy in Fig. 5.14. When we diffuse isotope Cu in pure Cu or isotope of Ni in Ni, we obtain the tracer diffusion coefficient of $D^*_{Cu/Cu}$ and $D^*_{Ni/Ni}$, respectively. They are shown on the vertical coordinates of pure Cu and pure Ni in Fig. 5.14. The intrinsic diffusivity of D_A and D_B are functions of alloy composition. They can be solved by knowing the marker velocity and interdiffusion coefficient, as discussed in the previous section.

Note that the interdiffusion coefficient determines how fast A and B mix, i.e. how fast A and B diffuse in the concentration gradient. There is a chemical force because of homogenization. The chemical effect is contained in the intrinsic diffusion coefficients of D_A and D_B:

$$D_A = D^*_A \left(1 + \frac{\partial \ln \gamma_A}{\partial \ln C_A} \right)$$ (5.39)

where γ_A is the activity coefficient.

5.4   Analysis of growth of a solid precipitate

In the derivation of Fick's second law or the continuity equation in Chapter 4, we assume that in the small cubic box, the concentration can change. This is true when we consider the diffusion in a gas phase or the diffusion of solute atoms in a solid phase. But in many solid-state kinetic problems, we consider the growth of a pure solid phase or the growth of a precipitate; there is negligible concentration change in the pure solid phase or the precipitate. When fluxes of atoms come to them, what will occur is a growth in size rather than change in concentration. In other words, we do not consider dC/dt, rather, we consider dx/dt or dr/dt. This is already shown in the growth of a planar precipitate in Section 5.2. Typically the following procedures are taken to solve the problem.

(1) Use the Fick's second law to set up the diffusion equation, and choose the coordinates and the initial and boundary conditions.
(2) Solve the diffusion equation to obtain the concentration profile. Then, a steady state is often assumed so we have C as a function of x or r.
(3) Use Fick's first law to obtain the atomic flux arriving at the growth front.
(4) By considering conservation of mass or volume, the growth equation is obtained.
(5) Check the dimension of the solution to see if it is correct.

We illustrate step (4) below and steps (1)–(3) and (5) in the next section. Let us consider the growth of a spherical particle by the diffusion in the spherical coordinate. The flux arriving at the particle surface is J and the radius of the particle is r. Since the total number of atoms arriving at the particle surface in a period of Δt is $N = JA\Delta t$, we have added the following volume to the precipitate,

$$V = \Omega N = \Omega J 4\pi r^2 \Delta t$$

where Ω is atomic volume. On the other hand, the volume of the sphere is

$$V' = \frac{4}{3}\pi r^3 \text{ and } dV' = 4\pi r^2 dr$$

The volume change, dV', should be equal to the added volume, V, in Δt. It will thicken the precipitate by dr or add a shell of volume of $4\pi r^2 dr$ to the precipitate in time dt; thus, we have

$$4\pi r^2 dr = \Omega J 4\pi r^2 dt$$

so the growth rate of the particle is $dr/dt = \Omega J$. A more detailed presentation of the growth is given in the next section.

5.4.1 Ham's model of growth of a spherical precipitate (C_r is constant)

This is the classic model of growth of a precipitate under diffusion-controlled kinetics. In Fig. 5.15, we depict a spherical particle of radius r. The solute in the large sphere of radius r_0 will contribute to the growth of the precipitate. Let R be the variable. The diffusion equation in spherical coordination, assuming a steady state, is

$$\frac{\partial^2 C}{\partial R^2} + \frac{2}{R}\frac{\partial C}{\partial R} = 0$$

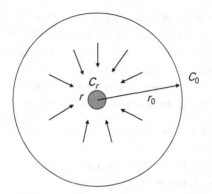

Fig. 5.15   Schematic diagram of growth of a precipitate under diffusion-controlled kinetics. We depict a spherical particle of radius r. The solute in the large sphere of radius r_0 will contribute to the growth of the precipitate.

The solution is

$$C = \frac{b}{R} + d \qquad (5.40)$$

The boundary conditions are

$$\text{At } R = r_0, C = C_0, \text{ we have } C_0 = \frac{b}{r_0} + d \qquad (5.41)$$

$$\text{At } R = r, C = C_r, \text{ we have } C_r = \frac{b}{r} + d \qquad (5.42)$$

Now, if we take the difference between the last two equations, we have

$$C_r - C_0 = b\left(\frac{1}{r} - \frac{1}{r_0}\right) = b\frac{r_0 - r}{rr_0} \cong \frac{b}{r} \quad \text{where } r_0 \gg r \qquad (5.43)$$

The approximation of $r_0 \gg r$ is an important assumption. It means that precipitates are far apart, and it also means that the volume fraction of the precipitates is very small. Note that if we take the volume fraction, f, the ratio of volume of the precipitate particles to the volume of the diffusion field, or the ratio of the total volume of the precipitated phase to the total volume of the matrix, to be

$$f = \frac{(4\pi/3)r^3}{(4\pi/3)r_0^3} = \frac{r^3}{r_0^3} \to 0 \qquad (5.44)$$

it is a very small value: $f \to 0$. (This is a very important assumption in the Lifshitz–Slezov–Wagner (LSW) theory of ripening to be discussed in the next section.)

We have $b = r(C_r - C_0)$. Substituting b into Eq. (5.42), we have

$$C_r = \frac{r(C_r - C_0)}{r} + d \tag{5.45}$$

We have $d = C_0$, and Eq. (5.40) becomes

$$C(R) = \frac{(C_r - C_0)r}{R} + C_0 \tag{5.46}$$

This is the solution of the diffusion equation. Therefore,

$$\frac{dC}{dR} = -\frac{(C_r - C_0)r}{R^2}$$

At the particle/matrix interface for a particle of radius r, or $R = r$, we have

$$\frac{dC}{dR} = -\frac{C_r - C_0}{r} \tag{5.47}$$

Then, the flux of atoms arriving at the interface is

$$J = +D\frac{\partial C}{\partial R} = \frac{D(C_0 - C_r)}{r}, \text{ at } R = r \tag{5.48}$$

Note that when $C_r > C_0$, $J < 0$, the net flux is toward the particle, and it grows. When $C_r < C_0$, $J > 0$, the flux leaves the particle, so the particle dissolves.

In the case of growth, if Ω is atomic volume, a volume is added to the spherical particle in time dt,

$$\Omega J A dt = \Omega J 4\pi r^2 dt = 4\pi r^2 dr$$

where the last term is the increment of a spherical shell due to the growth. Hence

$$\frac{dr}{dt} = \Omega J = \frac{\Omega D(C_0 - C_r)}{r} \tag{5.49}$$

By integration and assuming that $r = 0$ when $t = 0$,

$$r^2 = 2\Omega D(C_0 - C_r)t \tag{5.50}$$

Note here that if we follow Ham's approach [5] and take C_r as a constant, it is not a function of r. (If the precipitate is very small, C_r will be a function of r as required by the Gibbs–Thomson equation, to be discussed later.) From the above equation, we see that $r \cong t^{1/2}$ and $r^3 \cong t^{3/2}$. Or, we have

$$r^3 = [2\Omega D(C_0 - C_r)t]^{3/2} \tag{5.51}$$

5.4.2   Mean-field consideration

We consider the average lost concentration in the matrix, $\Delta \overline{C} = C_0 - \overline{C}$, due to the formation of the precipitate, where the average concentration in the matrix is $\overline{C}$, which can be regarded as the "mean-field" concentration (this is the starting point of mean-field theory). In the beginning, the average concentration is C_0, but it changes to $\overline{C}$ when the precipitate grows.

Let $1/\Omega = C_p$ be the concentration in the solid precipitate. We have by mass balance

$$\frac{4\pi}{3}r_0^3(C_0 - \overline{C}) = \frac{4\pi}{3}r^3\frac{1}{\Omega} = \frac{4\pi}{3\Omega}[2\Omega D(C_0 - C_r)t]^{3/2} \tag{5.52}$$

$$\overline{C} = C_0 - \left[\frac{2D(C_0 - C_r)\Omega^{1/3}}{r_0^2}t\right]^{3/2} = C_0 - \left[\frac{2Bt}{3}\right]^{3/2} \tag{5.53}$$

where $B \equiv \frac{3D(C_0 - C_r)}{C_p^{1/3}r_0^2}$.

Note that the above equation is the same as Eq. (1.36) in Chapter 1 in Shewmon's book *Diffusion in Solids* [5].

We can derive the last equation in a slightly different way. The growth of the precipitate reduces the concentration in the matrix. The amount of solute atoms which diffuse to the precipitate in time Δt is $J(r)4\pi r^2 \Delta t$. The number of atoms should be equal to the reduction of the average concentration in the volume of the sphere of diffusion of r_0. Hence, if we take the average concentration in the matrix to be $\overline{C}$,

$$\frac{4\pi r_0^3}{3}\Delta\overline{C} = J(r)4\pi r^2 \Delta t$$

Or, we have

$$\frac{\Delta\overline{C}}{\Delta t} = \frac{3}{4\pi r_0^3}4\pi r^2 J(r) = -\frac{3D}{r_0^3}(C_0 - C_r)r \tag{5.54}$$

The conservation of mass requires that

$$\frac{4\pi}{3}r_0^3(C_0 - \overline{C}) = \frac{4\pi}{3}r^3 C_p \tag{5.55}$$

where C_p is the concentration of solute in the solid precipitate and $C_p = 1/\Omega$. Hence,

$$r = r_0\left(\frac{C_0 - \overline{C}}{C_p}\right)^{1/3} \tag{5.56}$$

By substituting r into the rate equation in the above, we have

$$\frac{\Delta\overline{C}}{\Delta t} = -\frac{3D}{r_0^2}(C_0 - C_r)\frac{1}{C_p^{1/3}}(C_0 - \overline{C})^{1/3} \tag{5.57}$$

Let

$$B \equiv \frac{3D(C_0 - C_r)}{C_p^{1/3} r_0^2}$$

We have

$$\frac{d\overline{C}}{dt} = -B(C_0 - \overline{C})^{1/3}$$

By integration, we obtain

$$-\frac{3}{2}(C_0 - \overline{C})^{2/3} = -Bt + \beta$$

at $t = 0$, $C_0 = \overline{C}$, so $\beta = 0$.

Thus, we have the solution,

$$\overline{C} = C_0 - \left(\frac{2Bt}{3}\right)^{3/2} \tag{5.58}$$

which is the same as what we obtained before. Hence, we have $C_0 - \overline{C} \cong t^{3/2}$ for a three-dimensional growth.

$$\text{Let } \overline{C} = C_0 \left[1 - \left(\frac{2Bt}{3C_0^{2/3}}\right)^{3/2}\right] = C_0 \left[1 - \left(\frac{t}{\tau}\right)^{3/2}\right] = C_0 \exp\left[-\left(\frac{t}{\tau}\right)^{3/2}\right] \tag{5.59}$$

if we assume that $t \ll \tau$, where

$$\tau = \frac{C_p^{1/3} r_0^2 C_0^{2/3}}{2D(C_0 - C_r)} \cong \frac{r_0^2}{2D}\left(\frac{C_p}{C_0}\right)^{1/3} \tag{5.60}$$

Usually, D, C_p, C_0 are known; we can design the experiment to control the growth of the precipitate.

5.4.3 Growth of a spherical nanoparticle by ripening

In the above analysis of the growth of a large-size precipitate, we can assume the equilibrium concentration C_r to be constant, independent of the size of the precipitate. This assumption is not true when the particle is small, for example, in the nanoscale. We must consider the Gibbs–Thomson potential and allow C_r to be a function of r. When we have a large number of particles, there is a distribution of size. There will be ripening action among them. To analyze the ripening among particles having a size distribution, we shall use the concept of mean field, which can be regarded as the average concentration of all the particles. We assume that there is a critical-size r^* of a particle that is in equilibrium with the mean field. Then, the ripening of any particle can be analyzed against the critical size particle. For those that are larger than the critical size, they will grow. For those that are smaller, they will shrink. Below, we shall first develop the Gibbs–Thomson potential.

Consider a small sphere with radius r and surface energy per unit area γ. The surface energy exerts a compressive pressure on the sphere because it tends to shrink to reduce the surface energy. The pressure equals

$$p = \frac{F}{A} = \frac{dE/dr}{A} = \frac{d4\pi r^2 \gamma / dr}{4\pi r^2} = \frac{8\pi r \gamma}{4\pi r^2} = \frac{2\gamma}{r} \tag{5.61}$$

If we multiple p by atomic volume Ω, we have the chemical potential

$$\mu_r = \frac{2\gamma\Omega}{r} \tag{5.62}$$

This is called the Gibbs–Thomson potential owing to the small curvature of the surface. Note that it is not just the potential of the surface atoms of the precipitate, but of all the atoms in the precipitate. We see that for a flat surface $r = \infty$, $\mu_\infty = 0$ so we have

$$u_r - \mu_\infty = \frac{2\gamma\Omega}{r} \tag{5.63}$$

In the following, we shall apply this potential to determine the effect of curvature on solubility and then ripening among a set of particles of varying size.

We consider an alloy of $\alpha = A(B)$, where B is solute in solvent A. At a given low temperature, B will precipitate out. We consider a precipitate of B with a radius r. The solubility of B surrounding the particle is taken to be $X_{B,r}$. To relate the solubility to the Gibbs–Thomson potential, we have the chemical potential of B as a function of its radius as

$$\mu_{B,r} - \mu_{B,\infty} = \frac{2\gamma\Omega}{r} \tag{5.64}$$

where γ is the interfacial energy between the precipitate and the matrix. If we define the standard state of B as pure B with $r = \infty$, we have

$$\mu_{B,r} = \mu_{B,\infty} + RT \ln a_B \tag{5.65}$$

where a_B is the activity. According to Henry's law

$$a_B = KX_{B,r}$$

where $X_{B,r}$ is the solubility of B surrounding a precipitate of radius r. At $r = \infty$,

$$\mu_{B,\infty} = \mu_{B,\infty} + RT \ln a_B$$

It implies that $RT \ln a_B = 0$, or $a_B = 1$. So $K = 1/X_{B,\infty}$.
  Therefore,

$$\mu_{B,r} = \mu_{B,\infty} + RT \ln \frac{X_{B,r}}{X_{B,\infty}} \tag{5.66}$$

Hence,

$$\ln \frac{X_{B,r}}{X_{B,\infty}} = \frac{\mu_{B,r} - \mu_{B,\infty}}{RT} = \frac{2\gamma\Omega}{rRT}$$

Or, if we consider kT per atom instead of RT per mole, we have

$$X_{B,r} = X_{B,\infty} \exp\left(\frac{2\gamma\Omega}{rkT}\right) \tag{5.67}$$

The solubility of B around a spherical particle of B of radius r is given by the above equation. When $r = \infty$, the exponential equals unity. So $X_{B,r}$ goes up when r goes down. Now we replace $X_{B,r}$ by C_r and $X_{B,\infty}$ by C_∞, which is the equilibrium concentration on a flat surface; we have

$$C_r = C_\infty \exp\left(\frac{2\gamma\Omega}{rkT}\right) \tag{5.68}$$

If $2\gamma\Omega \ll rkT$, we have

$$C_r = C_\infty\left(1 + \frac{2\gamma\Omega}{rkT}\right)$$

$$C_r - C_\infty = \frac{2\gamma\Omega C_\infty}{rkT} = \frac{\alpha}{r} \tag{5.69}$$

where $\alpha = \frac{2\gamma\Omega}{kT} C_\infty$. So

$$C_r = C_\infty + \frac{\alpha}{r} \tag{5.70}$$

Thus, we obtain the very important result that C_r is not a constant, but a function of r. Now we substitute C_r into the growth equation of

$$\frac{dr}{dt} = \Omega J = \frac{\Omega D(C_0 - C_r)}{r}$$

We have

$$\frac{dr}{dt} = \frac{\Omega D}{r} \left(C_0 - C_\infty - \frac{\alpha}{r} \right) \tag{5.71}$$

Note that $C_0 - C_\infty > 0$ always. We can define a critical radius r^* such that

$$C_0 - C_\infty = \frac{\alpha}{r^*}$$

We can regard the concentration which is in equilibrium with r^* as the "mean-field" concentration. In considering the ripening of any particle of radius r, large or small, we just consider the ripening of this particle against the critical particle of r^*, or against the mean field. Then we have

$$\frac{dr}{dt} = \frac{\alpha \Omega D}{r} \left(\frac{1}{r^*} - \frac{1}{r} \right) \tag{5.72}$$

The parameter r^* is defined such that

$$r > r^*, \frac{dr}{dt} > 0 \quad \text{The particle is growing.}$$

$$r < r^*, \frac{dr}{dt} < 0 \quad \text{The particle is dissolving.}$$

$$r = r^*, \frac{dr}{dt} = 0 \quad \text{The particle is in a state of metastable equilibrium.}$$

It has a concentration $\overline{C}$ at the interface, or $C_r^* = \overline{C}$.

In ripening, the larger particles grow at the expense of the mean field and the smaller particles will shrink with respect to the mean field. The mean-field concentration will decrease with time. It will approach a dynamic equilibrium distribution of size of the particles. The distribution function can be obtained by solving the continuity equation in size space as given by the LSW theory of ripening. Knowing dr/dt, it is the beginning of the LSW theory of ripening [7, 8].

References

[1] D. Turnbull, "Phase changes," *Solid State Physics* **3** (1965) 225.
[2] J. W. Christian, *The Theory of Transformation in Metals and Alloys* (Pergamon Press, New York, 1965).
[3] D. A. Porter and K. E. Easterling, *Phase Transformations in Metals and Alloys* (Chapman and Hall, London, 1992).
[4] P. G. Shewmon, *Transformations in Metals* (Indo American Books, Delhi, 2006).
[5] P. G. Shewmon, *Diffusion in Solids*, 2nd edn (TMS, Warrendale, PA, 1989).
[6] A. P. Sutton and R. W. Balluffi, *Interfaces in Crystalline Materials* (Oxford University Press, Oxford, 1995).

[7] V. V. Slezov, Chapter 4 in *Kinetics of First-order Phase Transitions* (Wiley-VCH, Weinheim, 2009).

[8] A. M. Gusak and K. N. Tu, "Kinetic theory of flux-driven ripening," *Phys. Rev.* **B66**, (2002) 115403.

Problems

5.1 What is lattice shift in interdiffusion?

5.2 Define an interdiffusion coefficient, intrinsic diffusion coefficient, and tracer diffusion coefficient.

5.3 What is the Matano interface in interdiffusion?

5.4 In interdiffusion, what is the direction of marker motion? Is it moving in the direction as the faster diffusing species?

6 Elastic stress and strain in thin films

6.1 Introduction

Thin films are not used as structural parts in electronic devices to carry mechanical loads. Nevertheless, stress or strain does exist commonly in thin films as a result of constraints imposed by their substrates. A thin film and its substrate generally have different thermal expansion coefficients, so stress is produced during temperature excursion in deposition and annealing. Stress in thin films is known to cause serious yield and reliability problems in microelectronic devices. Ni thin film is known to have a high tensile stress deposited by e-gun or by sputtering at room temperature. In epitaxially grown silicon or a silicon-germanium layer, stress can affect the mobility of the carriers, and stress is introduced in devices for the purpose. In this chapter, we shall discuss the nature of biaxial stress in thin films, and the measurement of biaxial stress in thin films using the wafer-bending method. The chemical potential in a stressed solid that affects atomic diffusion and the time-dependent response of a solid to applied stresses in creep or stress-migration will be covered in Chapter 14.

A piece of solid is under stress when its atoms are displaced from their equilibrium positions by a force [1–6]. The displacement is governed by the interatomic potential. It is well known that the potential ϕ and the internal force F ($F = -\partial\phi/\partial r$) between two atoms as a function of interatomic distance generally obey the schematic relations shown in Fig. 3.1(a) and Fig. 3.1(b). When we consider an applied external force instead, we define

$$F_{ex} = +\frac{\partial\phi}{\partial r} \tag{6.1}$$

where we have changed the sign to be positive compared to the internal force between two atoms. An external tensile force tends to lengthen the solid and in turn to increase the interatomic distance. On the basis of the sign convention given in Chapter 3, Section 3.2, a force which increases the interatomic distance is positive, and hence the external tensile force (or stress) is positive. An external compressive force (or stress) which tends to shorten the solid is negative. The interatomic potential, the external force, and the sign of the force are shown schematically in Fig. 6.1(a) to (c), respectively.

Clearly, Fig. 6.1(b) is an inverted diagram of Fig. 3.1(b). We define the point F_{max} to be the maximum force which corresponds to the dissociation distance r_D. The maximum tensile force needed to pull the solid apart is F_{max}, because the force needed to increase the

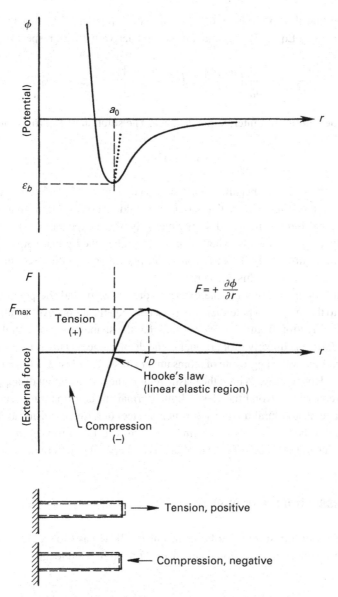

Fig. 6.1 (a) Interatomic potential function plotted against interatomic distance. The dotted curve shows the anharmonicity of atomic vibration. (b) Applied force plotted against atomic displacement. (c) The direction and sign of applied force by convention.

interatomic distance beyond r_D is less than F_{max}. We can regard F_{max} as the theoretical strength of the solid. To calculate F_{max}, we require that

$$\frac{\partial^2 \phi}{\partial r^2} = 0 \text{ at } r = r_D$$

If we assume that the solid obeys the Lennard-Jones potential and that the potential function is given by Eq. (3.7), we obtain its second derivative with respect to r,

$$\frac{\partial^2 \phi}{\partial r^2} = \varepsilon_b \frac{12}{a_0^2} \left(\frac{a_0}{r}\right)^8 \left[13\left(\frac{a_0}{r}\right)^6 - 7\right] = 0 \quad \text{at } r = r_D \qquad (6.2)$$

where a_0 is the equilibrium interatomic distance. The solution of Eq. (6.2) shows that

$$r_D = 1.11 a_0$$

Theoretically, the solid can be stretched (strained) by about 11% before it breaks. Furthermore, if stretched just below that strain, it would return to the original condition when the external force is removed. Experimentally, these expectations are not true at all. Most polycrystalline metals, whether or not they obey the Lennard-Jones potential, have an elastic limit of only 0.2%; beyond that, plastic deformation sets in. We shall consider elastic behavior in this chapter.

At the equilibrium position a_0, the external force is zero, and the potential energy corresponds to the minimum potential energy ε_b between the atoms. At a small displacement in either direction from a_0, if we assume that the shape of the potential energy is parabolic, the force is linearly proportional to the displacement. This is the origin of the elastic behavior in a solid aggregate of atoms under stress. The elastic behavior observed is described by Hooke's law. Within the elastic region, the displacement disappears when the force is removed. Beyond the elastic limit, permanent deformation occurs. In permanent damage, a structural ductile solid such as steel deforms by dislocation motion, but a brittle solid such as glass will deform by fracture via crack propagation. The major difference is due to the nature of chemical bonds and crystal structure in these solids.

6.2   Elastic stress–strain relationship

Consider a piece of thin solid film of dimensions $l \times W \times t$ as shown in Fig. 6.2. If we apply a force F to the cross-sectional area $A = Wt$ to stretch the length of the film l by Δl, we have

$$\frac{F}{A} = Y\frac{\Delta l}{l} \text{ or } \sigma = Y\varepsilon \qquad (6.3)$$

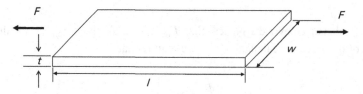

Fig. 6.2   Piece of thin film of dimensions $l \times w \times t$ under tension.

where $\sigma = F/A$ and $\varepsilon = \Delta l/l$ are stress and strain, respectively, and Y is Young's modulus. This is Hooke's law. In addition

$$\frac{\Delta t}{t} = \frac{\Delta W}{W} = -\nu\frac{\Delta l}{l} \tag{6.4}$$

where ν is Poisson's ratio. This ratio is a positive number for almost all materials and is about or less than one-half. Ideally, if we consider a cylindrical sample of radius r and length l under tension, we have the following relation under the assumption of constant volume before and after the tension, Eq. (3.22) in Chapter 3,

$$\pi r^2 l = \pi (r + dr)^2 (l + dl)$$

By ignoring the higher-order terms, we obtain

$$\frac{dr}{r} = -\frac{1}{2}\frac{dl}{l}$$

Under the theoretical assumption of constant volume in elastic deformation, the Poisson ratio is 0.5. Notice that there is a negative sign before ν which means that while we stretch l, both W and t shrink. An easy way to recognize Poisson's ratio is to observe the change in lattice parameter by X-ray diffraction in the direction normal to the tensile stress. For example, take a single-crystal film of cubic crystal structure and stretch it by bending, as shown in Fig. 6.3. The cubic unit cell deforms into a tetragonal cell (dashed rectangle), assuming that the unit cell volume is constant. The interplanar spacing normal to the substrate surface decreases. This decrease is measured by the shift in angle of the X-ray diffraction.

Different materials have different values of Y and ν. The elastic behavior of a polycrystalline material is characterized by just these two parameters. Sometimes other parameters such as shear modulus and bulk modulus are given, but they are interrelated. An example is given in the following discussion.

We consider shear strain and illustrate it with the sketch shown in Fig. 6.4(a). It is a schematic diagram of the cross-sectional view of a Si chip joined by two solder joints to a ceramic pad in the flip-chip packaging scheme. During operation, the device will experience a temperature rise of $\sim 100\,°\mathrm{C}$. Since Si expands more than the ceramic, the solder joints experience a shear strain. The strain is actually cyclic because the device is

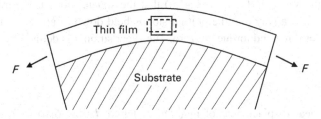

Fig. 6.3   Thin film is stretched by bending the substrate.

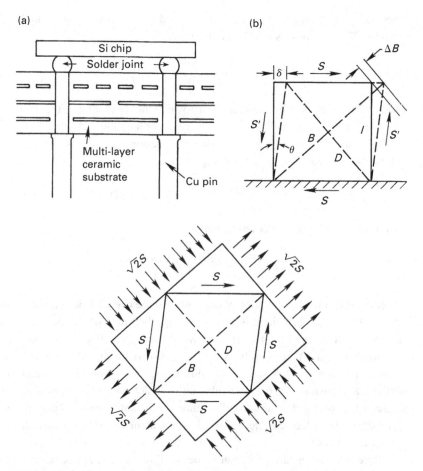

Fig. 6.4   (a) Schematic diagram of a Si chip solder-joined to a ceramic substrate. (b) A square is sheared by a pair of shear forces S. (c) The shear stress is converted to a combination of tensile and compressive stresses.

being turned on and off frequently. The cyclic strain has been found to cause reliability failure of low-cycle fatigue of the solder joints. Furthermore, we can imagine that if we increase the chip size, the solder joints at the edges of the chip will have a greater shear strain. Therefore, we cannot increase the chip size arbitrarily. Shear strain is a critical factor limiting the yield of the device and also the size of the Si chip.

By shear strain, we mean (see Fig. 6.4(b)) that the square block is deformed by a pair of forces in such a way that the bottom side is held down to prevent rotation. At equilibrium the net force and torque are zero. The shear strain θ is defined by

$$\theta = \frac{\delta}{l} \qquad (6.5)$$

where δ is the shear displacement of material of length l, Fig. 6.4(b). To relate the strain to the shear stress, we translate the shear stress to a combination of tensile and compressive stresses on the rectangular block as shown in Fig. 6.4(c). The tensile force is equal to $\sqrt{2}S$, yet the length and hence the area is also increased by a factor of $\sqrt{2}$. Therefore, the tensile stress is S/A, which is the same magnitude as the shear stress. The compressive stress is the same except for the sign. Consider the stress–strain relation along the diagonal of the rectangular block in Fig. 6.4(c); we have

$$\frac{\Delta B}{B} = \frac{1}{Y}\frac{S}{A} + \frac{-\nu}{Y}\left(\frac{-S}{A}\right) = \frac{1+\nu}{Y}\frac{S}{A} \qquad (6.6)$$

On the other hand, we have defined

$$\theta = \frac{\delta}{l} = \frac{\sqrt{2}\Delta B}{l} = \frac{2\Delta B}{B}$$

Then

$$\theta = 2\left(\frac{1+\nu}{Y}\right)s \qquad (6.7)$$

where $s = S/A$ is the shear stress. If we define the shear modulus as

$$\mu = \frac{s}{\theta}$$

then we can express the shear modulus in terms of Young's modulus and Poisson's ratio,

$$\mu = \frac{Y}{2(1+\nu)} \qquad (6.8)$$

Shear stress can be regarded as a pair of normal stresses, and vice versa. The relationship between the elastic constants for single crystals is given in Appendix D.

6.3 Strain energy

It is important to estimate the magnitude of the energy involved in elastic strain. Consider the case at the elastic limit. The elastic energy is given by

$$E_{elastic} = \int \sigma \cdot d\varepsilon = \frac{1}{2}Y\varepsilon^2 \qquad (6.9)$$

To estimate the elastic energy, we take values of Young's modulus from Table 3.4 in Chapter 3, and choose one of the stiffest materials, steel, with $Y = 2.0 \times 10^{12}$ dyne/cm^2 and 8.4×10^{22} atoms in 1 cm^3. When we take $\varepsilon = 0.2\%$, we have

$$E_{elastic} = \frac{1}{2}Y\varepsilon^2 = 4 \times 10^6 \frac{\text{dyne}}{\text{cm}^2} \cong 3 \times 10^{-5} \frac{\text{eV}}{\text{atom}}$$

The value of elastic energy obtained is three to four orders of magnitude smaller than the typical chemical energy, say the binding energy in Au, which is about 0.47 eV/atom as given in Chapter 3. We can also conclude from Fig. 6.1(a) that elastic energy is small. The potential energy corresponding to a strain of 0.2% is very close to the binding energy, so the difference is very small. For this reason, we can approximate the potential energy to have a parabolic relation with displacement within the elastic region, so $\phi = \frac{1}{2}(kr^2)$.

This is the reason that in measuring interdiffusion coefficients in a bulk diffusion couple or in silicide formation between a metal thin film and Si wafer, the part of the driving force due to stress on activation energy of motion is ignored. Therefore, in chemical reactions such as compound formation, the effect of elastic strain energy or stress effect is often ignored. On the other hand, the interdiffusion and reaction or IMC formation can lead to the bending of a very thick substrate. For example, the deposit of a thin film of 200 nm Al on a Si wafer of 200 μm (8 mils), just the interfacial adhesion between two layers of atoms across the interface can lead to the bending of the Si wafer. In the reaction of a Ni film of 200 nm with a Si wafer of 200 μm, the silicide formation can also lead to bending of the wafer.

Strain energy, although small, is important in cases where solids are near equilibrium. At equilibrium the forces are balanced, so any small additional force will be able to tilt the balance and to affect the equilibrium. Strain energy due to epitaxial misfit can stabilize metastable phases. Furthermore, in an epitaxial structure where the dislocation slip system is non-operative or the nucleation of dislocations is difficult, the elastic limit can be greatly extended (for instance, up to a few percent), so that the strain energy can be two orders of magnitude greater. But when a solid is far away from equilibrium, the strain energy tends to be unimportant in most kinetic processes.

Elastic strain energy, while small, is reversible, so the elastic energy can perform work in creep as a function of time. The elastic strain energy can affect vacancy formation energy, in turn the vacancy concentration, but is negligible in affecting vacancy motion energy, so it does affect atomic diffusion as in creep to be discussed in Chapter 14. In comparison, plastic strain energy can be much higher, yet it is a waste energy, like joule heating, so it cannot be used to perform work. Work has been done already by the plastic deformation so the energy is spent and cannot be stored for renewable use.

6.4   Biaxial stress in thin films

On a planar substrate, the stress experienced by the thin film due to differential thermal expansion is biaxial. As shown in Fig. 6.5, the stresses act along the two principal axes in the plane of the film, but there is no stress in the direction normal to the free surface of the film, and yet there is strain in the normal direction. Another example of biaxial stress is in the stresses on the surface of a balloon. The in-plane stress results in strain normal to the balloon surface and becomes thinner and thinner as it expands.

To express the biaxial stress, we start with a three-dimensional isotropic cubic structure as shown in Fig. 6.6. The linear dimensions in the $x, y,$ and z axes are $l, w,$ and t,

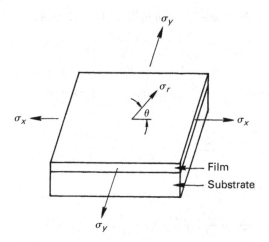

Fig. 6.5   Biaxial stress in a thin film deposited on a rigid substrate.

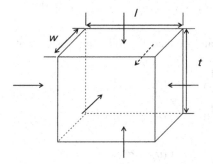

Fig. 6.6   Three-dimensional isotropic cubic of size l, w, t.

respectively. Now we consider the cubic structure under hydrostatic compression. We shall apply the compression in the x, y, and z directions in sequence. First, we apply the pressure p in the x-direction, so

$$p = -Y \frac{\Delta l_1}{l}$$

We have strain in the x-direction of

$$\frac{\Delta l_1}{l} = -\frac{p}{Y}$$

Second, we apply the compression in the y-direction, and we have

$$\frac{\Delta w}{w} = -\frac{p}{Y}$$

Owing to Poisson's effect, we have tensile strain in the x-direction given by

$$\frac{\Delta l_2}{l} = +v\frac{p}{Y} = -v\frac{\Delta w}{w}$$

Third, we apply the compression in the z-direction, and we have tensile strain in the x-direction given by

$$\frac{\Delta l_3}{l} = +v\frac{p}{Y}$$

The total strain in the x-direction is given by

$$\frac{\Delta l}{l} = \frac{\Delta l_1}{l} + \frac{\Delta l_2}{l} + \frac{\Delta l_3}{l} = -\frac{p}{Y}(1 - 2v)$$

or

$$\varepsilon_x = -\left(\frac{\sigma_x}{Y} - v\frac{\sigma_y}{Y} - v\frac{\sigma_z}{Y}\right) = -\frac{1}{Y}[\sigma_x - v(\sigma_y + \sigma_z)]$$

Now we change the stress to tension and we have the following basic equations:

$$\varepsilon_x = \frac{1}{Y}[\sigma_x - v(\sigma_y + \sigma_z)]$$

$$\varepsilon_y = \frac{1}{Y}[\sigma_y - v(\sigma_x + \sigma_z)] \qquad (6.10)$$

$$\varepsilon_z = \frac{1}{Y}[\sigma_z - v(\sigma_x + \sigma_y)]$$

In the thin-film biaxial stress state, we assume that there is tensile stress within the plane of the film (x and y) but no stress in the z-direction ($\sigma_z = 0$). Therefore,

$$\varepsilon_x = \frac{1}{Y}(\sigma_x - v\sigma_y)$$

$$\varepsilon_y = \frac{1}{Y}(\sigma_y - v\sigma_x)$$

$$\varepsilon_z = \frac{-v}{Y}(\sigma_x + \sigma_y)$$

From these equations, we have

$$\varepsilon_x + \varepsilon_y = \frac{1 - v}{Y}(\sigma_x + \sigma_y)$$

and

$$\varepsilon_z = \frac{-v}{1 - v}(\varepsilon_x + \varepsilon_y)$$

In two-dimensional isotropic systems where $\varepsilon_x = \varepsilon_y$,

$$\varepsilon_z = -\frac{2v}{1-v}\varepsilon_x \tag{6.11}$$

$$\sigma_x = \left(\frac{Y}{1-v}\right)\varepsilon_x \tag{6.12}$$

Later, we shall apply the relations above to obtain the Stoney's equation of the stress of a thin film on a substrate.

With thin films on a circular substrate, it is convenient to use cylindrical rather than Cartesian coordinates, and we have

$$\sigma_r = \sigma_x \cos^2\theta + \sigma_y \sin^2\theta + 2\tau_{xy}\sin\theta\cos\theta \tag{6.13}$$
$$\sigma_\theta = \sigma_x \sin^2\theta + \sigma_y \cos^2\theta - 2\tau_{xy}\sin\theta\cos\theta$$

where τ_{xy} is the shear stress. If there is no curl in the stress field ($\sigma_\theta = 0$), we obtain

$$\sigma_r = \sigma_x + \sigma_y$$

Similarly

$$\varepsilon_r = \varepsilon_x + \varepsilon_y$$

Then

$$\varepsilon_r = \frac{1-v}{Y}\sigma_r \text{ and } \varepsilon_z = -\frac{v}{Y}\sigma_r \tag{6.14}$$

These relationships are useful, for example, when we consider the growth of a circular hillock in thin films under a compressive stress.

The biaxial stress discussed above assumes that the substrate is rigid. We shall now consider the bending of the substrate under biaxial stress when the substrate is not rigid. The stress–strain relationship is given by Stoney's equation to be discussed below.

6.5　Stoney's equation of biaxial stress in thin films

We begin the analysis by assuming that the film thickness t_f is much less than that of the substrate thickness t_s, so the neutral plane where there is no stress can be taken to be at the middle of the substrate. In Fig. 6.7, we enlarge one end of the substrate to show the neutral plane, the stress distribution in the film and in the substrate, and the corresponding forces and moments. At equilibrium, the moment produced by the stress in the film must equal that produced by the stress in the substrate; see Fig. 6.7(b). Since

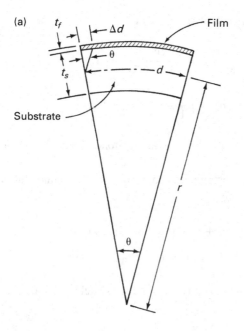

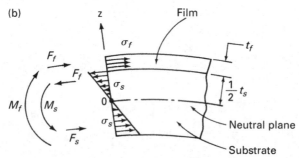

Fig. 6.7   (a) Cross-sectional view of a thin film under compression on a bent substrate. (b) Schematic diagram showing the stress distribution in film and substrate and the corresponding forces and bending moments.

we have assumed that the film thickness is thin, the stress σ_f is uniform across the film thickness. The moment M_f (force times perpendicular distance) due to the force in the film with respect to the neutral plane is

$$M_f = \sigma_f W t_f \frac{t_s}{2} \tag{6.15}$$

where W is the width of film normal to t_f. To calculate the moment of the substrate, we first obtain the geometrical relation

$$\frac{d}{r} = \frac{\Delta d}{t_s/2} \tag{6.16}$$

and so

$$\frac{1}{r} = \frac{\Delta d}{t_s d/2} = \frac{\varepsilon_{max}}{t_s/2} \tag{6.17}$$

where r is the radius of curvature of the substrate measured from the neutral plane, d is an arbitrary length of the substrate measured at the neutral plane, and $\Delta d/d = \varepsilon_{max}$ is the strain measured at the outer surfaces of the substrate. In the substrate, the elastic strain is zero at the neutral plane, yet it increases linearly with distance z measured from the neutral plane (i.e. it obeys Hooke's law and increases linearly with stress), so that

$$\frac{\varepsilon_s(z)}{z} = \frac{\varepsilon_{max}}{t_s/2} = \frac{1}{r} \tag{6.18}$$

where $\varepsilon_s(z)$ is the strain in a plane which is parallel to the neutral plane and is at a distance of z from the neutral plane. Then, by assuming a state of biaxial stress in the substrate, we have from Eq. (6.12) that

$$\sigma_s(z) = \left(\frac{Y}{1-v}\right)_s \varepsilon_s(z) = \left(\frac{Y}{1-v}\right)_s \frac{z}{r} \tag{6.19}$$

Therefore, the moment produced by the stress in the substrate is

$$M_f = W \int_{-t_s/2}^{t_s/2} z\sigma(z)dz = W \int_{-t_s/2}^{t_s/2} \left(\frac{Y}{1-v}\right)_s \frac{z^2}{r} dz = \left(\frac{Y}{1-v}\right)_s \frac{Wt_s^3}{12r} \tag{6.20}$$

By equating M_s to M_f, we have Stoney's equation

$$\sigma_f = \left(\frac{Y}{1-v}\right)_s \frac{t_s^2}{6rt_f} \tag{6.21}$$

where the subscripts f and s refer to film and substrate, respectively. Eq. (6.21) shows that by measuring the curvature and the thicknesses of the film and the substrate, and by knowing Young's modulus and Poisson's ratio of the substrate [7], we can determine the biaxial stress in the film. The curvature can be measured by laser interference or by stylus profiling.

Equation (6.21) has been applied to measure surface stress during epitaxial growth of a film on a substrate. The pseudomorphic (commensurate) growth as shown in Fig. 6.8 induces a stress between the film and the substrate. When the substrate is sufficiently thin, the misfit stress can bend the substrate as discussed. In the extreme case of one monolayer pseudomorphic growth of Ge on a 0.1 mm-thick (001) Si strip, the bending is large enough to be detected by laser reflection [8]. In essence, the force on the cross-section wt_f of a

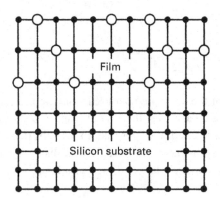

Fig. 6.8   Schematic diagram of the epitaxial structure of Ge_xSi_{1-x} (Ge open circle) on a (100) silicon substrate as strained layer with distorted unit cell.

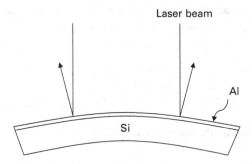

Fig. 6.9   Schematic diagram of the cross-section of an Al film on a bent Si wafer. The curvature can be measured from the reflection of a pair of parallel laser beams.

film is $F_f = \sigma_f W t_f$ or $F_f/W = \sigma_f t_f$. Rearranging Eq. (6.21), we have

$$r = \left(\frac{Y}{1-v}\right) \frac{t_s^2}{6(F_f/W)} \tag{6.22}$$

This shows that by measuring r (or determining r as a function of t_f) we determine F_f/W. The dimension of F_f/W is force per unit width of the film (i.e. it is a measure of surface stress). Recall the discussion of surface energy in Section 3.3 where we have shown that the surface energy and surface tension of a liquid have the same magnitude. This is not so for solids. Liquids cannot take shear stresses and the surface of a liquid cannot sustain a compressive stress along the surface, yet solids can. On the other hand, neither a solid nor liquid surface can have a normal stress.

For example, if we deposit an Al film on a Si wafer at liquid nitrogen temperature and bring them up to room temperature, the Si wafer bends because Al has a larger thermal expansion coefficient than Si. Their cross-section with a concave curvature is sketched in Fig. 6.9. The Al film is constrained by the substrate, assuming a very good adhesion, and is under compression. The compressive stress in the Al film can be determined by measuring the curvature of the Si wafer. The stress in Al will change as a function of temperature and the change can be measured by using the following analysis.

6.6 Measurement of thermal stress in Al thin films

Stoney's equation shows that by measuring the curvature of the substrate, we can determine the biaxial stress in the thin film, whether it is tensile or compression, when we know the thickness of the film, the thickness of the substrate and the mechanical modulus of the substrate. The key is to measure the curvature. This can be done by using laser reflection from a curved surface of a bending beam or a freestanding circular wafer. Fig. 6.9 depicts the measurement of the curvature of a 4 in freestanding Si wafer by the reflection of a pair of parallel laser beams from the curved Si wafer surface. The relative displacement of the pair of reflected beams will allow us to determine the curvature.

We consider below the deposition of an Al thin film on a Si wafer at room temperature and followed by heating the film/substrate to 400 °C, holding the sample there for a while and then cooling down the sample to room temperature. Fig. 6.10 depicts the change in curvature in the temperature cycling. Fig. 6.10(a) depicts an Al film deposited and kept on a Si wafer at room temperature. We assume that there is no thermal stress, so the sample of film/substrate is flat and has no curvature. Upon heating to 400 °C, the Al expands more than the Si substrate, so the wafer bends with a curvature concave downward, as shown in Fig. 6.10(b) and the Al film is under compression. The compression is because the Si wafer will exert a torque to compress the Al film when the wafer tries to bend back to reduce the curvature. If we keep the sample at 400 °C for a while, the compressive stress in the Al will disappear due to stress relaxation in Al by atomic diffusion. We can calculate the lattice diffusivity of Al at 400 °C, and find that the diffusion distance is more than 200 nm, which is the Al film thickness, in 5 min, so the stress can be relaxed by atomic diffusion and rearrangement. The curvature of the wafer will become flat, as shown in Fig. 6.10(c).

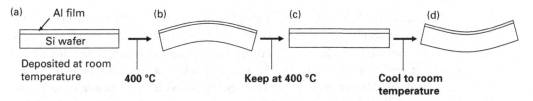

Fig. 6.10 Schematic diagram of the cross-section of an Al film deposited on a Si wafer upon temperature excursion from room temperature to 400 °C and back to room temperature. (a) The sample is flat in room-temperature deposition. (2) The Al film is under compression when the sample is heated to 400 °C; the curvature is concave downward. (c) When the sample is kept at 400 °C for 5 min, the compressive stress in Al is relaxed and the sample is flat. (d) When the sample is cooled to room temperature, the Al film is under tension and the sample is concave upward.

In Fig. 6.10(b), we assume an Al grain of the size of 1000 λ under compression at 400 °C, where λ is interatomic distance or the thickness of one atomic plane. We assume that the compressive strain is 0.1%. In other words, the strain is 1 λ. Then, if we can remove one atomic plane normal to the compressive stress, the stress will be relaxed. To do so, we imagine a dislocation loop in the grain. If a vacancy comes from the free surface and diffuses to the edge of the loop and exchange position with the atom at the end, the dislocation will climb one atomic step. By a successive diffusion of vacancies to the dislocation, we can remove an atomic plane and relax the stress. This is the mechanism of creep.

In the schematic diagram shown in Fig. 6.10(b), it assumes a microstructure of columnar grains. If we assume that there is no surface oxide so that the source of vacancies on the surface can function fully, the vacancies needed in diffusion creep are available. A uniform relaxation over the entire film can occur in the sense that a uniform layer of materials can be transported to the surface. The film will become slightly thicker and be stress-free. On the other hand, since Al has a protective oxide and we should consider the case where the Al film is oxidized, then the source of vacancies at the Al surface is not fully operative. The compressive stress may break the oxide at a few weaker spots and the relaxation behavior will be localized or non-uniform, and hillock or whisker will form which tends to involve a long-range atomic diffusion driven by stress gradient. Stress migration or diffusion creep will be covered in Chapter 14.

Then, when we cool down the sample to room temperature, a tensile stress will build up in the Al film. The curvature of the wafer will be concave upward as shown in Fig. 6.10(d). At room temperature, atomic diffusion in Al is very slow, so not much stress relaxation will occur.

Fig. 6.11 is an experimental measurement of the stress evolution of an Al film on a Si wafer. In the initial stage, the stress was nearly zero and upon heating, a compressive stress was found to increase with temperature. Above 100 °C; the stress ceased to increase owing to relaxation and the compressive stress stayed nearly constant until 350 °C, above that the stress decreased because of fast stress relaxation. When the film was kept at 400 °C, the stress was almost completely relaxed. Then cooling occurred and tensile stress was found to increase with decreasing temperature. During cooling, the tensile stress kept increasing. When it was back to room temperature, the film had a very high tensile stress.

The stress evolution and the corresponding microstructure change in the Al film has been a subject of much study. During the heating stage, the compressive stress has led to hillock formation on the Al surface. During the cooling stage, the tensile stress can lead to void formation. The void formation has been a serious reliability issue in Al thin-film interconnect technology. The stress-induced atomic diffusion, or "stress-migration" as it is called, will be covered in Chapter 14. In Chapter 4, we discussed atomic diffusion in fcc metals as requiring vacancy. We must consider not only how the stress can affect the equilibrium vacancy concentration, but also what are the sources and sinks of vacancy in stress-migration for both stress generation and stress relaxation, and in turn, how vacancies are created and annihilated in the lattice of the Al film.

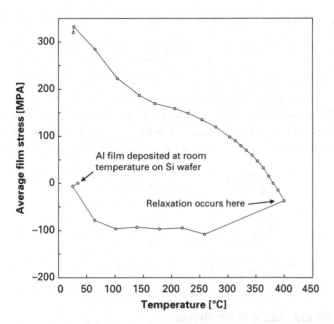

Fig. 6.11   Plot of stress against temperature of an Al film deposited on a Si wafer. The stress is calculated from the curvature of the wafer by using Stoney's equation.

Another case of thermal stress in thin films is a Pb thin film on a Si substrate. The coefficients of thermal expansion of Pb and Si are $29.5 \times 10^{-6}/°C$ and $2.6 \times 10^{-6}/°C$, respectively. We deposit Pb on Si at room temperature and then cool them to 4.2 K, where the Pb becomes superconducting. The sample experiences a temperature drop of about 300 K, and the net change in linear dimension is 0.86% for Pb. While the Pb tries to shrink, the Si substrate restricts it from doing so; hence in cooling, the Pb is under tension. The tensile stress will be relaxed to some extent because of the yielding of the Pb film. Upon heating the sample back to room temperature, the Pb tends to expand and again it is restricted by the Si substrate. The Pb is under compression upon heating. Since room temperature is about half of the melting point of Pb, atomic diffusion is substantial. The Pb film will release its compressive stress partly by atomic diffusion; hence, hillock formation occurs. This is a well-known phenomenon in Josephson junction devices where Pb has been used as electrodes and has experienced the temperature cycling between room temperature and 4.2 K. Hillock formation causes rupture of the ultra-thin oxide layer used for junction tunneling and the device fails.

6.7   Application of Stoney's equation to thermal expansion measurement

We can rewrite thin-film biaxial stress σ_f as

$$\sigma_f = \left(\frac{Y}{1-v}\right)_f \varepsilon_f = \left(\frac{Y}{1-v}\right)_f \Delta\alpha\,\Delta T$$

or we can write

$$\frac{d\sigma_f}{dT} = \left(\frac{Y}{1-v}\right)_f (\alpha_f - \alpha_s)$$

where α_f and α_s are thermal expansion coefficients of the thin film and its substrate, respectively. By measuring the slope of σ_f versus temperature, we obtain $d\sigma_f/dT$. Thus, if we use two kinds of substrate for the same thin film, e.g. we deposit Al film on a wafer of Si and on a wafer of Ge and then measure the slopes, we have two equations:

$$\left(\frac{d\sigma_f}{dT}\right)_{Si} = \left(\frac{Y}{1-v}\right)_f (\alpha_f - \alpha_{Si}) \qquad (6.23)$$

$$\left(\frac{d\sigma_f}{dT}\right)_{Ge} = \left(\frac{Y}{1-v}\right)_f (\alpha_f - \alpha_{Ge})$$

We can solve for the two unknowns, $[Y/(1-v)]_f$ and α_f.

6.8   Anharmonicity and thermal expansion

The thermal expansion coefficient is an intrinsic property of elements. Indeed, the large differences in thermal expansion coefficients between different kinds of material (metals, semiconductors, and insulators) may be one of the limiting factors in growing high-quality epitaxial structures when a composite of materials is used.

    If the bottom of an interatomic potential well of an atom is parabolic, the atom's displacement will be linearly proportional to the driving force. The atom undergoes a harmonic oscillation so that its average position does not change and there is no thermal expansion upon heating. This is the same as a simple pendulum whose mean position remains unchanged in oscillation. But in reality, the interatomic potential well is not parabolic; the Lennard-Jones potential shows that resistance to compression is stronger than tension. Thus, thermal vibration tends to drive atoms apart. The stronger the vibration or the larger the vibration amplitude, the greater the separation. This leads to thermal expansion by the anharmonicity of the atomic potential. The dotted line in Fig. 6.1(a) depicts the increase of a_0 with energy or temperature. Since we can also change a_0 by stress and the change is described by Young's modulus, the thermal expansion coefficient and Young's modulus are related. Recall that in Section 3.7 the relation between Young's modulus and interatomic potential was discussed. The equation of state of solids which relates changes among pressure, temperature, and volume is given by Grüneisen's equation (see Mott and Jones) [5].

6.9   The origin of intrinsic stress in thin films

There are intrinsic and extrinsic stresses in a thin film. The origin of extrinsic stress in a thin film comes mainly from thermal stress or externally applied stress. The effect of adhesion of the film to its substrate is important. Thermal stress can be introduced in a thin film due to differential thermal expansion between the film and its substrate, due to lattice misfit with its substrate, or due to chemical reaction with its substrate when the IMC formed is coherent to the film but has a slight lattice misfit. The origin of intrinsic stress is not as clear as the extrinsic stress.

Metallic thin films often have thermal stress at room temperature when it has been deposited at a temperature higher or lower than room temperature. However, if the thin film was deposited and kept at room temperature, we expect no thermal stress, yet a stress is commonly found; this is called the "intrinsic stress." For example, Ni thin films deposited by e-beam evaporation or sputtering at room temperature have had a very high tensile stress. When the thickness of the Ni film is over 300 nm, it tends to pill off from the substrate.

During the heterogeneous nucleation stage of a thin film on a substrate, the nucleus is under compression due to the Gibbs–Thomson effect. When nuclei coalesce, the stress will change from compression to tension. This is because if it is assumed that the grain size in the film is directly proportional to the film thickness, as the nuclei coalesce and the film thickens, grain size increases, and in turn the number of grain boundaries per unit area will decrease. Since the structure of the grain boundary contains free volume, the removal of a grain boundary means that the free volume has to be absorbed into the film if the film adheres to the substrate. A tensile stress will be generated in the film [9–11]. In the later stage of film deposition, when the deposited atoms can diffuse into grain boundaries in the film, especially by oxygen atoms, compressive stress will be generated.

6.10 Elastic energy of a misfit dislocation

In the heteroepitaxial growth of a film on a substrate, such as a SiGe film on Si, strain will exist in the film when the thickness of the film is much less than that of the substrate. Beyond a critical thickness, interfacial misfit dislocations are introduced to release the strain in the film. Hence, the elastic energy of an array of misfit edge dislocations at the interface between a film and its substrate is of interest. We show in Fig. 6.12 a schematic atomic picture of an edge dislocation. We can regard it as a misfit dislocation in which the crystal structure and atom of the film and the substrate are the same.

To form misfit dislocations, we assume a simple cubic structure for the film and substrate, having lattice parameters a_f and a_s, respectively. We further assume that

$$(n + 1)a_f = na_s \tag{6.24}$$

and we define the misfit

$$f = \frac{a_s - a_f}{a_s} \tag{6.25}$$

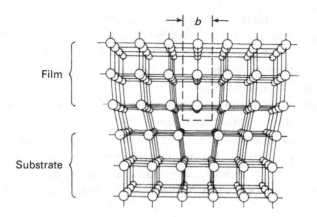

Fig. 6.12   Edge-type misfit dislocation created by inserting an extra half sheet of atoms.

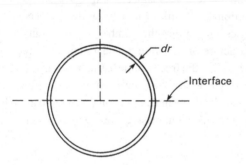

Fig. 6.13   Ring of radius r and width dr surrounding an interfacial misfit dislocations.

Hence, we have the spacing between neighboring misfit dislocations to be

$$na_s = \frac{a_f}{f} \approx \frac{b}{f} \tag{6.26}$$

where b ($\approx a_f$) is the Burgers vector of the misfit dislocation. Clearly, to fit the film epitaxially without misfit dislocations on the substrate, we need to stretch the film by a strain of $\Delta l / l = 1/n$. Since the strain energy increases with the film thickness, a thick film is energetically favorable to reduce the strain by introducing misfit dislocations. We imagine that we make cuts in the film at a spacing of na_s insert a single sheet of atoms in each cut as indicated by the broken lines in Fig. 6.12, and rejoin the atoms to form the misfit dislocations.

   The elastic energy involved in forming a straight dislocation (an edge dislocation) has been covered in many textbooks. For a detailed discussion, see Chapter 2 in Hirth and Lothe [2]. Here, we shall present only a simple analysis. We consider in Fig. 6.13 a ring surrounding a misfit dislocation. We make a cut at the top of the ring, open the cut, and insert a sheet of atoms of width b. The strain and stress introduced into the ring are, respectively,

$$\varepsilon = \frac{b}{2\pi r} \tag{6.27}$$

$$\sigma = \frac{\mu b}{2\pi r}$$

where μ is the shear modulus. The elastic energy per unit length of the dislocation is

$$E_d = \int_{r_1}^{r_2} \frac{1}{2} \frac{\mu b^2}{(2\pi r)^2} 2\pi r \, dr = \frac{\mu b^2}{4\pi} \int_{r_1}^{r_2} \frac{dr}{r} = \frac{\mu b^2}{4\pi} \ln \frac{r_2}{r_1} \tag{6.28}$$

Note that in a rigorous analysis for an edge dislocation, we must divide Eq. (6.28) by a factor of $(l - v)$ where v is Poisson's ratio. Now, what are r_1 and r_2? When we make the cut in the film, the cut should open up automatically since the film is under tension, so that it does not cause much strain to insert a sheet of atoms into the cut. Although the elastic field of a dislocation is long-range and proportional to l/r, the elastic field produced by an array of misfit dislocations extends only to a distance of $na_s/2$. The strain fields produced by neighboring dislocations cancel each other and we can take $r_2 = nb/2$. Then, for r_1 it is clear that the dislocation core is a singularity, so that we have to take $r_1 > 0$ in the integration in order to avoid the singularity. Typically, $r_1 \approx b$. Treatment of the actual dislocation core is a longstanding problem. The core energy is not infinite, because we know that the energy is finite for the case of a small-angle grain boundary which consists of a set of dislocations. This in turn means that the elastic field in the core of the dislocation does not go to infinity. In any case, the interatomic distances between atoms in the core are finite.

If we take $r_2 = nb/2$ and $r_1 = b$, we have

$$E_d = \frac{\mu b^2}{4\pi(1 - v)} \ln \frac{n}{2} \approx \frac{1}{2} \mu b^b \tag{6.29}$$

for a typical misfit of 0.1% or less, and $n = 10^3$ to 10^4. Using Al as an example, we have $\mu \approx 2 \times 10^{11}$ dyne/cm^2 and $b = 2.5 \times 10^{-8}$ cm.

Then,

$$E_d = \frac{1}{2} \times 2 \times 10^{11} \times \left(2.5 \times 10^{-8}\right)^2$$

$$= 6.25 \times 10^{-5} \text{ erg/cm of dislocation}$$

If the misfit is 0.1%, then there is a dislocation in every 1000 lattice spacing ($\sim 2.5 \times 10^{-5}$ cm) or an (areal) energy density of 2.5 erg/cm^2. For comparison purposes, it is useful to evaluate this energy in eV/atom. A thin film of Al, thickness t (cm), contains $t \cdot 6 \times 10^{22}$ atom/cm^2. A typical film of 100 nm then contains 6×10^{17} atom/cm^2. The dislocation energy averaged over all atoms in the film reduces to

$$E_d = \frac{(2.5\,\text{erg/cm}^2)(6.25 \times 10^{11}\,\text{eV/erg})}{6 \times 10^{17}\,\text{atom/cm}^2}$$

or a value of 2.6×10^{-6} eV/atom. This value is comparable to the strain energy in a stressed solid discussed in Section 6.2. In both cases the stored energy is basically the displacement from an equilibrium position in an elastic solid. While the average energy/atom is small, the displacements and stored energy of atoms at the dislocation core can be large, of order 1 eV/atom. Thus, dislocations are rarely created by random thermal fluctuations. Stresses during thermal and mechanical processing are usually the cause of excess dislocations.

References

[1] R. P. Feynman, R. B. Leighton and M. Sands, *The Feynman Lectures on Physics* (Vol. II, Ch. 38) (Addison-Wesley, Reading, MA, 1964).

[2] J. P. Hirth and J. Lothe, *Theory of Dislocations* (McGraw-Hill, New York, 1969).

[3] R. W. Hoffman, "Nanomechanics of thin films: emphasis: tensile properties," in *Physics of Thin Films* (Vol. 3, p. 211), eds G. Hass and R. E. Thun (Academic Press, New York, 1964).

[4] H. B. Huntington, "The elastic constants of crystals," in *Solid State Physics*, Vol. 7, eds F. Seitz and D. Turnbull (Academic Press, New York, 1958).

[5] N. F. Mott and H. Jones, *The Theory of the Properties of Metals and Alloys* (Dover, New York, 1958).

[6] A. S. Nowick and B. S. Berry, *Anelastic Relaxation in Crystalline Solids* (Academic Press, New York, 1972).

[7] M. Murakami and A. Segmiiller, in *Analytical Techniques for Thin Films*, eds K. N. Tu and R. Rosenberg, Vol. 27 in *Treatises on Materials Science and Technology* (Academic Press, Boston, 1988).

[8] A. J. Schell-Sorokin and R. M. Tromp, "Mechanical stresses in (sub)monolayer epitaxial films," *Phys. Rev. Lett.* **64** (1990), 1039.

[9] P. Chaudhari, "Grain growth and stress relief in thin films," *J. of Vac. Sci. Tech.* **9** (1972), 520.

[10] W. D. Nix, "Mechanical properties of thin films," *Metallurgical Transaction, A. Physical Metallurgy and Materials Science* **20** (1989), 2217–2245.

[11] F. Spaepen, "Interfaces and stress in thin films," *Acta Mat.* **48** (2000), 31–42.

Problems

6.1 Using the Lennard-Jones potential with $n = 8 \times 10^{22}$ atom/cm^3 and $\varepsilon_b = 0.6$ eV/atom, $a_0 = (1/n)^{1/3}$,

(a) calculate the maximum force F_{max};

(b) assume the solid is in the linear elastic region and calculate Young's modulus Y;

(c) Find the elastic energy $E_{elastic}$ at F_{max}.

6.2 A 1 μm-thick Al film is deposited without thermal stress on a 100 μm-thick Si wafer at a temperature 100 °C above ambient temperature. The wafer and film are allowed to cool to the ambient. Using the values provided in the table and assuming that $v = 0.272$ for Si:
(a) calculate the thermal strain and stress for $\Delta T = 100$ °C;
(b) calculate the radius of curvature.

	Expansion coefficient $\alpha(10^{-6}/°C)$	Young's modulus $Y(10^{11} \text{ N/m}^2)$
Al	24.6	0.7
Si	2.6	1.9

6.3 Consider Al and Cu at two-thirds their melting temperature T_m under a tensile strain of 0.2%. Using the data provided in the table, calculate for both materials:
(a) the concentration of vacancies;
(b) the enhancement of the vacancy concentration caused by the strain.

	$N(\times 10^{22}/\text{cm}^3)$	$Y(\times 10^{11}\text{N/m}^2)$	$\Delta H_f(\text{eV})$	$T_m(°C)$
Al	6.02	0.7	0.67	660
Cu	8.45	1.1	1.28	1028

6.4 For a 10^{-5} cm cubic grain of Cu held at two-thirds the melting temperature for 10 min, use the data in Table 3.1 and the Nabarro–Herring equation (see Section 14.3) to calculate the volume and number of atoms accumulated ($\varepsilon = 0.2\%$).

6.5 For Si, calculate the elastic energy per unit length of a misfit dislocation, and discuss whether dislocations would be formed by heating the crystal to 100 °C. Use the Si parameters, $\mu = 7.5 \times 10^{10}\text{N/m}^2$, $b = a/\sqrt{2}$, $a = 0.543$ nm, and $N = 5 \times 10^{22}/\text{cm}^3$.

6.6 A 200 μm diameter spherical balloon is made of 1 μm thick Al. At 20 °C and 760 torr, the balloon is fully inflated with no stress on the film. If the temperature is changed to 30 °C and the pressure is constant, will the balloon burst? (The thermal expansion coefficient of Al is $2.5 \times 10^{-5}°\text{C}$.)

6.7 A 3000 Å oxide film is deposited on a 500 μm-thick bare Si wafer that has a radius of curvature of 300 m. After deposition the radius is measured to be 200 m. A 6000 Å nitride film is now deposited on the oxide and the radius of curvature is measured to be 240 m. Calculate the dual film stress and the stress of the nitride film ($v_{Si} = 0.272$, $Y_{Si} = 1.0 \times 10^{12}$ dyne/cm²).

6.8 The stress in films on wafers can be determined from the amount of bow of the substrate. The bow can be measured by a surface profilometer with an 18 cm scan length.

(a) Prove that Eq. (6.21) in the text is equivalent to the equation shown here.

$$\sigma = \left(\frac{\delta}{3\rho^3}\right)\left(\frac{Y}{1-v}\right)\left(\frac{t_s^2}{t_f}\right)$$

where δ is the maximum bow height of the profileometer scan and ρ is half the scan length.

(b) Given a scan length of 5 cm and a bow of 20 000 Å, calculate the stress for an unknown film 2 μm-thick. The substrate is a 200 μm-thick Si wafer. ($Y/(1-v)$ for (100) Si $= 1.8 \times 1011$ N/m^2.)

7   Surface kinetic processes on thin films

7.1   Introduction

Surface kinetic processes deal with nucleation, growth, and ripening processes on a surface from the point of view of atomic absorption, desorption, and diffusion on the surface. The surface of a thin film, more specifically the surface of a single crystal, is the starting place for these processes. The single-crystal surface has a microscopic structure associated with crystallographic structure and symmetry and reconstruction, as well as a macroscopic structure associated with surface steps, kinks, and other surface defects [1–4]. Surface chemical reaction such as oxidation is not considered here.

In Chapter 3, there was a brief discussion of surface crystallographic structure. A ball-and-stick crystal model could depict a (100) surface of silicon, the cubic face of a diamond lattice, as a portion of a plane of infinite extent consisting of a square array of atoms. Each of these atoms has two unpaired electron bonds. On a (100) surface of silicon, the atoms displace laterally (as shown schematically in Fig. 3.12) to satisfy the bonding requirement. Such surfaces are called "reconstructed."

On a larger scale, single-crystal surfaces can contain terraces, steps, kinks, and other defects. As shown in Fig. 7.1, the step spacing is associated with the miscut of the crystal. The miscut angle is the angular difference between a crystal plane and the mechanical surface of the crystal established when an ingot is cut. Under production conditions this angle is about $0.1°$. The relationship shown in the figure relates the step spacing L_0 to the miscut angle. For monolayer steps of height h (for Si, $h = a/4 = 0.136$ nm where the lattice constant $a = 0.543$ nm) and miscut angle of $0.1°$, the step spacing L_0 is 77 nm. Steps are far from perfect. The atomic structure of steps has been revealed by scanning tunneling microscopy. Fig. 7.2 shows an array of steps with highly irregular step edges on (100) Si surface. Terrace size distribution is given in Appendix E.

Steps and terraces form because the low-index planes, such as the (100) plane, are more stable than higher (hkl) planes. Steps and kinks form high-energy binding sites as the source and sink of adatoms of deposited materials. It is the lateral displacement of the edges of the sheets of low-index planes that defines the growth of the surface. The step-mediated growth model will be explained in Section 7.5.

The surface layers of compound semiconductors must be further characterized by their atomic composition. For example, the (100) surface of GaAs contains either all Ga atoms (the "A" face) or all As atoms (the "B" face) of this A–B compound. Other surfaces can contain a mixture of atoms. The mono-atomicity of the (100) surface has

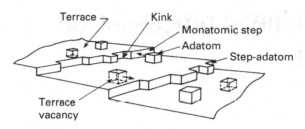

Fig. 7.1   Schematic diagram of surface steps, kinks, and terrace.

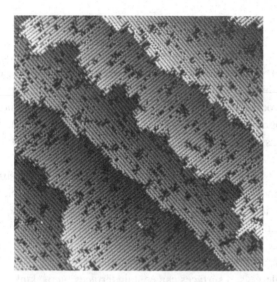

Fig. 7.2   Scanning tunneling microscopic images of the Si (100) surface. The average distance between steps is about 17 nm (see Fig. 1.5). (Courtesy of R. J. Hamers, U. K. Kohler, and J. E. Demuth, *J. of Vac. Sci. Tech.* **A8**, 195 (1990)).

made it the preferred growth face for GaAs epitaxy; the growth proceeds monolayer by monolayer in the sequence of A, B, A, B.

Several characteristics of surface kinetic processes are given below.

(1) Surface processes must take into account the equilibrium vapor pressure above the surface.

(2) Adatoms may absorb from or desorb to the vapor, depending on the vapor pressure. They can rearrange themselves on the surface even if local equilibrium with vapor exists.

(3) The surface has steps and terraces, and there are kinks along a step. The kinks are the sources and sinks of adatoms.

(4) If we dig an atom out of a flat surface and place the atom on the surface as an adatom, we leave a vacancy in the plane, so we create a pair of an adatom and a surface vacancy. However, it is important to note that such kinds of vacancy are

not required for surface diffusion. This is conceptually different from the vacancy model of lattice diffusion in crystalline solids as discussed in Chapter 4. Typically, we assume that surface diffusion on a close-packed plane occurs by the migration of adatoms. There is no need to create a surface vacancy on a free surface for the migration of an adatom.

(5) The surface may have a curvature consisting of uneven terraces or steps.

(6) If the surface is perfectly flat, we need to nucleate steps or islands for growth.

7.2 Adatoms on a surface

An atom impinging on a solid surface sees an array of binding sites or potential wells formed by the substrate atoms. There is a finite probability that the atom diffuses along the surface by hopping from a potential well to a neighboring potential well. There is also a probability of an atom escaping from the well to the vacuum (desorption). To characterize the surface diffusion and desorption, we assume that the surface atoms have a surface vibration frequency v_s ($\sim 10^{13}$ s^{-1}). The frequency of desorption is defined as

$$v_{des} = v_s \exp\left(-\frac{\Delta G_{des}}{kT}\right) \tag{7.1}$$

where ΔG_{des} is the change in free energy associated with desorbing one atom. As in all such formulae, the frequency can be thought of as the number of attempts (attempt frequency) multiplied by the probability of success given by the exponential factor, as shown in Eq. (7.1). The residence time τ_0 of an adatom on the surface before desorption occurs is

$$\tau_0 = \frac{1}{v_{des}} = \frac{1}{v_s}\exp\left(\frac{\Delta G_{des}}{kT}\right) \tag{7.2}$$

During the residence time, an adatom moves from one surface site to a neighboring site (surface diffusion) with a frequency

$$v_{diff} = v_s \exp\left(-\frac{\Delta G_s}{kT}\right) \tag{7.3}$$

where ΔG_s is the activation energy of motion needed to move to a neighboring site. The surface diffusivity D_s is defined as

$$D_s = \lambda^2 v_s \exp\left(-\frac{\Delta G_s}{kT}\right) \tag{7.4}$$

where λ is the jump distance between two neighboring surface sites. Desorption is much less likely than surface diffusion; the difference lies in their activation energies. For the

(111) surface of a fcc noble metal, we typically have ΔG_{des} and ΔG_s to be about 1 eV and 0.5 eV, respectively. Desorption involves breaking bonds while surface diffusion consists of motion only and the broken bonds are recovered.

Using the concepts of desorption and surface diffusion, we can describe the behavior of an adatom on a stepped surface where the width of terrace or the mean spacing between steps is L_0. Adatoms will not desorb if they arrive at a high binding energy site, i.e. a step, in less time than the residence time. For growth, we require that the adatom diffuses to a step before desorption. So, in order to arrive at the step before desorption

$$\sqrt{4D_s\tau_0} > \frac{L_0}{2} \tag{7.5}$$

where $L_0/2$ is the largest distance to some step. The diffusion time t_D for an adatom to reach the step is

$$t_D = \frac{L_0^2}{16D_s} \tag{7.6}$$

We define the sticking coefficient S_c to be the ratio of

$$S_c = \frac{\tau_0}{\tau_D} \tag{7.7}$$

by taking a "characteristic" diffusion time of L_0^2/D_s and $\tau_D = 16t_D$. For $S_c > 1$, an adatom will have sufficient time to diffuse to the step and become bound on the surface. For $S_c < 1$, desorption can occur. From Eq. (7.7),

$$S_c = \frac{\lambda^2}{L_0^2} \exp\left(\frac{\Delta G_{des} - \Delta G_s}{kT}\right) \tag{7.8}$$

For a noble metal surface with $\lambda^2 = 10^{-15}$ cm^2, $L_0 = 10^{-5}$ cm, $\Delta G_{des} = 1$ eV, $\Delta G_s = 0.5$ eV, and $T = 293$ K, we have S_c approximately equal to 10^3. This estimate shows that sticking is complete at room temperature for deposition on a noble metal surface. If we increase the substrate temperature to 600 K, we obtain $S_c \sim 10^{-1}$, the sticking is incomplete, and desorption dominates.

Thin-film growth is more complicated than a simple description of an adatom on a stepped surface. Deposited atoms can react with the surface to form new chemical compounds. Deposited atoms can interact with each other to form dimers and larger two-dimensional clusters called islands. Surface morphologies can change as the result of surface energy variations from one material to another. In epitaxial growth, strain energy in heteroepitaxial growth can be a driving factor for clustering and defect propagation. All of these phenomena have been discussed in many textbooks.

7.3 Equilibrium vapor pressure above a surface

The equilibrium density of adatoms, N_0, on a stepped surface in units of atom/cm^2 is given by

$$N_0 = N_s \exp\left(-\frac{W_s}{kT}\right) \tag{7.9}$$

where N_s is the number of atoms per unit area of surface, which is about 10^{15} atom/cm^2. W_s is the energy needed to remove an atom from a kink site to a stepped surface, but with no desorption from the surface.

At equilibrium, the desorption of adatoms from the stepped surface is given by

$$J_d = J_0 = \frac{N_0}{\tau_0} = N_s \exp\left(-\frac{W_s}{kT}\right) v_s \exp\left(-\frac{\Delta G_{des}}{kT}\right) = N_s v_s \exp\left(-\frac{W}{kT}\right) \tag{7.10}$$

where $\tau_0 = \frac{1}{v_s} \exp\left(\frac{\Delta G_{des}}{kT}\right)$ is the residence time of an adatom on a stepped surface as given before. $W = W_s + \Delta G_{des}$ is the activation energy of sublimation.

The activation energy of sublimation or of desorption can be measured by using thermo-desorption of a solid in an enclosed chamber. The vapor pressure can be measured as a function of temperature, so we can plot $\ln J_0$ versus $1/kT$ to obtain W.

For Si, the activation energy of sublimation W has been measured to be 4.5 eV/atom. Since W consists of W_s and ΔG_{des}, if we know one of them, we can estimate the other. Since each atom of Si has four bonds in the diamond structure, we can use the bond-breaking argument for the estimation:

$$W_s = \frac{3}{4} W$$

$$\Delta G_{des} = \frac{1}{4} W = \frac{4.5}{4} = 1.125 \text{ eV/atom}$$

Also, knowing W, we can estimate J_0 at a given temperature. Take $T = 1223$ K (950 °C); we have

$$J_0 = N_s v_s \exp\left(-\frac{W}{kT}\right)$$

$$= 10^{15} \times 10^{13} \exp\left(-\frac{4.5 \times 23\,000}{2.3 \times 2 \times 1223}\right) \approx 10^{28} \times 10^{-19} \approx 10^{10} \text{ atom/cm}^2\text{s}$$

Since on a cubic surface there are 10^{15} atom/cm^2, the desorption rate is about 10^{-5} atoms per unit area/s.

Table 7.1. Values of vapor pressure p_0 and flux density J_0 for Si*

$T(K)$	823	923	1023	1123	1223
P_0 (Pa)	2.5×10^{-16}	2.7×10^{-13}	7.1×10^{-11}	6.9×10^{-11}	3.2×10^{-7}
J_0 (1/cm^2 s)	4.4×10^2	4.3×10^5	1.1×10^8	1.0×10^{10}	4.6×10^{11}

* From Allen and Kasper (1998), p. 65 [7].

Table 7.2. Values of saturation σ at $J_{Si} = 2 \times 10^{15}/\text{cm}^2$ s*

$T(K)$	723	823	923	1023	1123	1223
σ	3×10^{16}	4.5×10^{12}	4.7×10^9	1.8×10^7	2×10^5	4.4×10^3

* From Allen and Kasper (1998), p. 65 [7].

In Chapter 2 we presented an equation to relate the partial pressure to the equilibrium desorption flux as given in Table 7.1:

$$J_0 = p_0 \sqrt{\frac{1}{2\pi mkT}}$$

In thin-film deposition, the typical deposition flux or rate is about 1 monolayer/s. In other words, we deposit about 60 monolayers/min, or we grow a film of thickness of about 20 nm/s. Thus, we take $J_c = 10^{15}$ atom/cm^2 s. For the deposition of Si at 1223 K, we have obtained in the above that $J_0 = 10^{10}$ atom/cm^2 s. Thus,

$$\frac{J_{Si}}{J_0} = \frac{J_c}{J_0} = 10^5 \gg 1$$

It indicates that the deposition occurs with a high supersaturation. We define the super-saturation coefficient, σ, below as listed in Table 7.2:

$$\sigma = \frac{J_c}{J_0} - 1 \tag{7.11}$$

7.4   Surface diffusion

In Chapter 4, the concept of vacancy-mediated lattice diffusion is given. The diffusion of an atom requires the presence of a vacancy in the nearest neighbor and the exchange of position with the vacancy. The diffusion of an atom to the right is equal to the diffusion of a vacancy to the left. The activation energy of diffusion of the atom consists of the activation energy of formation of a vacancy and the activation energy of motion of a vacancy. The pre-exponential factor of diffusivity includes the correlation factor, number of nearest neighbors, jump frequency, square of jump distance, and the entropy factor.

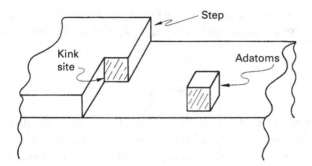

Fig. 7.3 Schematic diagram of breaking an atom from a kink site from a step on the surface and moving it to other parts of the surface to form an adatom. (see Fig. 6.4)

The concept of surface diffusion of an adatom on a surface is similar to lattice diffusion that the adatom has to jump to a nearest-neighboring vacant site, except that no vacancy is required on the surface because the adatom is surrounded by vacant sites. Thus, the activation energy of surface diffusion of an adatom does not contain the activation of formation of a surface vacancy.

Note that we can form a vacant site on a surface by digging an atom out of a flat surface and placing the atom on another part of the flat surface as an adatom. If we consider the (111) surface of fcc lattice, nine bonds need breaking in order to dig an atom out of the flat surface and it will regain three bonds when the adatom is made. The energy change in this process is more than that in breaking an atom from a kink site from a step on the surface as depicted in Fig. 7.3. Thus, the vacancy model is unfavorable. As presented in the last section, the equilibrium concentration of adatoms on a stepped surface is related to the energy of breaking an atom from the kinks on the surface, not to the formation of vacant sites in the flat surface.

Under equilibrium conditions, the diffusion of an adatom requires only the activation energy of motion. Yet, when there is a flux of adatoms driven away by a driving force, e.g. under an electric field in electromigration on the surface of a Cu interconnect, we need to supply adatoms to keep the equilibrium concentration, so we must consider the source of adatoms coming from the kink sites along a surface step. The equilibrium concentration of adatoms on a surface is given as

$$N_0 = N_s \exp\left(-\frac{W_s}{kT}\right)$$

where N_0 and N_s are the equilibrium concentration of adatoms and the number of atomic sites on the surface, respectively, and W_s is the amount of energy it takes to remove an atom from the kink site to the terrace of the surface, as depicted in Fig. 7.3. The surface diffusion flux is given as

$$J_s = N_0 \frac{D_s}{kT} F \qquad (7.12)$$

where F is the driving force of surface diffusion flux. Since the adatoms have a residence time τ_0 on the stepped surface of terrace width of $L_0/2$, we have defined surface diffusivity in Eq. (7.5):

$$\sqrt{4D_s\tau_0} > \frac{L_0}{2}$$

In surface diffusion of an adatom, when we compare it to lattice atom diffusion in fcc metals by vacancy mechanism, no formation of vacancy is considered and the correlation factor and the number of nearest neighbors are ignored. Instead, the formation energy of an adatom is considered. Therefore, in the flux equation of $J = C\langle v \rangle = CMF$, the concentration term is given by $C = N_0$, the adatom concentration on the surface as given in Eq. (7.9). On the other hand, if we take $C = N_s$, we need to include the adatom formation energy of into D_s as shown below:

$$J_s = N_s \exp\left(-\frac{W_s}{kT}\right)\frac{D_s}{kT}F = N_s\frac{D_{ss}}{kT}F$$

$$\text{where } D_{ss} = \lambda^2 v_s \exp\left(-\frac{\Delta G_s + W_s}{kT}\right) \tag{7.13}$$

The activation energy of the surface diffusion of adatoms is equal to the sum of activation energy of motion and the activation energy needed to remove an atom from a kink site to the flat surface. Comparing Eq. (7.4) and Eq. (7.13), the difference is in whether the diffusion of adatoms is in equilibrium state or in steady state.

On the stepped surface, the diffusion distance is given by

$$\lambda_s = \sqrt{D_s\tau_0} = \lambda \exp\left(\frac{\Delta G_{des} - \Delta G_s}{2kT}\right)$$

On a Si (001) surface, we have $\Delta G_{des} \cong 1.1$ eV/atom and $\Delta G_s \cong 0.5$ eV/atom. We can estimate the diffusion distance and the homoepitaxial growth rate as shown in Table 7.3.

Surface atomic structure tends to be anisotropic; often, there are valleys and the diffusion along a valley is faster than across a valley. Dimensionality also affects the migration distance in a fundamental way; that is, if $\langle x^2 \rangle = 4Dt$ for the one-dimensional case, then $\langle R^2 \rangle = \langle x^2 \rangle + \langle y^2 \rangle = 8Dt$ for a two-dimensional isotropic solid. The rms distance for two dimensions is then $\sqrt{2/3}$ smaller than that for three dimensions, for the same time and assuming an isotropic diffusion coefficient.

By far the most significant quantitative difference between surface and bulk diffusion is the aspect of vacancy formation, required for the bulk case and not required in the surface case. An inspection of Table 4.1 shows that the vacancy formation energy is at least comparable to the vacancy migration energy for many solids. For self-diffusion in Ge and Si, the vacancy formation energy is the dominant term. Since both the formation energy, ΔH_f, and migration energy, ΔH_m, enter the exponential, the difference in diffusion coefficient (with and without ΔH_f) can be enormous. For example, the ratio

Table 7.3. Growth parameters on (100) Si with $W = 4.5$ eV, $\lambda = a = 0.543$ nm, $\nu = 10^{13}$/s, $\Delta G_S = 0.5$ eV and $\Delta G_{des} = 1.1$ eV

	25 °C	300 °C	500 °C	700 °C
kT (eV)	0.0257	0.0494	0.0666	0.0838
D_s (cm^2/s)	5.2×10^{-11}	5.9×10^{-7}	8.1×10^{-6}	3.8×10^{-5}
τ_0 (s)	3.8×10^5	4.7×10^{-4}	1.4×10^{-6}	5.0×10^{-8}
λ_s (μm)	44.4	0.16	0.033	0.014

of the exponential factors in Si (Table 3.1), with and without the vacancy formation term, is

$$\frac{\exp{(0.4 \text{ eV}/kT)}}{\exp{(4.3 \text{ eV}/kT)}} \approx 10^{24}$$

at $T = 550$ °C. The pre-exponential factor for surface diffusion is relatively simple, since the statistics of vacancy formation are not involved. Thus, an estimate of the pre-exponential factor, $D = \lambda^2 \nu$ with $\lambda \cong 10^{-8}$ cm and $\nu = 10^{13}$/s yields $\lambda^2 \nu = 10^{-3}$ cm^2/s. This value is close (within a factor of 10) to the pre-exponential factor of many measured surface diffusion coefficients. An order-of-magnitude estimate for the ratio of the surface diffusion coefficient to the bulk diffusion coefficient for the Si/Si system is 10^{18} at $T = 550$ °C. More graphically $\sqrt{4D_{surf}t} = 1.1 \times 10^{-2}$ cm and $\sqrt{4D_{bulk}t} = 1.1 \times 10^{-11}$ cm for $T = 550$ °C, using Table 4.1 for the bulk diffusion coefficient and $D_{surf} = 10^{-3}e^{-0.4/kT}$ cm^2/s. Thus, there is extensive surface diffusion at 550 °C while the bulk diffusion is essentially non-existent at this temperature, i.e. $\sqrt{4D_{bulk}t}$ is less than an atom spacing.

7.5 Step-mediated growth in homoepitaxy

On a regularly stepped surface as shown in Fig. 7.4, we assume the following: that V is the rate of growth normal to the step surface; h is the step height; L is the width of each step; and v is the step velocity in the lateral direction (see Appendix E).

We also assume that the growth times in the vertical and lateral directions are the same, and we have

$$t = \frac{h}{V} = \frac{L}{v} \tag{7.14}$$

Thus, we have $V = vh/L = vhN_L$, where $N_L = 1/L$ is the number of steps per unit length. Since h and N_L are given for a cut surface, we can calculate V, the growth rate of the epitaxial thin film, provided that we know v, the lateral growth velocity.

To consider a model for the analysis of v, we depict in Fig. 7.5 a model of deposition in which the lateral diffusion of adatoms leads to the lateral growth. The net flux of atoms

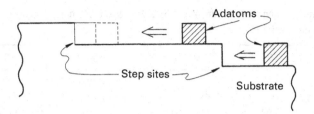

Fig. 7.4   Schematic diagram of step-mediated growth on a regularly stepped surface.

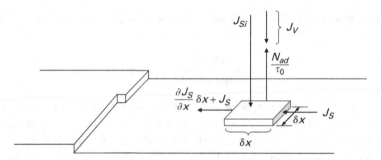

Fig. 7.5   Schematic diagram depicting a model of deposition in which the lateral diffusion of adatoms leads to the lateral growth.

being deposited on the stepped surface per unit area is

$$J_V = J_{Si} - \frac{N_{ad}}{\tau_0} \tag{7.15}$$

where J_V is the net flux of atoms deposited on the surface per unit area per unit time, J_{Si} is the incident flux of atoms by deposition, N_{ad} is the density of adatoms per unit area on the surface, which is much larger than N_0, the equilibrium adatoms on the surface in the deposition, and N_{ad}/τ_0 is the desorption rate.

Note that both J_V and $J_{Si} - N_{ad}/\tau_0$ are in units of number of atom/cm² s, but J_s as shown in Fig. 7.5 is the surface flux diffusing to the step per unit length of step, and it has a unit of number of atom/cm s.

According to Fick's first law, we have

$$J_s = -D_s \frac{\partial N_{ad}}{\partial x}$$

Now we consider the growth to be one-dimensional, then the number of atoms deposited by $(J_{Si} - N_{ad}/\tau_0)$ into the unit area of $\delta y \delta x$ must be equal to the net flux of atoms diffusing out from the unit area. The net flux is given by $(\partial J_s/\partial x)dx$, so we have

$$\left(\frac{\partial J_s}{\partial x} dx \right) dy = \left(J_{Si} - \frac{N_{ad}}{\tau_0} \right) dxdy$$

Thus,

$$\frac{\partial J_s}{\partial x} = J_{Si} - \frac{N_{ad}}{\tau_0}$$

We obtain the diffusion equation of Fick's second law below

$$-D_s \frac{\partial^2 N_{ad}}{\partial x^2} = J_{Si} - \frac{N_{ad}}{\tau_0} \tag{7.16}$$

Since we have defined the supersaturation coefficient, $\sigma = \frac{J_{Si}}{J_0} - 1$, we have $(\sigma + 1)J_0 = J_{Si}$, or $\tau_0 J_0(\sigma + 1) = \tau_0 J_{Si}$. Rearranging terms, we have

$$-D_s \tau_0 \frac{\partial^2 N_{ad}}{\partial x^2} = \tau_0 J_{Si} - N_{ad}$$

Substituting $D_s \tau_0 = (\lambda_s)^2$, we have the following diffusion equation of

$$-\lambda_s^2 \frac{\partial^2 N_{ad}}{\partial x^2} + N_{ad} = \tau_0 J_0(\sigma + 1) \tag{7.17}$$

Two kinds of solution can be given. The first solution is for a single step, in which

$$N_{ad} = N_0 \left\{ 1 + \sigma \left[1 - \exp\left(-\frac{x}{\lambda_s}\right) \right] \right\} \tag{7.18}$$

The second solution is for a periodic step of width L_0, and we have

$$N_{ad} = N_0 \left\{ 1 + \sigma \left[1 - \frac{\cosh\left(\dfrac{x}{\lambda_s}\right)}{\cosh\left(\dfrac{L_0}{2\lambda_s}\right)} \right] \right\} \tag{7.19}$$

Knowing the solution, we can use Fick's first law to obtain the flux arriving at the step, then we can calculate the step growth velocity by considering mass conservation in the growth. Fig. 7.6 depicts the arriving of flux J_s at the step and the growth of the step of length x on the surface. In other words, we consider the conservation of the number of atoms arriving at the front of the step and the number of atoms leading to the growth of xdy, where dy is the width as indicated in Fig. 7.6:

$$J_s dy t = -N_s(xdy) \tag{7.20}$$

The negative sign in the above equation is because the growth direction is opposite to the flux, J_s. Note that while we conserve the total number of atoms in the above equation using a two-dimensional model, we can multiply both sides by an atomic height and

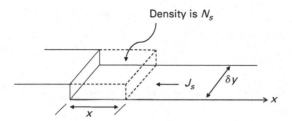

Density is N_s

Fig. 7.6   Schematic diagram depicting the arrival of flux J_s at the step and the growth of the step of length x on the surface.

obtain the conservation of atoms in volume in the growth. We need to calculate the flux of J_s arriving at the front of the step by using Fick's first law:

$$J_s = -D_s \frac{\partial N_{ad}}{\partial x} = -N_s \frac{dx}{dt} = -N_s v$$

Note that the dimension of the above equation is correct. Now we shall use the solution of the single step to obtain the flux in the above equation:

$$\frac{\partial N_{ad}}{\partial x} = \frac{\sigma N_0}{\lambda_s} \exp\left(-\frac{x}{\lambda_s}\right) = \frac{\sigma N_0}{\lambda_s} \quad \text{at } x = 0$$

Thus, the step velocity is obtained:

$$v = \frac{J_s}{N_s} = \frac{D_s \partial N_{ad}/\partial x}{N_s} = \frac{D_s \sigma N_0}{\lambda_s N_s} \tag{7.21}$$

$$= \frac{N_0 D_s}{\lambda_s N_s}\left(\frac{J_{Si} - J_0}{J_0}\right) \quad \text{(since } \sigma = (J_{Si}/J_0) - 1)$$

$$= \frac{\tau_0 D_s}{\lambda_s N_s}(J_{Si} - J_0) \quad \text{(since } J_0 = N_0/\tau_0)$$

$$= \frac{\lambda_s}{N_s} J_{Si} \quad \text{(since } D_s\tau_0 = (\lambda_s)^2 \text{ and } J_{Si} \gg J_0)$$

So the rate of normal growth is given as

$$V = vhN_L = \frac{\lambda_s J_{Si} h N_L}{N_s} \tag{7.22}$$

7.6   Deposition and growth of an amorphous thin film

In thin-film deposition, when the substrate is kept at a very low temperature, for example, at liquid nitrogen temperature, the surface diffusion is frozen. When a flux of vapor atoms in deposition reaching the surface, they are quenched and the atoms cannot move on the surface, not even a small amount of shifting of position, in order to obtain the low-energy position of a lattice site in a crystalline solid. Therefore, no nucleation of a crystalline nucleus occurs. An amorphous solid can be formed as in the case of the deposition of amorphous Si thin films. Under such conditions, the time taken to grow the amorphous film of one atomic layer thick is given by N_s/J_{Si}, thus the rate of amorphous thin-film growth is given by

$$R_{ND} = \frac{h}{N_s/J_{Si}} = \frac{J_{Si}h}{N_s} \tag{7.23}$$

where h is the thickness of one atomic layer and R_{ND} is the rate of no-diffusion growth. If we compare the step-mediated growth to the no-diffusion growth, we have

$$\frac{V}{R_{ND}} = \lambda_s N_L = \frac{\lambda_s}{L_0} = \frac{\text{diffusion-length}}{\text{step-width}}$$

We see that if we define a condensation coefficient which is greater than one as below,

$$\eta = \frac{\lambda_s}{L_0} = \lambda_s N_L > 1$$

the deposited atoms can diffuse to the steps and we can have stepwise growth.

7.7 Growth modes of homoepitaxy

As discussed in the last two sections, the ability to grow an epitaxial layer or an amorphous layer depends primarily on surface diffusion and on a variety of experimental factors: surface temperature, surface cleanliness, deposition rate, surface miscut, etc. The key factor is surface or substrate temperature. Fig. 7.7 depicts the different configurations of the deposited layers. At the lowest temperature (room temperature or below) impinging materials will generally be deposited as an amorphous layer. There is little surface diffusion at such a low temperature and the deposited atoms get trapped into a conglomerate of non-crystalline sites with no thermal energy for atomic rearrangement. Subsequent heat treatment is necessary for regrowth to form an epitaxial layer. At higher temperatures, the atoms have sufficient surface mobility to arrange themselves epitaxially on the surface. Epitaxial islands are formed. At yet higher temperatures, epitaxial growth occurs by lateral stepwise growth. We indicate the transition temperature in Fig. 7.7 with the understanding that it is a concept of convenience. In principle, step-mediated growth could occur at any temperature in a perfect vacuum if the deposition rates were slow enough.

The temperature dependence can be understood on a more quantitative basis by consideration of the surface diffusion coefficient. In general, such diffusion coefficients are not well known, but there has been enough work on a few systems that we can make some estimates. From the previous discussion on surface diffusion coefficients, we can

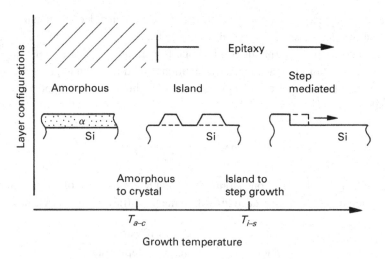

Fig. 7.7   Growth modes of deposited layers having of amorphous, epitaxial island and epitaxial layer versus growth temperature.

estimate the time for an atom to make a single jump. If D_s is written as

$$D_s = \lambda^2 v_s \exp\left(-\frac{\Delta G_s}{kT}\right)$$

the mean time for a single jump, τ_s, is given by

$$\lambda^2 = D_s \tau_s$$

$$\tau_s = \frac{1}{v_s} \exp\left(\frac{\Delta G_s}{kT}\right)$$

As usual, $v_s \cong 10^{13}$/s and we choose $\Delta G_s = 1.0$ eV/atom so at room temperature $\tau_s \cong 10^4$ s $\cong 2.8$ h. If the deposition rate is R monolayer/s, other atoms will fall in the vicinity of the initial deposited atom in a time $1/R$. We need time for at least one jump before other atoms impinge. If we require at least one jump before completion of a monolayer, this implies a growth rate of a monolayer every 2.8 h.

It is possible to perform epitaxy at such low temperatures and low rates. In practice, such slow growth is limited by the quality of the vacuum system. As we discussed in Chapter 2, even at a pressure of 1×10^{-10} torr, a full monolayer worth of residue gases (O_2, H_2, N_2, etc.) impinges on the sample every 10^4 s. If only a small fraction of impinging gas atoms stick, they contaminate the surface and ruin the epitaxial growth of a high-purity film.

Because of the exponential dependence on temperature of the diffusion coefficient, the relevant deposition time changes dramatically with temperature. For example, at 252 °C (525 K), τ_s is approximately 4×10^{-4} s, and under typical conditions there are many possible jumps with deposition rates of 1–10 monolayer/s. In this case the growth is pictured as the assembly of two-dimensional epitaxial growth. A value of $\Delta G_s = 0.5$ eV/atom implies $\tau_s \cong 3 \times 10^{-5}$ s, rather than 10^4 s, a change of nine orders of magnitude. Nevertheless, the general $\Delta G_s/kT$ scaling is correct and practical limits are set by experimental conditions.

High-binding-energy locations near surface steps are favored sites for deposited atoms. If atoms have sufficient diffusivity, they can diffuse to a step and reach the high-temperature regime associated with step-mediated growth.

One type of experiment that gives information on growth modes is the dynamic electron diffraction known as reflection high-energy electron diffraction (RHEED). In this technique, electrons are diffracted from the first monolayer of an ordered surface during growth, and the diffracted intensity indicates the perfection of a long-range order on the surface. One of the advantages of RHEED is that the diffraction intensity can be measured during growth. The two-dimensional island-growth mode gives a characteristic RHEED signal in the form of an oscillating intensity, with one oscillation corresponding to the deposition of one monolayer of material. The maximum in the diffraction intensity occurs when the layer is complete while the minimum occurs at $^1/_2$ monolayer coverage, corresponding to the maximum disorder in the sense of a random array of two-dimensional epitaxial growth. As the substrate temperature is increased, the RHEED oscillations decay and eventually vanish, signaling the onset of step-mediated growth. In this latter growth mode, the surface appears as a series of flat terraces at all times, and essentially yields constant diffraction intensity.

In summary, homoepitaxy is associated with three different growth modes depending primarily on the substrate temperature. At low temperatures, there is an amorphous layer deposition which requires subsequent heat treatment for epitaxial regrowth. At intermediate temperatures, structure epitaxy occurs via two-dimensional clusters or island formation. We shall discuss the nucleation event of these clusters in the next section. At higher temperatures, epitaxy via the step-mediated growth occurs.

7.8 Homogeneous nucleation of a surface disc

In the classical theory of nucleation, it is known that the rate of homogeneous nucleation is much less than that of heterogeneous nucleation. The event of nucleation in real materials systems is generally heterogeneous. In the following, we shall consider the homogeneous nucleation of a surface disk on a flat surface, and we shall see that energetically it is difficult for such an event to happen. This is why we have step-mediated growth on a cut surface so that the steps are given for epitaxial growth without the need of nucleation. Before we consider the energy in the nucleation, we need to consider the vapor pressure above a small radius cluster. As stated at the beginning of the chapter, any surface kinetic process must be linked to the vapor pressure on the surface.

Figure 7.8(a) and (b) depict the nucleation of a circular disk on a flat surface and the schematic cross-sectional view of a single atomic layer of disk on the surface, respectively. Note that there is no interface between the surface disk and the substrate as shown

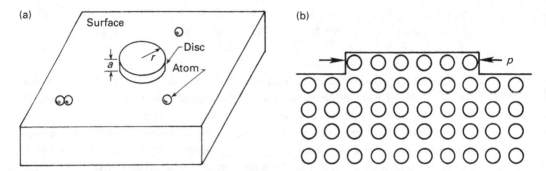

Fig. 7.8   (a) Schematic diagram of the nucleation of a circular disk on a flat surface. (b) Schematic cross-sectional view of a single atomic disk layer on the surface.

in Fig. 7.8(b). Thus, we can regard the nucleation of such a disk as homogeneous nucleation. The nucleation requires the formation of the circumference of the disk, so it is the energy barrier of the nucleation event. Before we consider the energies involved in the nucleation, we must first consider the change in equilibrium conditions of the surface when a disk exists.

We assume that the equilibrium pressure on the flat surface is p_0. At equilibrium, the fluxes condensing on or desorbing from the surface are equal. In vacuum, the condensing flux can be given as

$$J_c = Cv = nv_a$$

where $C = n(= p_0/kT)$ is the density of atom in the vapor, or the number of atoms per unit volume of the vapor, and v_a is the root mean square velocity of the atoms in the vapor. For desorption flux, we express

$$J_d = J_0 = N_0 v_{des} = N_0 v_s \exp\left(-\frac{\Delta G_{des}}{kT}\right)$$

where $J_d = J_0$ is the equilibrium desorption flux, N_0 is the number of adatoms per unit area of the planar surface, v_{des} is the desorption frequency, v_s is the surface vibration frequency, and ΔG_{des} is the activation energy of desorption. At equilibrium, we have $J_c = J_d$, or $N_0 v_{des} = nv_a$.

To nucleate a disk on the flat surface, we need supersaturation, $J_c \gg J_d$. The question is, how much supersaturation is needed? First, we consider the energy change in the nucleation of a disk. The surface energy of the circumference is given as

$$E_d = 2\pi r a \gamma \tag{7.24}$$

where a is the atomic layer thickness and γ is the surface energy per unit area of the circumference. The surface energy of the circumference exerts a pressure on the disk, which is

$$p_s = \frac{1}{A}\frac{dE_d}{dr} = \frac{2\pi a\gamma}{2\pi ar} = \frac{\gamma}{r} \tag{7.25}$$

where A is the area of the circumference of the disk. Under the pressure, the energy of each atom in the disk is increased by the amount

$$p_s\Omega = \frac{\gamma}{r}\Omega \tag{7.26}$$

Recall that this is the chemical potential energy increase due to the formation of the circumference surface of the disk. Because of the increase in energy, the atoms in the disk can sublimate more easily,

$$J_d^* = N_0 v_s \exp\left(-\frac{\Delta G_{des}}{kT} + \frac{p_s\Omega}{kT}\right) \tag{7.27}$$

$$\frac{J_d^*}{J_0} = \exp\left(\frac{p_s\Omega}{kT}\right) = \exp\left(\frac{\gamma\Omega}{rkT}\right) = \frac{nv}{n_0v} = \frac{p_s}{p_0} \tag{7.28}$$

The last equation is the Gibbs–Thompson equation for the vapor pressure above a cylindrical disk of radius r. It means if we keep the disk under the equilibrium pressure of p_0, since $p_s > p_0$, the disk tends to evaporate and shrink, so the disk will reduce its radius and will continue to shrink until it disappears. However, if we want to keep the disk stable or to grow the disk, we must increase the pressure, the supersaturation, or the condensation rate. What is the supersaturation needed so that it becomes stable? Or what is the supersaturation needed to nucleate a new disk of the stable critical size on the surface?

The energy change in nucleating the disk is given as

$$\Delta E_{disc} = 2\pi ra\gamma - \pi r^2 a\Delta E_s \tag{7.29}$$

where ΔE_s is the latent heat of condensation per unit volume, and γ is the surface energy per unit area of the circumference of the disk. We define r_{crit} such that at $r = r_{crit}$,

$$\frac{d\Delta E_{disc}}{dr} = 0$$

Thus, $2\pi a\gamma - 2\pi ra\Delta E_s = 0$, and

$$r_{crit} = \frac{\gamma}{\Delta E_s} \tag{7.30}$$

The net energy change in nucleating the critical disk is

$$\Delta E_{crit} = \pi r_{crit} a\gamma \tag{7.31}$$

The critical disk at r_{crit} is metastable since ΔE_{crit} is a maximum in the plot of ΔE versus r. Any slight deviation from r_{crit} will lead to a decrease of energy. Now we consider the vapor pressure on the critical disk, we have

$$\frac{p_{crit}}{p_0} = \exp\left(\frac{\gamma\Omega}{r_{crit}kT}\right)$$

Thus,

$$r_{crit} = \frac{\gamma\Omega}{kT\ln(p_{crit}/p_0)}$$

Then,

$$\Delta E_{crit} = \pi r_{crit} a\gamma = \frac{\pi\gamma^2 a\Omega}{kT\ln(p_{crit}/p_0)} = \frac{\pi\gamma^2 a^4}{kT\ln(p_{crit}/p_0)} \tag{7.32}$$

In the last step of the above equation, we have taken $\Omega = a^3$. Knowing the critical energy or activation energy of formation of the critical disk, we can calculate the probability of nucleation of the disk or the density of nucleus, i.e. the number of nuclei per unit area:

$$N_{crit} = J_c\tau_0 \exp\left(-\frac{\Delta E_{crit}}{kT}\right) \tag{7.33}$$

In the above equation, recall that J_c is the condensation flux, having a unit of number of atoms per unit area per unit time, and τ_0 is the residence time of adatoms on the surface. Therefore, we have

$$N_{crit} = J_c\tau_0 \exp\left[-\frac{\pi(\gamma a^2)^2}{(kT)^2\ln(p_{crit}/p_0)}\right] \tag{7.34}$$

On the basis of the last equation, we can estimate the nucleation rate and we shall show that it is extremely low. For the condensation flux, we shall take $J_c = 10^{15}$ atom/cm^2 s, which means that we deposit 1 monolayer/s, or we deposit 100 nm-thick film in 5 min, which is a typical rate of deposition.

For the residence time, recall that

$$\tau_0 = \frac{1}{\nu_s}\exp\left(\frac{\Delta G_{des}}{kT}\right)$$

We take $\Delta G_{des} = 1.1$ eV, $T = 1223$ K, which means that $kT = 0.1$ eV, then

$$\tau_0 = \frac{1}{10^{13}}\exp\left(\frac{1.1}{0.1}\right) = 10^{-13}10^{11/2.3} = 10^{-8}\text{s}$$

For the activation energy, we take the surface energy per atom of Si, $\gamma a^2 = 0.6\,\mathrm{eV/atom}$ and $kT = 0.1$ eV at 1223 K, we have

$$\Delta E_{crit} = \frac{\pi (\gamma a^2)^2}{kT \ln(p_{crit}/p_0)} = \frac{3.14 \times 3.6}{\ln(p_{crit}/p_0)} \tag{7.35}$$

If we give

$$p_{crit}/p_0 = 10, \quad \ln 10 \ \ = 2.3, \quad \Delta E_{crit} = 4.9\mathrm{eV}$$

$$p_{crit}/p_0 = 100, \quad \ln 100 \ = 4.6 \quad \Delta E_{crit} = 2.5\mathrm{eV}$$

$$p_{crit}/p_0 = 1000, \quad \ln 1000 = 6.9 \quad \Delta E_{crit} = 1.6\mathrm{eV}$$

If we input the above values into N_{crit}, for the case of $p_{crit}/p_0 = 1000$, we find that

$$N_{crit} = J_c \tau_0 \exp\left(-\frac{\Delta E_{crit}}{kT}\right) = 10^{15} \times 10^{-8} \times \exp\left(-\frac{1.6}{0.1}\right) = 10^7 \times 10^{-7} = 1$$

This means that we have one nucleus on a unit area of 1 cm^2 even under the super-saturation of 1000, which is indeed a very low rate of nucleation. Clearly, heterogeneous nucleation will take over if it can occur.

7.9 Mass transport on a patterned surface

7.9.1 Early stage of diffusion on a patterned surface

A square wave-type periodic structure on a surface can be produced by patterning and etching or by imprinting, as shown in Fig. 7.9(a). Upon annealing, the periodic structure will decay in magnitude by surface diffusion. Such structure has been used to study surface diffusion. The driving force is the reduction in surface area and surface energy. In the following, we consider two stages of the diffusion. The first is the very beginning of the process in which the sharp corners in the patterned structure become rounded. The second is the later stage of the diffusion when the curvature of the periodic waves becomes very small. The middle stage of diffusion has been covered in Section 5.3.1.

In Fig. 7.9(b), we depict the cross-section of a square wave with rounded corners. In annealing, the radius of the corners will increase. We try to find out the rate of change of the radius. On the rounded corner surface or the surface of a cylindrical surface, the chemical potential is given by

$$\mu_r = \frac{\gamma}{r}\Omega$$

where γ is the surface energy per unit area of the cylindrical surface, r is the radius, and Ω is atomic volume. We can also write

$$\mu_r - \mu_\infty = \frac{\gamma}{r}\Omega$$

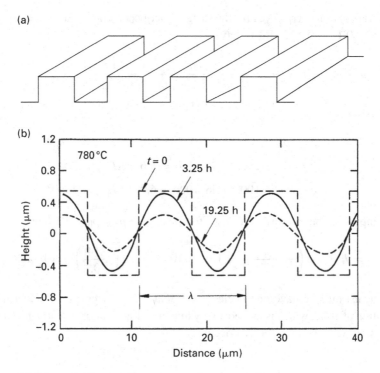

Fig. 7.9   (a) Schematic diagram of a square wave-type periodic structure on a surface. (b) Schematic diagram of the cross-section of a square wave with rounded corners. In annealing, the radius of the corners will increase.

where μ_∞ is the chemical potential of a flat surface and we can let $\mu_\infty = 0$. Thus, the chemical potential difference between the upper and the lower corners is

$$\Delta\mu = \mu_r - \mu_{-r} = \frac{2\gamma}{r}\Omega$$

Thus, the force driving the atoms to diffuse from the upper to the lower corners is given by

$$F = -\frac{\Delta\mu}{\Delta x} = -\frac{2\gamma}{rh}\Omega$$

where h is the height of the patterned step. Then the flux of atoms diffusing from the upper to the lower corners is

$$J = C_s M_s F = C_s \frac{D_s}{kT}\left(-\frac{2\gamma}{rh}\Omega\right) = -\frac{2C_s D_s \gamma \Omega}{kTrh} \tag{7.36}$$

where C_s is the concentration of atoms per unit area, D_s is surface diffusivity, and J is the surface flux, number of atom/cm s.

The flux, departing from the upper corner, will round off more of the upper corner, and when it arrives at the lower corner, will fill up the lower corner; in other words, both r and $-r$ will increase. Let us consider the growth in the lower corner and assume a unit width, so the number of atoms in the volume of the lower corner is

$$\frac{1}{\Omega}\left(r^2 - \frac{1}{4}\pi r^2\right)$$

If we assume that no flux is leaving the lower corner, all arriving fluxes of atoms will grow in the corner and increase its radius. Then, the time-rate change of the radius of the lower corner will be equal to the arriving flux:

$$\frac{d}{dt}\left[\frac{1}{\Omega}\left(r^2 - \frac{1}{4}\pi r^2\right)\right] = \frac{2C_sD_s\gamma\Omega}{kTrh} \tag{7.37}$$

$$\left(1 - \frac{\pi}{4}\right)\frac{kThr^2 dr}{C_sD_s\gamma\Omega^2} = dt$$

$$r^3 \approx \frac{3C_sD_s\gamma\Omega^2}{(1 - (1/4)\pi)kTh}t \tag{7.38}$$

We can check the dimension in the above equation to see if it is correct, and we have

$$\text{cm}^3 = \frac{(\text{atom}/\text{cm}^3)(\text{cm}^2/\text{s})(\text{eV}/\text{cm}^2)(\text{cm}^3)^2}{\text{eV}(\text{cm})}\text{s}$$

and the dimension checks.

7.9.2 Later stage of mass transport on a patterned structure

In a later stage of mass transport on the surface of the patterned structure, we assume that the curvature is large so that the slope variation on the curved surface is small, as depicted in Fig. 7.10(a) and (b). The curvature is given as [5]

$$\frac{1}{r} = \frac{\frac{d^2z}{dx^2}}{[1 + (dz/dx)^2]^{3/2}} \tag{7.39}$$

If we assume that $dz/dx \ll 1$, in other words, when r is very large, we take

$$\frac{1}{r} = \frac{d^2z}{dx^2} \tag{7.40}$$

On the curved surface, we can take the chemical potential to be

$$\mu_r = \frac{\gamma}{r}\Omega = \gamma\Omega\frac{d^2z}{dx^2}$$

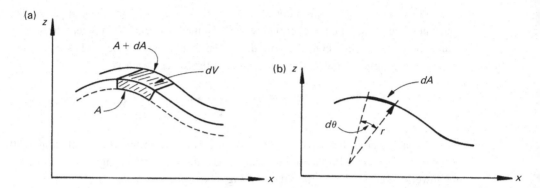

Fig. 7.10   (a) Surface profile showing volume element dV. (b) Schematic of a surface profile where $r = dS/d\theta$.

The driving force is

$$F = -\frac{d\mu_r}{dx} = -\gamma\Omega\frac{d^3z}{dx^3}$$

The diffusing flux will be

$$J = C_sM_sF = C_s\frac{D_s}{kT}\left(-\gamma\Omega\frac{d^3z}{dx^3}\right) \tag{7.41}$$

When we apply Fick's second law to the above flux in the one-dimensional case, we have

$$-\frac{\partial C}{\partial t} = \frac{\partial J_x}{\partial x} = -\frac{C_sD_s\gamma\Omega}{kT}\frac{\partial^4z}{\partial x^4} \tag{7.42}$$

We cannot solve the above equation until we change $\partial C/\partial t$ to $\partial z/\partial t$. To do so, we consider the change of the surface contour by surface diffusion. Mass transport will lower the hills and will fill up the valleys. It means a change in height in the z-direction.

Now we consider an area of ΔL by ΔW on the curved surface. The total number of atoms, Q, entering this area of $\Delta L\Delta W$ per unit time is equal to the flux divergence in this area. Here, we have ignored the unit height above the area, so it is the same as considering the divergence in a volume.

$$(\Delta L\Delta W)\frac{\partial J_x}{\partial x} = \frac{\partial Q}{\partial t}$$

The volume of Q atoms is equal to $Q\Omega$. When we add this volume to the area $\Delta L\Delta W$, it leads to the growth in the z-direction, $Q\Omega = z\Delta L\Delta W$, as shown in Fig. 7.11. Thus,

$$\frac{\partial Q}{\partial t} = \frac{1}{\Omega}\frac{\partial z}{\partial t}\Delta L\Delta W = -\frac{\partial J_x}{\partial x}\Delta L\Delta W$$

And we have

$$\frac{1}{\Omega}\frac{\partial z}{\partial t} = -\frac{\partial J_x}{\partial x} = \frac{C_sD_s\gamma\Omega}{kT}\frac{\partial^4z}{\partial x^4}$$

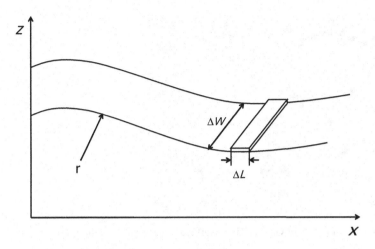

Fig. 7.11 Schematic diagram of adding the growth of a small volume to the area $\Delta L \Delta W$, it leads to the growth in the z-direction, $Q\Omega = z\Delta L \Delta W$.

Finally, we obtain a fourth-order diffusion equation:

$$\frac{\partial z}{\partial t} = \frac{C_s D_s \gamma \Omega^2}{kT} \frac{\partial^4 z}{\partial x} \tag{7.43}$$

Note that a similar equation was obtained for the cusp formation at the triple point where a grain boundary meets a free surface. If we check the dimension of the equation, it is correct as shown below:

$$\frac{cm}{s} = \frac{(atom/cm^2)(cm^2/s)(eV/cm^2)(cm^3)^2}{eV} \frac{cm}{cm^4}$$

The solution of the differential equation is

$$z(x,t) = \sum_{n=0}^{\infty} A_n \exp\left(-\frac{t}{\tau_n}\right) \sin\left(\frac{2n\pi x}{\lambda}\right) \tag{7.44}$$

where $\tau_n = \left(\frac{C_s D_s \gamma \Omega^2}{kT}\right)^{-1} \left(\frac{2n\pi}{\lambda}\right)^{-4}$ and λ is the period of the function, so $\tau_n \cong (\lambda/n)^4$. Thus, the small period function decays very fast.

7.10 Ripening of a hemispherical particle on a surface

Thin-film deposition often results in the formation of a distribution of discrete three-dimensional nuclei or particles when the thin film does not wet the substrate. Annealing of the deposited film results in particle growth. Large nuclei grow larger at the expense of smaller ones. When the particles are very close, they can agglomerate by coalescence

Fig. 7.12   SEM image of clusters of Ga on a GaAs surface heated to 660 °C for 5 min.

to reduce the total surface-to-volume ratio of the deposit at constant volume. When there is a distribution of size of particles and when the particles are far apart, they tend to grow and dissolve through a single-atom process of diffusion known as ripening. In this section we describe the kinetics of ripening on a surface. Energy minimization represents the driving force in ripening, while diffusion as outlined below describes the mechanism and time-dependence of the process [6]. Fig. 7.12 shows an image of clusters of Ga on a GaAs surface heated to 660 °C for 5 min.

In a distribution of particle clusters as shown in Fig. 7.12, the ripening process is a many-body problem. The nearest neighbors of a particle may have sizes both bigger and smaller than the particle itself. Thus, it is difficult to describe the kinetics of ripening, since both growth and dissolution of the particle can occur. To overcome this problem, the mean-field approach was developed. It is assumed that a mean concentration exists among all the particles. The mean concentration is the equilibrium particle concentration of the mean size, $\bar{r}$. Any particle which has a size larger than the mean-size will grow and any particle smaller than the mean size will shrink. The kinetics of ripening of any particle can be described with respect to the mean-size particle. For simplicity we take the particles to be hemispherical of radius r and the wetting angle is constant and kept at 90°. The Gibbs–Thomson equation now represents an enhanced concentration of adatoms surrounding the particle of radius r. A particle of radius r is in thermodynamic equilibrium with the adatom concentration N_r:

$$N_r = N_0 \exp\left(\frac{2\gamma\Omega}{rkT}\right) \tag{7.45}$$

where N_0 is the adatom concentration corresponding to the vapor pressure for the planar surface. Note that the units of N_0 and N_r are number of atoms per cm^2.

Similarly, we imagine a mean-size particle, and we have

$$N_m = N_0 \exp\left(\frac{2\gamma\Omega}{\bar{r}kT}\right) \tag{7.46}$$

where N_m is the adatom concentration corresponding to the mean-size particle, and we have $N_0 < N_m$.

Below we shall consider the growth of a particle in the mean field, so the size of the particle is larger than the mean size. The mechanism of particle growth is taken to be surface diffusion. We shall solve the steady-state diffusion equation to obtain the growth rate as a function of time. Note that the steady state is an approximation here, as the mean size is actually increasing with time and the mean-field concentration is decreasing with time. However, the rates involved are sufficiently slow to allow a steady-state approximation.

The continuity equation for the concentration N of a diffusing species corresponding to a two-dimensional equation in cylindrical coordinates R and θ is given as

$$\frac{\partial N}{\partial t} = \frac{1}{R}\frac{\partial}{\partial R}\left(RD_s\frac{\partial N}{\partial R}\right) + \frac{D_s}{R^2}\frac{\partial^2 N}{\partial \theta^2} \tag{7.47}$$

where $N(R)$ is a local concentration of adatom/cm^2, and D_s is the surface diffusion coefficient and is assumed to be independent of concentration. Next, we shall consider a steady-state solution with the cylindrical coordinates and we further assume a cylindrical symmetry, so we can ignore the term in θ, giving

$$\frac{1}{R}\frac{d}{dR}\left[RD_s\frac{dN}{dR}\right] = 0 \tag{7.48}$$

The solution to this equation is

$$N(R) = K_1 \ln R + K_2 \tag{7.49}$$

where K_1 and K_2 are arbitrary constants to be determined by given boundary conditions. We impose the following boundary conditions:

$$N(R) = N_r \quad \text{at } R = r$$

$$N(R) = N_m \quad \text{at } R = Lr$$

where r is the radius of the particle and L is a multiplier of the radius r and measures the distance (in units of r) in which the enhanced vapor concentration corresponding to the mean field. Typically, we take $L \sim 10$. Then, the solution becomes

$$N(R) = \frac{N_r \ln(Lr/R) - N_m \ln(r/R)}{\ln(L)} \tag{7.50}$$

In a diffusion problem we use Fick's first law to obtain the flux of adatoms coming to the particle:

$$J_s = -D_s\frac{dN}{dR}$$

where the unit of J_s is the number of adatom/cm s. This is the flux of atoms per unit length (rather than per unit area) per unit time because the problem is two-dimensional.

The total number of atoms diffusing to the particle of circumference $2\pi r$ is J, where

$$J = -2\pi r D_s \left.\frac{dN}{dR}\right|_{R=r} = \frac{2\pi D_s}{\ln(L)}(N_r - N_m)$$

On the basis of Gibbs–Thomson equations of N_r and N_m, and using a small argument expansion for small particle size, we have

$$J = \frac{2\pi D_s}{\ln(L)} N_0 \frac{2\gamma\Omega}{kT}\left(\frac{1}{r} - \frac{1}{\bar{r}}\right) \qquad (7.51)$$

By considering the conservation of mass, the hemispherical particle will grow upon the arrival of the total number of atoms of J. The number of atoms in a hemispherical particle of r is Q where

$$Q = \left(\frac{2}{3}\pi r^3\right)\frac{1}{\Omega}$$

where Ω is atomic volume. Then

$$\frac{dQ}{dt} = \frac{2\pi r^2}{\Omega}\frac{dr}{dt}$$

By conservation of mass or the number of atoms, it must be equal to J, which is the divergence of J_s. So we have

$$\frac{r^2 dr}{\Omega dt} = \frac{D_s}{\ln(L)} N_0 \frac{2\gamma\Omega}{kT}\left(\frac{1}{r} - \frac{1}{\bar{r}}\right) \qquad (7.52)$$

The units in the above equation are correct, since we have

$$\frac{cm^3}{cm^3 s} = \frac{cm^2}{s}\frac{1}{cm^2}\frac{eV}{cm^2}\frac{cm^3}{eV}\frac{1}{cm}$$

However, the solution is non-trivial (similar to LSW theory), so we will not solve it here. Experimentally, it was found that for surface diffusion-limited growth the larger hemispherical particles form with the fourth power of the radius growing linearly with time, i.e. $r^4 = Kt$. Fig. 7.13 shows the experimental data of ripening of Sn particles on a Si (111) surface at 525 K, in which the time-dependence agreement with r^4 is better than with r^3. However, in LSW theory, the kinetic of ripening has r^3 dependence on time.

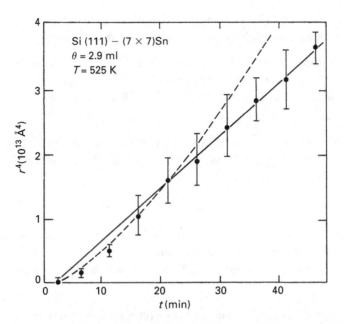

Fig. 7.13 Experimental data of ripening of Sn particles on Si (111) surface at 525 K, in which the time-dependence agreement with r^4 is better than with r^3.

References

[1] G. A. Somorjai, *Chemistry in Two Dimensions: Surfaces* (Cornell University Press, Ithaca, NY, 1981)

[2] A. Zangwill, *Physics of Surfaces* (Cambridge University Press, Cambridge, 1988).

[3] C. Ratsch, M. F. Gyure, R. E. Caflisch, F. Gibou, M. Petersen, M. Kang, J. Garcia and D. D. Vvedensky, "Level-set method for island dynamics in epitaxial growth," *Phys. Rev.* **B 65** (2002) 195403 .

[4] C. Ratsch, A. P. Seitsonen and M. Scheffler, "Strain-dependence of surface diffusion: Ag on Ag(111) and Pt(111)," *Phys. Rev.* **B 55** (1997) 6750 .

[5] W. W. Mullins, "Solid surface morphologies governed by capillarity," in *Metals Surfaces* (ASM, Metal Park, Ohio, 1963).

[6] M. Zinke-Allmang, L. C. Feldman and S. Nakahara, "Role of Ostwald ripening in islanding processes," *Appl. Phys. Lett.* **51** (1987) 975.

[7] F. Allen and E. Kasper, "Models of silicon growth and dopan incorporation," Chapter 4 in *Silicon Molecular Beam Epitaxy* (Vol. I, p. 65), eds E. Kasper and J. C. Bean (CRC Press, Boca Raton, FL, 1988).

Problems

7.1 What is the source of adatoms on a surface?

7.2 In desorption, what is the residence time of an adatom on a surface?

7.3   In Fig. 7.8, we consider the homogeneous nucleation of a circular disk on a surface. Why is it a homogeneous nucleation, not a heterogeneous nucleation?

7.4   A 1 cm^2 clean surface is covered with 4×10^{10} hemispherical tantalum clusters with equal numbers that are either 10 or 30 nm in diameter. If ripening allows the large clusters to consume all of the small clusters, what will the reduction of the total surface energy be if $\gamma_{Ta} = 2890$ erg/cm^2?

7.5   Find the surface energy E_s of two hemispherical Ti clusters ($\gamma = 1650$ erg/cm^2) of radius $= 40$ nm. How much energy will be gained if they coalesce? Ignore the contribution from the interface.

7.6   What process has (time)$^{1/3}$, (time)$^{1/2}$, or (time)^1 dependence?

7.7   Given that at time $t = 1200$ s, a cluster has radius $r = 1$ μm at temperature $T = 800$ K, at what time did the cluster radius $r = 0.5$ μm at 800 K? Assume that the activation energy for diffusion $= 1.5$ eV/atom and the pre-exponential factor is 0.1 cm^2/s.

7.8   A hemispherical cluster is in the process of ripening. Calculate the time at which the pressure over the cluster will be 2.5 times the equilibrium pressure. $N_{eq} = 10^{15}$ cm^{-2}, $\gamma = 2890$ erg/cm^2, $\Omega = 2.96 \times 10^{-23}$ cm^3, $T = 500$ K, $Lr = 10^{-6}$ cm, $D_s = 10^{-10}$ cm^2/s.

7.9   On a (100) Si ($a = 0.543$ nm), what is the surface density of atom/cm^3, and what miscut angle along a [100] direction is required to produce a step length of 10^{-5} cm and of 10^{-6} cm?

7.10   A simple cubic ($a = 0.3$ nm) substrate has been cut along the (100) at a 3° angle to create a stepped surface. If homoepitaxial growth is attempted at 600 °C, will the diffusion length be greater than the distance between steps? Use $\Delta G_{des} = 1.0$ eV/atom, $\Delta G_s = 0.5$ eV/atom, and $v_s = 10^{13}$/s.

7.11   We want to carry out SI MBE at a deposition rate of 1 monolayer/s with step-mediated growth on a sample where steps are separated by 50 nm. What growth temperature and what vacuum level are required if $\Delta G_s = 1.0$ eV/atom or if $\Delta G_s = 0.5$ eV/atom?

7.12   For diffusion-limited growth on (100) Si at 500 °C with a spacing between steps of 50 nm, 6.8×10^{14}/cm^2 surface sites, and $s = 0.033$ μm, what is the growth rate R for $J_{Si} = 10^{15}$ atom/cm^2 s? What is the condensation coefficient for the single-step growth case?

7.13   For a substrate with a simple cubic lattice ($a = 0.3$ nm) homoepitaxial growth with $J = 5 \times 10^{15}$ atom/cm^2 s results in a growth rate $R = 3$ monolayer/s. If the path length $\lambda_s = 15$ nm, estimate the average step length.

7.14 In step-mediated growth in the high-temperature growth regime, Si atoms have a step velocity v (along the step) of 5×10^{-7} cm/s. If the substrate was created by making a 0.69° miscut along the (100) plane, how long does it take to grow 1 μm of Si? Calculate the step velocity for a Si flux of 5×10^{15} atom/cm^2 s at 650 °C on (100) Si. Use values from Table 7.3.

8   Interdiffusion and reaction in thin films

8.1   Introduction

Modern microelectronic semiconductor devices use layered thin films on Si wafers. Interdiffusion and reaction between two neighboring thin-film layers has been a technological issue from the point of view of yield and reliability of the devices. On a piece of Si chip the size of a fingernail, there are now more than several hundred millions of FETs, each of them with source, drain, and gate contacts. These contacts are typically made of silicides, which are IMCs of metal and Si, and each of the contacts must be the same or have the same electrical properties. Hence, in manufacturing VLSI circuits on a Si chip, the formation of silicide contacts and gates has been a critical processing step. So the controlled formation of silicide by depositing and reacting a thin metal film on a Si substrate has been a very active area of study [1–4]. What is unique in the thin-film interfacial reaction is the requirement of "single-phase formation" [5]. This means that we need to form a specific single-silicide phase in all the contacts and gates. Why it is unique is because of the difference between the reaction in bulk diffusion couples and that in thin-film couples. Multiple phases are formed simultaneously in the bulk couples, but multiple phases are formed sequentially one-by-one in thin films, so we can have single-phase instead of multiple-phase formation in thin-film reactions.

When we anneal a bulk-diffusion couple at a very high temperature and for a very long time, kinetics will not be a limiting factor and the system can approach the thermodynamic equilibrium reaction condition as closely as possible. It should have formed all the IMC phases shown in the equilibrium binary phase diagram of the diffusion couple. Indeed, this is true, except that sometimes we may find one or two of the intermetallic phases to be missing. However, the missing phenomenon is much more serious and common in reactions in a thin-film couple. Typically, only one of the intermetallic phases forms and all the rest are missing. This is called the "single-phase formation." In single-phase formation, there are two fundamental questions. The first is: why is it different from the bulk case? The second is: how can we predict which is the first phase to form from all the phases present in the binary-phase diagram?

We shall use the diffusion couple of Au and Al as an example. A bulk couple of Au and Al were interdiffused at 460 °C for 100 min, and five IMCs were found as

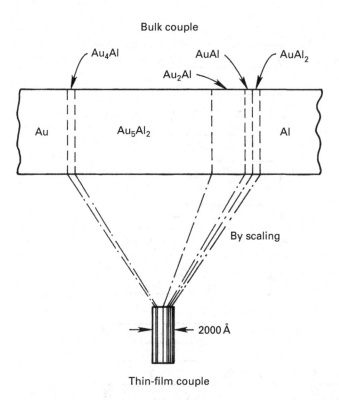

Bulk couple

Thin-film couple

Fig. 8.1   Schematic diagram of a bulk couple of Au and Al, interdiffused at 460 °C for 100 min; five IMCs were found.

sketched in Fig. 8.1. They are formed in the correct composition order as expected from the binary-phase diagram of AuAl. However, if the couple is annealed at a different temperature such as 200 °C, $AuAl_2$ and AuAl are absent. Nevertheless, multiple phases are formed in the bulk couple. An interesting question is: if we scale the thickness down from a bulk-diffusion couple to a thin-film couple as shown in Fig. 8.1, and anneal it at the same temperature, do we have the same five-phase or three-phase formation as the bulk couple, provided that the thickness of each of the phases is proportionally thinner? This does not occur in thin-film reactions because we cannot scale down the thickness and expect to find the same reaction products.

When we annealed a bimetallic Au/Al thin film at 200 °C, there is only one phase of Au_2Al formed between the Au and Al. The other phases do not form; rather, they will form sequentially one-by-one as shown in Fig. 8.2. When there is more Au than Au_2Al in the bimetallic thin film, the second phase formed will be more Au-rich. When there is more Al than Au_2Al in the bimetallic thin film, the second phase to form will be more Al-rich. In the following, we shall present another example of silicide formation in thin-film reactions.

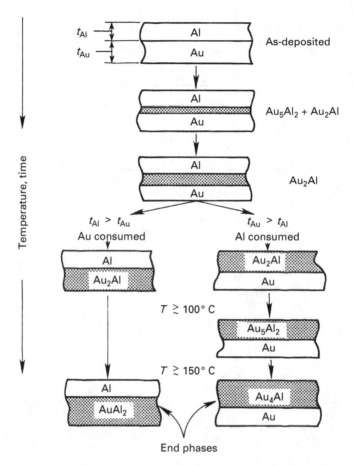

Fig. 8.2   When a bimetallic Au/Al thin film is annealed at 200 °C, there is only one phase of Au_2Al formed between the Au and Al. The other phases do not form; rather, they will form sequentially one-by-one as shown.

8.2   Silicide formation

8.2.1   Sequential Ni silicide formation

We shall use silicide formation in the Ni/Si system as the example. Recall that single-phase formation of a specific silicide on Si to serve as ohmic contacts and gates in FET devices has been a very important technological issue. The millions or even billions of silicide contacts and gates on a Si chip must be the same in order to offer the same physical properties such as resistance and Schottky barrier height. For example, we cannot have a contact that consists of a mixture of silicide phases, because it will affect the rate of transport of charges across the contact. Therefore, the device application demands single-phase formation, which in principle is against thermodynamics. For example, thermodynamic equilibrium requires the phase of $NiSi_2$ to form in a sample of a Ni thin film deposited on a Si wafer, according to the NiSi binary-phase diagram. This is because it is the most Si-rich phase. Yet when a thin film of Ni reacts on a Si wafer, the Ni_2Si phase always forms first between the Ni and Si. Furthermore, from the viewpoint of resistivity, it is NiSi that has the lowest resistivity among all the NiSi silicide phases. Therefore, device manufacturing wants to have a single-phase NiSi to form in every contact. Yet, according to the principle of thermodynamics, we cannot have it! Thus, a manufacturing process based on kinetics rather than thermodynamics has to be developed. Below, we shall compare Ni silicide formation in bulk samples and in thin-film samples.

An optical cross-sectional image of a bulk couple of Ni and Si after annealing at 850 °C for 8 h is shown in Fig. 8.3. The polished cross-section shows the formation of four compounds between the Ni and Si; they are Ni_3Si, Ni_5Si_2, Ni_2Si, Ni_3Si_2, and NiSi [6].

Similar to bulk couples, we can obtain multiple-phase formation in the lateral reaction in thin-film samples, wherein the distance of interdiffusion has been greatly increased. This is shown in Fig. 8.4 by a schematic diagram of a Ni film deposited on half of a Si stripe. Upon annealing at 750 °C for 20 min, five Ni-silicide phases were formed. Figure 8.5 shows a bright-field TEM image of the lateral growth of multiple phases in the NiSi sample. We can identify Ni_3Si, Ni_5Si_2, Ni_2Si, Ni_3Si_2, and NiSi in the sample. Four of them are the same as those observed in the bulk couple as shown in Fig. 8.3. By

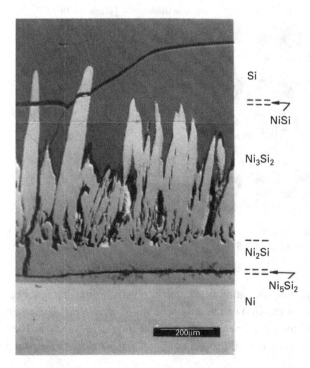

Fig. 8.3   Optical cross-sectional image of a bulk couple of Ni and Si after annealing at 850 °C for 8 h. The polished cross-section shows the formation of four compounds of NiSi between the Ni and Si.

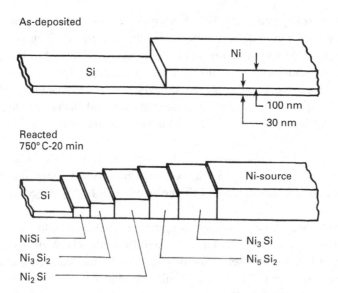

Fig. 8.4   Schematic diagram showing a Ni/Si lateral diffusion couple before and after annealing at 750 °C for 20 min. (Courtesy of C. H. Chen *et al.*, 1985) [6].

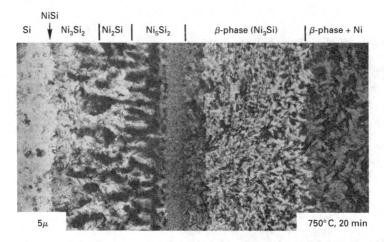

Fig. 8.5   Bright-field TEM image of a Ni/Si lateral diffusion couple after annealing at 750 °C for 20 min. (Courtesy of C. H. Chen *et al.*, 1985) [6].

measuring the compound formation in shorter intervals of time, the sequence of phase formation in the lateral reaction can be followed and shown in Fig. 8.6, where a plot of width (thickness) of compound phases against annealing time is shown. The first phase formed is Ni_2Si, and it reaches a thickness of about 20 μm before the second NiSi phase appears.

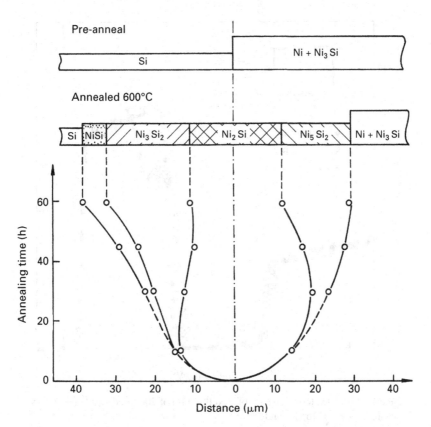

Fig. 8.6 Length of individual phases versus annealing time at 600 °C for a Ni/Si lateral diffusion couple. (Courtesy of Zheng, PhD thesis, Cornell University, 1985) [11].

Figure 8.7 shows the Rutherford backscattering spectra of a Ni thin film of 200 nm deposited on a Si wafer and annealed at 250 °C for 1 h and 4 h. The Ni_2Si formed, and the spectra showed a step in the signal. By measuring the step height of the Ni and the Si, the ratio of the step heights confirmed the composition of the phase to be Ni_2Si. When the annealing was extended from 1 h to 4 h, the width of the step increased, indicating the thickening or growth of the phase. The phase can be identified to be Ni_2Si by X-ray diffraction using a glancing incidence Seeman–Bohlin X-ray diffractometer. The diffraction spectrum is shown and indexed in Fig. 8.8.

The above results show that when a Ni film reacts with a Si wafer, the first phase formed is Ni_2Si. When all the Ni is consumed, then NiSi forms, and after all the Ni_2Si is consumed, $NiSi_2$ will form between NiSi and Si, as shown in Fig. 8.9. It shows a sequential formation of phases from Ni_2Si to NiSi to $NiSi_2$. On the other hand, if we deposit a thin film of Si on a thick Ni substrate, the first phase formed is still Ni_2Si, but the subsequent phases are different; they are Ni-rich silicides, as shown in Fig. 8.9.

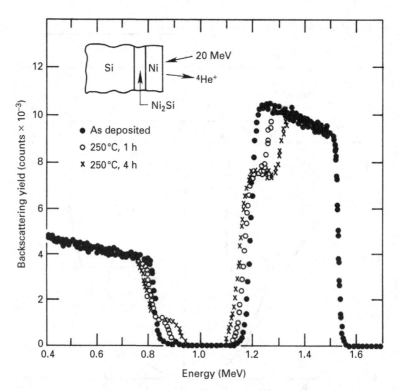

Fig. 8.7   Rutherford backscattering spectra of a Ni thin film of 200 nm deposited on a Si wafer and annealed at 250 °C for 1 h and 4 h.

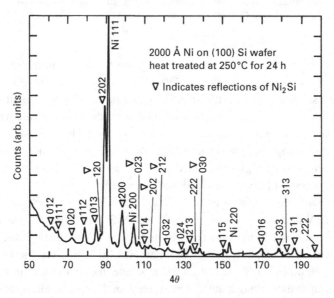

Fig. 8.8   X-ray diffraction spectrum of Ni_2Si by using a glancing incidence Seeman–Bohlin X-ray diffractometer of a 200 nm Ni film on Si after annealing at 250 °C for 24 h.

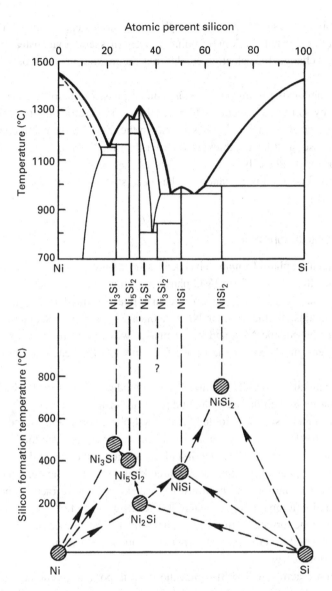

Fig. 8.9   Sequential phase formation in a NiSi thin-film system.

Recall the case of a thin film of Ni on a Si wafer, where the overall composition of the whole sample is very Si-rich. According to thermodynamics, the phase in equilibrium with the Si substrate should be $NiSi_2$. Yet whether the overall composition is Si-rich or Ni-rich, the first phase formed is always Ni_2Si. How we can predict the first-phase formation has been a very challenging question. There have been many attempts to use binary-phase diagrams and to apply a selection condition such as maximum free energy change to predict the first-phase formation. However, one of the difficulties in carrying out such an attempt is that it cannot predict the first-phase formation to be an amorphous phase, yet certain first-phase formations can be amorphous such as in the Ti-Si case. From the kinetic point of view, it seems that instead of maximum free energy change, the selection condition is the maximum "rate" of free energy change.

On the phenomenon of missing phases and/or single-phase formation, it can be verified experimentally by using cross-sectional high-resolution TEM lattice imaging [7, 8]. We can resolve whether or not a certain phase exists in the thin-film diffusion couple. The existence of a phase must have a thickness greater than the linear dimension of its unit cell, so it can be resolved easily.

Lately, silicide formation in Si nanowires has been studied. It will not be covered in this chapter [9, 10].

8.2.2   First phase in silicide formation

Besides Ni, sequential phase formation has been widely studied in the reaction between Si and other metallic films to form IMC phases of silicide. For example, in the reaction between Si and Pt, Pd, Ni, or Co, the first phase formed is M_2Si, where M represents the metals. In the case of Ni film on a Si wafer, the sequential formation is that after Ni_2Si, the NiSi and $NiSi_2$ phases will follow to form one by one. The same happens in Co-Si. In the case of Pt and Pd, the reaction stops at PtSi and PdSi.

For transition metals such as Ti, the first phase formed was determined to be TiSi. Yet sometimes, amorphous TiSi has been the first phase to form.

For refractory metals such as Mo and W, the first-phase formation has been $MoSi_2$ and WSi_2, respectively. Since they are the most Si-rich phase, it is also the last phase.

For rare-earth metals, such as Dy, the first phase formed is the disilicide.

Table 8.1 lists the phase formation of the three kinds of metal silicide. Note that for the noble and near-noble metals, the transition metals, and the refractory metals, the first-phase formation temperature is around 200 °C, 400 °C, and 600 °C, respectively. Actually, Pd_2Si can be formed around 100 °C; it has the lowest temperature of formation on Si wafers. Table 8.2 lists the silicide formation sequence of various metals on Si and their free formation energy.

In a metal film reacting with a Si wafer, the most important kinetic step is how to break the covalent bonds in the single Si crystal. The formation temperature indicates that the silicide formation mechanism of various metals is not the same because of the difference in formation temperature.

Experimentally, it is of interest to prepare a Ni/Si thin-film sample and age at 400 °C for a short time to form a structure of NiSi/Si, followed by a thin-film Ni deposition to form the Ni/NiSi/Si structure, as shown in Fig. 8.10. Then we anneal the sample at 250 °C and might expect the growth of NiSi, since NiSi exists already. Yet what was found was that Ni_2Si formed between the Ni and NiSi, and the growth of Ni_2Si occurs at the expense of NiSi, as shown in Fig. 8.10. When there is unreacted Ni and Si and when the temperature is around 250 °C, Ni_2Si is the first phase to form.

Table 8.1. Comparison of the three transition metal silicide classes*

Characteristics	Near-noble metal (Ni, Pd, Pt, Co, ...)	Refractory metal (W, Mo, V, Ta, ...)	Rare-earth metal (Eu, Gd, Dy, Er, ...)
First phase formed	M_2Si	MSi_2	MSi_2
Formation temperature	$\sim 200\,^\circ C$	$\sim 600\,^\circ C$	$\sim 350\,^\circ C$
Growth rate	$X^2\,\alpha\,t$	$X\,\alpha\,t$	?
Activation energy of growth	$1.1 \sim 1.5$ eV	> 2.5 eV	?
Dominent species	Metal	Si	Si
Barrier height to n-Si	$0.66 \sim 0.93$ eV	$0.52 \sim 0.68$ eV	~ 0.40 eV
Resistivity	$20 \sim 100\mu\Omega\text{-cm}$	$13 \sim 1000\mu\Omega\text{-cm}$	$100 \sim 300\mu\Omega\text{-cm}$

* R. D. Thompson and K. N. Tu, *Thin Solid Films* **53** (1982), 4372.

Table 8.2. Free energy of formation ΔH of silicides*

Silicide	ΔH kcal g-atom^{-1}	Silicide	ΔH kcal g-atom^{-1}	Silicide	ΔH kcal g-atom^{-1}
Mg_2Si	6.2	Ti_5Si_3	17.3	V_3Si	6.5
		$TiSi$	15.5	V_5Si_3	11.8
$FeSi$	8.8	$TiSi_2$	10.7	VSi_2	24.3
Fe_2Si	6.2				
		Zr_2Si	16.7	Nb_5Si_3	10.9
Co_2Si	9.2	Zr_5Si_3	18.3	$NbSi_2$	10.7
$CoSi$	12	$ZrSi$	18.5, 17.7		
$CoSi_2$	8.2	$ZrSi_2$	12.9, 11.9	Ta_5Si_3	9.5
				$TaSi2$	8.7, 9.3
Ni_2Si	11.2, 10.5	$HfSi$			
$NiSi$	10.3	Hf_5i_2		Cr_3Si	7.5
				Cr_5Si_3	8
Pd_2Si	6.9			$CrSi$	7.5
$PdSi$	6.9			$CrSi_2$	7.7
				Mo_3Si	5.6
Pt_2Si	6.9			Mo_5Si_3	8.5
$PtSi$	7.9			$MoSi_2$	8.7, 10.5
$RhSi$	8.1			W_5Si_3	5
				WSi_2	7.3

* J. M. Poate, K. N. Tu and J. W. Mayer, eds, *Thin Films: Interdiffusion and Reactions*, Wiley-Interscience, New York (1978).

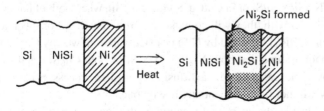

Fig. 8.10   Formation of Ni_2Si between NiSi and Ni.

8.3   Kinetics of interfacial-reaction-controlled growth in thin-film reactions

In Chapter 5, Section 5.2.2, we presented Kidson's analysis of a diffusion-controlled growth of a single-layer phase. The most important kinetic property of diffusion-controlled growth is that its growth rate is inversely proportional to its thickness. Thus, it cannot disappear because when its thickness reduces to zero, its growth rate will become infinitive. If we have two layers in growth competition and both of them are diffusion-controlled, we cannot get rid of any one of them in order to achieve single-phase growth. To overcome this problem, we shall introduce interfacial-reaction-controlled growth.

Figure 8.11 depicts the growth of a layered IMC phase between two pure elements, for example, the growth of Ni_2Si between Ni and Si. We represent Ni, Ni_2Si, and Si by $A_\alpha B$, $A_\beta B$, and $A_\gamma B$, respectively. The thickness of Ni_2Si is x_β and the position of its interface with Ni and Si is defined by $x_{\alpha\beta}$ and $x_{\beta\gamma}$, respectively. Across the interfaces, there is an abrupt change in concentration. In Fig. 8.11, the concentration change of Ni across the interfaces is shown. In a diffusion-controlled growth of x_β, the concentrations at its interfaces are assumed to have the equilibrium values, represented by the broken curve in the x_β layer in Fig. 8.11(a). In an interfacial-reaction-controlled growth, the concentration at its interface is assumed to be non-equilibrium, represented by the solid curve.

To consider a diffusion-controlled growth of a layered phase of x_β in Fig. 8.11(a), we have shown in Section 5.2.2 that we can use Fick's first law of diffusion in one dimension to represent the fluxes, as shown in Fig. 8.11(b). We have obtained the width of the β phase w_β or x_β to be (Eq. 5.13)

$$w_\beta = x_{\beta\gamma} - x_{\alpha\beta} = (A_{\beta\gamma} - A_{\alpha\beta})\sqrt{t} = B\sqrt{t} \tag{8.1}$$

which shows that the β phase has a parabolic rate or diffusion-controlled growth.

A fundamental nature of a diffusion-controlled layer growth is that its velocity of growth is inversely proportional to its thickness. As the thickness w approaches 0,

$$\lim \frac{dw}{dt} = \frac{B}{w} \to \infty$$

The growth rate will approach infinity, or the chemical potential gradient to drive the growth will approach infinity. Therefore, in a multilayered structure of silicide formation, for example, $Ni/Ni_2Si/NiSi/Si$ or in $Cu/Cu_3Sn/Cu_6Sn_5/Sn$, when both of the IMCs have diffusion-controlled growth, they will co-exist and grow together. For this reason, in a sequential growth of Ni_2Si followed by NiSi or in Cu_6Sn_5 followed by Cu_3Sn in Cu/Sn reaction, we cannot assume that both Ni_2Si and NiSi or both Cu_6Sn_5 and Cu_3Sn can nucleate and grow at the same time by a diffusion-controlled process, because they will co-exist and they will not form sequentially one by one.

Next, we shall consider the interfacial-reaction-controlled growth in which the growth rate is linear with time, or the rate is constant and finite and independent of thickness.

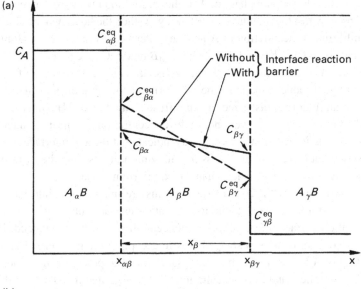

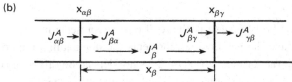

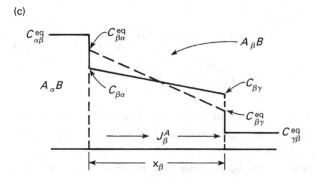

Fig. 8.11   (a) Schematic of concentration profile in interdiffusion. (b) Flux diagram. (c) Schematic of the flux and the interfacial concentration.

Note that the linear growth rate cannot keep on going forever; when the layer grows to a certain thickness, diffusion across the thicker and thicker layer will be rate-limiting and the growth will change to diffusion-controlled or its time dependence will change from linear to parabolic.

To formulate interfacial-reaction-controlled growth, we assume that the concentrations at the interfaces are not the equilibrium values as shown in Fig. 8.11(a). Physically, we consider for the moment that $A_\beta B$ is a liquid solution and is dissolving A atoms from $A_\alpha B$, which is a pure phase of A. If the dissolution rate is extremely high and only limited by how fast the A atoms can diffuse away, then the liquid will be able to maintain the equilibrium concentration of A near the interface, even though A atoms are being drained away to the other end through the $A_\beta B$ phase. On the other hand, if the process of breaking A atoms from the $A_\alpha B$ surface is slow or the dissolution of A is slow, the liquid will not be able to maintain the equilibrium concentration at the interface because whenever an A atom is dissolved, it can diffuse away quickly. The interfacial-reaction-controlled process is the slower one, so the concentration of A near the interface will be undersaturated; it leads to $C_{\alpha\beta} < C_{\alpha\beta}^{eq}$, as indicated at the $x_{\alpha\beta}$ interface in Fig. 8.11(a). We assume that there is sluggishness in removing A atoms from the surface of $A_\alpha B$, so that the concentration $C_{\beta\alpha}$ is less than the equilibrium value.

At the other end of the $A_\beta B$ phase, A atoms are incorporated into the $A_\gamma B$ surface for the growth of the latter. If the incorporation can take place as soon as the atoms arrive at the interface, the equilibrium concentration will be maintained. However, at the $x_{\beta\gamma}$ interface, if there is sluggishness in accepting the incoming A atoms, there is a build-up of A atoms. It becomes supersaturated. The process is therefore interfacial-reaction-controlled, and the concentration of $C_{\beta\gamma}$ is greater than the equilibrium value, and we have $C_{\beta\gamma} > C_{\beta\gamma}^{eq}$.

In Fig. 8.11(a), the broken curve in the $A_\beta B$ phase represents the equilibrium concentration gradient, and the solid curve represents the non-equilibrium concentration gradient. The growth rate of the $A_\beta B$ phase does not depend on diffusion across itself, but rather on the interfacial-reaction processes at the two interfaces.

Now we consider the interface $x_{\alpha\beta}$ which is moving with a velocity $v = dx_{\alpha\beta}/dt$,

$$(C_{\alpha\beta}^{eq} - C_{\beta\alpha})\frac{dx_{\alpha\beta}}{dt} = J_{\alpha\beta}^A - J_{\beta\alpha}^A = \left(-\overline{D_\alpha}\frac{dC_\alpha^A}{dx}\right) - \left(-\overline{D_\beta}\frac{dC_\beta^A}{dx}\right)$$

$$= \overline{D_\beta}\frac{dC_\beta^A}{dx} = -J_\beta^A$$

(8.2)

where J is atomic flux having the unit of number of atom/cm^2 s; D is atomic diffusivity, cm^2/s; C is concentration, number of atom/cm^3; x is length, cm; and v is the velocity of the moving interface, cm/s.

The term $J_{\alpha\beta}^A$ goes to zero because we have assumed that $A_\alpha B$ is pure A, so the concentration of A is flat and its gradient is zero. The last equality in the last equation is the definition of a flux equation and it shows that $J_{\beta\alpha}^A = J_\beta^A$. In the compound of $A_\beta B$, we can assume a linear concentration gradient that

$$\frac{dC_\beta^A}{dx} = \frac{C_{\beta\alpha} - C_{\beta\gamma}}{x_\beta}$$

Therefore, we have

$$(C_{\alpha\beta}^A - C_{\beta\alpha})\frac{dx_{\alpha\beta}}{dt} = \overline{D_\beta}\frac{C_{\beta\alpha} - C_{\beta\gamma}}{x_\beta} = -J_\beta^A$$

(8.3)

If we consider the flux from the viewpoint of a reaction-controlled process, we have

$$J_\beta^A = (C_{\beta\alpha}^{eq} - C_{\beta\alpha})K_{\beta\alpha} \tag{8.4}$$

where $K_{\beta\alpha}$ is defined as the interfacial-reaction coefficient of the $x_{\alpha\beta}$ interface. It has the unit of velocity, cm/s, and it infers the rate of removal of A atoms from the $A_\alpha B$ surface. If there is no interface sluggishness, $C_{\beta\alpha}$ will approach $C_{\beta\alpha}^{eq}$. Because of sluggishness, however, the actual concentration at the interface is lower than the equilibrium value, so $K_{\beta\alpha}$ is a measure of the actual flux J_β^A leaving the interface with respect to the concentration change at the interface. The physical meaning of the interfacial-reaction coefficient K (velocity) is that of interfacial mobility.

Similarly, at the $x_{\beta\gamma}$ interface, we have

$$J_\beta^A = (C_{\beta\gamma} - C_{\beta\gamma}^{eq})K_{\beta\gamma} \tag{8.5}$$

From Eq. (8.4) and Eq. (8.5), we have, respectively,

$$\frac{J_\beta^A}{K_{\beta\alpha}} = C_{\beta\alpha}^{eq} - C_{\beta\alpha}$$

$$\frac{J_\beta^A}{K_{\beta\gamma}} = C_{\beta\gamma} - C_{\beta\gamma}^{eq}$$

By adding the last two equations, we have

$$J_\beta^A \left(\frac{1}{K_{\beta\alpha}} + \frac{1}{K_{\beta\gamma}} \right) = (C_{\beta\gamma} - C_{\beta\alpha}) + (C_{\beta\alpha}^{eq} - C_{\beta\gamma}^{eq})$$

Let

$$\frac{1}{K_\beta^{eff}} = \frac{1}{K_{\beta\alpha}} + \frac{1}{K_{\beta\gamma}}$$

Since we have from Eq. (8.3)

$$C_{\beta\gamma} - C_{\beta\alpha} = -\frac{J_\beta^A x_\beta}{D_\beta}$$

if we define

$$\Delta C_\beta^{eq} = C_{\beta\alpha}^{eq} - C_{\beta\gamma}^{eq}$$

we obtain

$$J_\beta^A = \frac{\Delta C_\beta^{eq} K_\beta^{eff}}{\left(1 + \dfrac{x_\beta K_\beta^{eff}}{D_\beta} \right)} \tag{8.6}$$

Now to calculate the thickening rate of $A_\beta B$, we take

$$\frac{dx_\beta}{dt} = \frac{d}{dt}(x_{\beta\gamma} - x_{\alpha\beta}) = \left(\frac{1}{C_{\beta\gamma} - C_{\gamma\beta}^{eq}} - \frac{1}{C_{\alpha\beta}^{eq} - C_{\beta\alpha}}\right) J_\beta^A = G_\beta J_\beta^A \qquad (8.7)$$

By substituting Eq. (8.6) into the last equation, we have

$$\frac{dx_\beta}{dt} = \frac{G_\beta \Delta C_\beta^{eq} K_\beta^{eff}}{1 + x_\beta K_\beta^{eff}/D_\beta} \qquad (8.8)$$

where we recall that $\frac{1}{K_\beta^{eff}}$ is the effective interfacial-reaction coefficient of the β-phase. If we define a "changeover" thickness of

$$x_\beta^* = \frac{D_\beta}{K_\beta^{eff}} \qquad (8.9)$$

$$\frac{dx_\beta}{dt} = \frac{G_\beta \Delta C_\beta^{eq} K_\beta^{eff}}{1 + \dfrac{x_\beta}{x_\beta^*}}$$

For a large changeover thickness, or $x_\beta/x_\beta^* \ll 1$, i.e. $\overline{D_\beta} \gg K_\beta^{eff}$, under the condition that the interdiffusion coefficient is much larger than the effective interfacial-reaction coefficient, we obtain

$$\frac{dx_\beta}{dt} = G_\beta \Delta C_\beta K_\beta^{eff} \qquad \text{or } x_\beta \propto t \qquad (8.10)$$

The process is interfacial-reaction-controlled, and the growth rate is constant. For a small changeover thickness, $x_\beta/x_\beta^* \gg 1$, i.e. $\overline{D_\beta} \ll K_\beta^{eff}$, the process is diffusion-controlled. Then

$$\frac{dx_\beta}{dt} = G_\beta \Delta C_\beta^{eq} \frac{\overline{D_\beta}}{x_\beta} \qquad \text{or } (x_\beta)^2 \propto t \qquad (8.11)$$

The above demonstrates the well-known relationship that in an interfacial-reaction-controlled growth, the layer thickness is linearly proportional to time, but in a diffusion-controlled growth, the layer thickness is proportional to the square root of time. Furthermore, a reaction-controlled growth will always change over to a diffusion-controlled growth when the layer thickness has grown sufficiently large, as shown in Fig. 8.12.

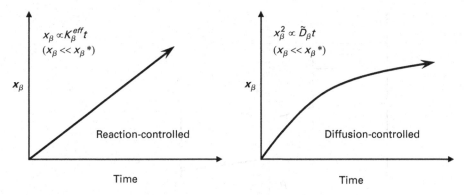

Fig. 8.12   Layer width x_β versus time for (a) reaction-controlled, and (b) diffusion-controlled growth.

8.4   Kinetics of competitive growth of two-layered phases

Recall that an interfacial-reaction-controlled growth has a constant growth rate, independent of layer thickness. Thus, a phase which is very thin and has a slow interfacial-reaction-controlled growth can be consumed by a neighboring layer which has a faster interfacial-reaction-controlled growth.

In considering the competition of growth between two co-existing phases, we may have three combinations. They are:

(1) both are diffusion-controlled;
(2) both are interfacial-reaction-controlled; and
(3) one is diffusion-controlled and the other is interfacial-reaction-controlled.

The first two cases are simple, and they have been discussed before. We shall analyze the kinetics of the third case.

In Fig. 8.13, we consider the competing growth of two-layered phases of $A_\beta B$ and $A_\gamma B$ between A and B. We assume that the growth of $A_\beta B$ is interfacial-reaction-controlled (i.e. $x_\beta \ll x_\beta^*$) and has a velocity v_1. The growth of $A_\gamma B$ is diffusion-controlled (i.e. $x_\gamma \gg x_\gamma^*$) and has a velocity v_2. When their thickness is small, the magnitude of v_2 can be quite large due to the inverse dependence on layer thickness, so we can assume that $v_2 \gg v_1$ and the rapid growth of $A_\gamma B$ can consume all of $A_\beta B$. We have single-phase growth. Quantitatively, we have

$$J_\beta^A \cong \Delta C_\beta^{eq} K_\beta^{eff}$$

$$J_\gamma^A = \frac{\Delta C_\gamma^{eq} \tilde{D}_\gamma}{x_\gamma}$$

Then the flux ratio is

$$\frac{J_\beta^A}{J_\gamma^A} = \frac{\Delta C_\beta^{eq} K_\beta^{eff}}{\Delta C_\gamma^{eq} \tilde{D}_\gamma} x_\gamma \tag{8.12}$$

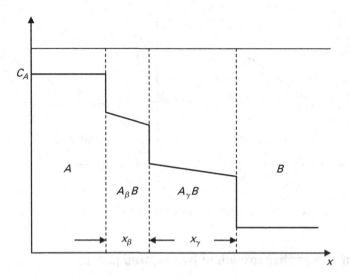

Fig. 8.13   Competing growth of two layered phases of $A_\beta B$ and $A_\gamma B$ between A and B.

When the above ratio is small, which means that $J_\beta^A \ll J_\gamma^A$, the flux in the diffusion-controlled growth of $A_\gamma B$ is very large (when it is very thin), so its growth rate is very fast and it can be much larger than the constant growth rate of $A_\beta B$. If the thickness of the latter is small, it can be taken over or consumed by the rapid growth of $A_\gamma B$.

We can rewrite the last equation as

$$x_\gamma^{crit} = \frac{\Delta C_\gamma^{eq} \overline{D_\gamma}}{\Delta C_\beta^{eq} K_\beta^{eff}} \frac{J_\beta^A}{J_\gamma^A} \tag{8.13}$$

The physical meaning of x_γ^{crit} is that when the thickness of x_γ is below the critical thickness, it will be able to consume x_β and achieve the single-phase formation. The critical thickness is estimated to be of the order of magnitude of microns and is verified by the lateral growth in thin films as discussed before. Since most thin-film diffusion couples have thicknesses in hundreds of nm, we observe only a single-phase growth if the above criteria are satisfied. The conclusion is that by combining diffusion-controlled growth and interfacial-reaction-controlled growth, we can explain the phenomena of single-phase growth in thin-film reactions.

8.5   Marker analysis in intermetallic compound formation

One of the key kinetic questions in silicide formation is: which is the dominant diffusion species? As in Darken's analysis of interdiffusion in bulk-diffusion couples, we need marker analysis to determine the faster or the dominant diffusing species in thin-film reactions. Because thin films are too thin to use Mo wires as markers, implanted inert gas atoms such as Ar and Xe were used as markers. The inert gas atoms were implanted

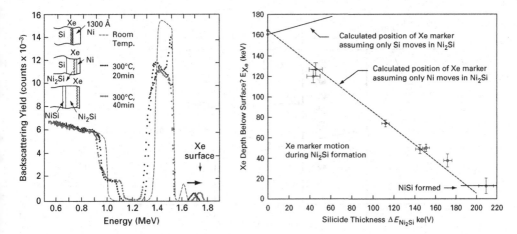

Fig. 8.14 (a) Rutherford backscattering spectra from a Xe marker-implanted sample before and after the formation of Ni₂Si at 300 °C for 20 and 40 min. After silicide formation, the Xe marker is buried with the silicide and displaced toward the surface. (b) The amounts of the displacement of the Xe marker versus the silicide thickness are shown.

into the Si surface and the Si were annealed to temperatures above 600 °C to reorder the implanted layer and remove the major amount of implantation-induced defect structures. During this annealing, it has been shown for the case of Xe that Xe bubbles 5–10 nm in diameter are formed in the Si. Then the Si surface was etched to remove any oxide or hydrocarbon layers. Following the surface-cleaning step, a metal layer such as Ni is deposited on the Si surface for silicide formation. Fig. 8.14(a) shows the Rutherford backscattering spectra from a Xe marker-implanted Ni-Si sample before and after the formation of Ni₂Si at 300 °C for 20 and 40 min. After silicide formation, the Xe marker is buried within the silicide and displaced toward the surface. In Fig. 8.14(b), the amounts of the displacement of the Xe marker versus the silicide thickness are shown. These results show that interfacial drag of the Xe bubbles did not occur and that Ni is the dominant diffusing species.

If Si is the moving species, when the advancing silicide front reaches the implanted marker, the marker will move in the opposite direction of the faster-moving species and will move deeper into the sample with the interface. The marker can be dragged by the interface. To determine Si as the faster-moving species, it is then necessary for implanted markers to move into the deposited metal layer. Then, during silicide growth, as the silicide–metal interface advances to the marker, if the marker becomes buried, its position will be shifted deeper into the samples, indicating that Si is the dominant diffusing species.

To determine the intrinsic diffusivity of Ni and Si in the silicide growth, besides marker motion, we need to know the chemical interdiffusion coefficient. The latter can be assumed to obey the growth equation of

$$x_i^2 = 4\overline{D}_i t \tag{8.14}$$

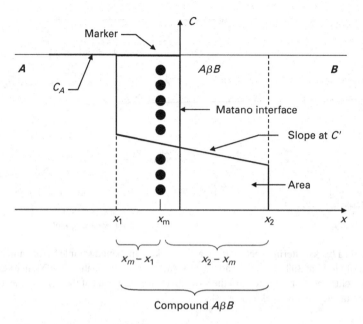

Fig. 8.15   Schematic diagram of a single-compound formation in a thin-film reaction, showing the Matano interface and marker displacement.

where x_i is the thickness of the IMC, and $\overline{D_i}$ is the interdiffusion coefficient. To analyze the marker motion, we depict in Fig. 8.15 a schematic diagram of single-compound formation in a thin-film reaction, showing the Matano interface and marker displacement. If the marker is placed at the original interface between A and B, each B atom passing the marker will form a molecule of $A\beta B$, and each time βB atoms pass the marker, they will form a molecule of $A\beta B$ on the other side. If we assume the molar volume of A and B to be the same, we have

$$\frac{J_B}{\beta J_A} = \frac{x_m - x_1}{x_2 - x_m} \tag{8.15}$$

From the flux equations,

$$J_A = -D_\beta^A \frac{\partial C_A}{\partial x}$$

$$J_B = -D_\beta^B \frac{\partial C_B}{\partial x}$$

we have

$$\frac{D_\beta^B}{D_\beta^A} = \frac{\beta(x_m - x_1)}{x_2 - x_m} \tag{8.16}$$

Therefore, from measurement of the marker displacement x_m and knowledge of the composition of $A_\beta B$, we can determine the ratio of the intrinsic diffusion coefficients from the last equation. From the knowledge of $(x_2 - x_1)$, we can find the interdiffusion coefficient using Eq. (8.14). An example of marker analysis of the formation of Al_3Ti IMC in a bimetallic thin-film couple of Al and Ti has been reported by J. Tardy and K. N. Tu, *Phys. Rev. B.*, **32**, 2070 in 1985.

8.6   Reaction of a monolayer of metal and a Si wafer

The most important kinetic question in silicide formation is how to break the Si covalent bonds in Si so that Si atoms can be removed from the Si surface to react with metal atoms. It was puzzling to find that Pd and Ni can react with Si to form Pd_2Si at 100 °C and Ni_2Si at 250 °C, respectively. They are low-temperature reactions when compared to the transformation temperature of 550 °C which is needed in order to transform amorphous Si to crystalline Si, the temperature of 800 °C is needed for the oxidation of Si, and the doping temperature of around 1000 °C is needed to dope B or P in Si. However, not all metals can react with Si at low temperatures. As shown in Table 8.1, while noble and near-noble metals react with Si at around 200 °C, transition metals such as Ti will react with Si beyond 400 °C, and refractory metals such as W will react with Si at 800 °C.

The noble and near-noble metals have a unique kinetic behavior in that they diffuse interstitially in Si. For example, Ni can diffuse across a Si wafer rather quickly. Also, if we deposit an Au thin film of 10 nm on a Si wafer, if the sample is kept at room temperature for a few days, the gold color disappears because the Au has diffused and dissolved into the Si. When the interstitial diffusion is combined with the finding that Ni is the dominant diffusing species during Ni_2Si formation, a plausible mechanism of low-temperature reaction on Si can be given as below. At the Ni_2Si/Si interface, some Ni atoms will diffuse interstitially into Si and lead to charge transfer from the covalent Si-Si bond to metallic NiSi bonds, so that the Si-Si bond is unsaturated and can be broken more easily. Without interstitials, other metals will take a much higher silicide formation temperature.

A monolayer of Pd atoms has been deposited on a (001) Si surface and the interaction between the Pd and Si was studied by using ultraviolet photoemission spectroscopy (UPS). It was found that the Pd atoms did not stay on the Si surface; rather, they sank into Si and formed interstitials in the Si.

References

[1] J. W. Mayer, J. M. Poate and K. N. Tu, Thin films and solid-phase reactions, *Science*, **190** (1975), 228–234.

[2] K. N. Tu and J. W. Mayer, "Silicide Formation," Ch. 10 of *Thin Films: Interdiffusion and Reactions*, eds J. M. Poate, K. N. Tu and J. W. Mayer (Wiley-Interscience, New York, 1978).

[3] Marc-A. Nicolet and S. S. Lau, "Formation and characterization of transition-metal silicides," in *VLSI Electronics*, Vol. 6, eds N. G. Einspruch and G. B. Larrabee (Academic Press, New York, 1983).

[4] L. J. Chen and K. N. Tu, "Epitaxial growth of transition-metal silicides on silicon," *Materials Science Reports* **6** (1991), 53–140.

[5] U. Goesele and K. N. Tu, "Growth kinetics of planar binary diffusion couples: thin film case versus bulk cases", *J. Appl. Phys.* **53** (1982), 3252.

[6] S. H. Chen, L. R. Zheng, C. B. Carter and J. W. Mayer, "Transmission electron microscopy studies on the lateral growth of nickel silicides", *J. Appl. Phys.* **57** (1985), 258.

[7] H. Foell, P. S. Ho and K. N. Tu, "Cross-sectional TEM of silicon–silicide interfaces", *J. Appl. Phys.* **52** (1981), 250.

[8] R. T. Tung and J. L. Batstone, "Control of pinholes in epitaxial CoSi2 layers on Si(111)", *Appl. Phys. Lett.* **52** (1988), 648.

[9] Kuo-Chang Lu, Wen-Wei Wu, Han-Wei Wu, Carey M. Tanner, Jane P. Chang, Lih J. Chen and K. N. Tu, "In-situ control of atomic-scale Si layer with huge strain in the nano-heterostructure NiSi/Si/NiSi through point contact reaction," *Nano Letters* **7**:8 (2007), 2389–94, s.

[10] Y. C. Chou, W. W. Wu, L. J. Chen and K. N. Tu, "Homogeneous nucleation of epitaxial CoSi2 and NiSi in Si nanowires," *Nano Lett.* **9** (2009), 2337–42.

Problems

8.1   Material A reacts with material B to form the compound A_2B. Markers were placed at the A–B interface before reaction. After annealing at 250 °C for 15 min, the sample was examined; the compound had a thickness of 80 nm and the markers were 12.3 nm from the A/A_2B interface. Calculate the diffusion coefficient of each constituent, A and B, in A_2B.

8.2   A diffusion couple is formed by joining alloy 1 ($A_{50}B_{50}$) and alloy 2 (pure B) and heating for $t_1 = 40$ minutes. Assume that $D_A = D_B = D_{interdiff.} = 3.04 \times 10^{-7}$ cm^2/s, which is constant independent of composition.

(a) What is the concentration of A at a distance $x = 0.2$ cm and $x = -0.2$ cm?
(b) For a longer annealing time t_2, the same concentration of A as in (a) above at $x = 0.2$ cm was found at $x = 0.4$ cm. What is t_2?

8.3   A compound A_2B is formed by heating for 10 000 s. Measured from x_1 ($x_1 = 0$), we found that $x_m = 40$ nm and $x_2 = 200$ nm. Calculate the interdiffusion coefficient, and the ratio of intrinsic diffusivities of D_A/D_B.

8.4   A Pb-Pb (50 atomic %) In bulk diffusion couple is produced with Al marker wires. The grain size in the couple is large, so that the effect of grain boundary diffusion can be ignored. After 118 h anneal at 173 °C, the concentration profile A_2 of In (see the figure below) was obtained using an energy dispersive X-ray analysis unit in a SEM. The initial concentration profile of In before interdiffusion was the curve A1. Use the Matano–Boltzmann analysis to determine the interdiffusion coefficient D at the Matano interface. Use Darken's analysis to obtain the intrinsic diffusivities at In and Pb. The marker velocity can be assumed to follow $v = x_m/2t$. The markers have moved 67 μm into the alloy. The Matano interface has an In concentration of 0.41.

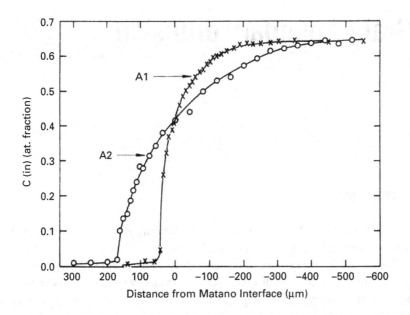

8.5 A phase $A_\beta B$ ($\beta = 3$) grows between the pure A and the $A_\gamma B$ ($\gamma = 1.5$) phase in a lateral diffusion couple where A is the dominant moving species. Given that the interdiffusion coefficient is 1.27×10^{-11} cm^2/s, $\Delta C_\beta^{eq} = 2$ atomic%, $K_\beta^{eff} = 2.79 \times 10^{-7}$ cm/s, and $\Omega =$ atomic volume $= 18 \times 10^{-24}$ cm^3/atom,

(a) find the changeover thickness of the $A_\beta B$ phase;
(b) find the $A_\beta B$ growth rate at the changeover thickness.

What is the flux of A through the $A_\beta B$ phase?

8.6 The growth of the single phase δ of Al$_3$Pd$_2$ between Al and Pd follows the diffusion-limited regime with $x_\delta^2 = K_\delta t$, where $K_\delta = 3.3 \times 10^{-12}$ cm^2/s and 250 °C. The concentration drop $\Delta C_\delta^{eq} = 3.6$ atomic%, and the concentration ratio $G_\delta = 4.17$ and $\Omega = 10 \times 10^{-24}$ cm^3. Calculate the interdiffusion coefficient.

8.7 In the text, we have studied the case of competing growth of two layered compound phases where one layer exhibits diffusion-controlled growth and the other layer shows interfacial-reaction-controlled growth. Describe what happens in the cases when

(a) both layers exhibit diffusion-controlled growth;
(b) both layers exhibit interfacial-reaction-controlled growth.

9 Grain-boundary diffusion

9.1 Introduction

Most of the metallic thin films used in microelectronic devices are polycrystalline rather than monocrystalline. Grain-boundary diffusion is of concern. It has caused two very well-known failure modes in Si devices: electromigration in Al interconnects and Al-penetration into Si through diffusion barriers. In electromigration, it is known that voids are formed at the triple point of grain boundaries and extend out along grain boundaries. The topic of electromigration and electromigration-induced failure will be covered in the next two chapters. On Al-penetration, Al forms a pit in Si and short in p-n junction, and also Si precipitates decorate the Al grain boundaries.

In general, atomic diffusion along grain boundaries is faster than in the bulk of grains. This assumes that atoms have a lower activation energy of motion in the boundary and also that vacancy formation within a grain boundary is easier because of the excess volume in the boundary [1–5]. The behavior of grain-boundary diffusion is shown by examining a few experiments.

(1) By comparing tracer diffusivities of Ag* in single crystal Ag and polycrystalline Ag, it has been found that the diffusivity is faster in the polycrystalline Ag at temperatures below 750 °C.

(2) Radioactive tracers deposited on a bulk bicrystal show deeper penetration along the grain boundary in auto-radiography images.

(3) We compare thin-film reactions in two sets of Pb/Ag/Au samples: one was epitaxially grown on rock salt and the other was deposited on fused quartz to grow polycrystalline grains. The latter but not the former showed Pb_2Au compound foundation at 200 °C. In the former case the Ag layer was a single crystal layer grown on Au/NaCl and it became a diffusion barrier to prevent the reaction between Pb and Au, yet the Ag and Au layers on fused quartz were polycrystalline and grain-boundary diffusion occurred in the Ag and allowed Pb_2Au formation.

Grain boundaries are rapid paths of atomic diffusion. Many studies have been devoted to the understanding of the subject. However, if grain-boundary diffusion is compared to the lattice diffusion discussed in Chapter 4, we find that at the present time our understanding of grain-boundary diffusion is phenomenological. We have yet to form an atomistic picture of grain-boundary diffusion. This is because the atomic structure and atomic positions of an arbitrary grain boundary are not known. Without them the atomic jump frequency and jump distance will be ill-defined, so we can only have a crude and macroscopic analysis of grain-boundary diffusion, or a continuum approach.

We must point out that there have been very serious efforts undertaken to determine the grain-boundary structure. For low-angle tilt-type and twist-type grain boundaries, dislocation models have been presented and are quite successful. Hence, the problem of diffusion in low-angle grain boundaries can be reduced to diffusion in an individual dislocation, i.e. "pipe diffusion," provided that the dislocation cores in the grain boundary are far apart. In a later section, we shall describe the diffusion measurement in small-angle tilt-type grain boundaries. For the large-angle grain boundaries, there have been systematic efforts to develop the "coincidence site lattice" for modeling the energy and structure of the boundaries. At the same time, high-resolution transmission electron microscopy (TEM) has been used to observe the atomic images of grain boundaries, and periodic units of atom clusters or repeating kite structures in the grain boundaries have been found. For example, Fig. 9.1 shows a lattice image of a large-angle (100) tilt-type grain boundary in an Au thin film. Periodicity has also been detected by X-ray diffraction in the large-angle grain boundaries which have a high density of coincidence lattice sites. Still, the direct link to atomic jump process in grain-boundary diffusion is missing. For example, the basic concept in lattice diffusion is point defects; whether or not we can define a point defect in a grain boundary is unclear.

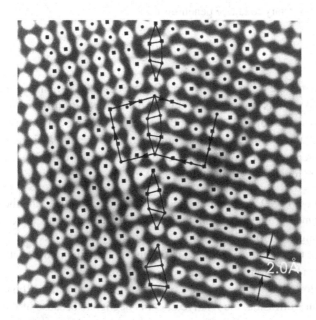

Fig. 9.1 High-resolution TEM image of a (100) tilt-type large-angle grain boundary in an Au thin film. The dots and squares indicate the atomic position in the two alternating stacking layers. The nearest interplanar distance is 0.202 nm. (Courtesy of W. Krakow, IBM T. J. Watson Research Center)

9.2   Comparison of grain-boundary and bulk diffusion

In order to have a real feeling about the relative magnitude of grain-boundary diffusion to lattice diffusion, the diffusion data of a radioactive tracer of Ag* in Ag is examined.

For self-diffusion in Ag, the diffusivities are

$$D_l = 0.67 \times e^{-\frac{1.95\,eV}{kT}} \text{ (lattice diffusion)} \tag{9.1}$$

$$D_b = 2.6 \times 10^{-2} e^{-\frac{0.8\,eV}{kT}} \text{ (grain-boundary diffusion)} \tag{9.2}$$

At 200 °C,

$$D_l = 0.67 \times e^{-\frac{1.95 \times 23000}{2.3 \times 2 \times 473}} \cong 10^{-21} \text{ cm}^2/\text{s}$$

$$D_b = 0.67 \times e^{-\frac{0.8 \times 23000}{2.3 \times 2 \times 473}} \cong 10^{-10} \text{ cm}^2/\text{s}$$

If the annealing time is 10^5 s (28 h or approximately one day), we estimate the diffusion distances:

$$x_l^2 \cong D_l t \cong 10^{-16} \text{ cm}^2; x_l \cong 1\text{Å}$$

$$x_b^2 \cong D_b t \cong 10^{-5} \text{ cm}^2; x_b \cong 30 \text{ microns}$$

The ratio of grain-boundary to lattice penetration is

$$\frac{x_b}{x_l} \cong 3 \times 10^5$$

We see a very large difference between D_l and D_b for Ag at 200 °C. There is negligible lattice diffusion in Ag at this temperature, yet grain boundary (GB) diffusion goes a long way. With the diffusivities at 400 °C and 800 °C, we list in Table 9.1 the values of D_l and D_b in Ag at the three temperatures for comparison. We see that D_l increases much more rapidly with temperature than D_b; the same is true for their ratio. We can classify their ratio into three regions as shown in Fig. 9.2.

(a) In the first region, Fig. 9.2(a), grain-boundary diffusion dominates and the penetration occurs primarily along the grain boundary. This region occurs typically in the temperature range of 150 to 300 °C.

(b) At intermediate temperatures, there is a noticeable penetration into the lattice of the grains adjacent to the grain boundary; see Fig. 9.2(b).

(c) In the last region, Fig. 9.2(c) at 800 °C, although grain-boundary diffusion is still faster than lattice diffusion, the relative effect is negligible because of the drain of atoms from the grain boundary into grains on both sides of the boundary. Therefore, the penetration along the grain boundary is slower and shallower if there was no drain into the adjacent grains.

Table 9.1. Comparison of D_l and D_b in Ag

	200 °C	400 °C	800 °C
D_l (cm^2/s)	10^{-21}	10^{-15}	10^{-10}
D_b (cm^2/s)	10^{-10}	10^{-8}	10^{-6}
D_l/D_b (cm^2/s)	10^{-11}	10^{-7}	10^{-4}

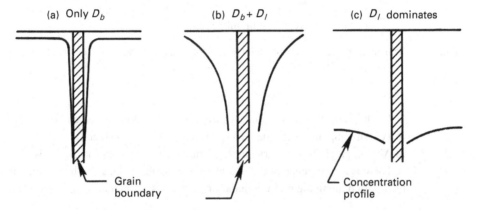

(a) Only D_b (b) $D_b + D_l$ (c) D_l dominates

Grain boundary   Concentration profile

Fig. 9.2 Schematic diagram of penetration profiles of concentration in three cases of combining grain-boundary and lattice diffusion where: (a) grain-boundary diffusion is dominant, (b) both are comparable, and (c) lattice diffusion is dominant.

The ratio of mass transport through the grain boundary and the grain is

$$R = \frac{J_l A_l}{J_b A_b} = \frac{D_l \pi r^2}{D_b 2\pi r\delta} \tag{9.3}$$

where r is the radius of a grain. Assuming that $r = 50$ nm (grain size of 100 nm in a film of 100 nm thickness), and $\delta = 0.5$ nm where δ is the grain-boundary width, we have at 200 °C

$$R = \frac{10^{-21} \times 5 \times 10^{-6}}{10^{-10} \times 2 \times 5 \times 5 \times 10^{-8}} \cong 10^{-9}$$

The mass transport in the grain boundary is much larger than that in the grains.

We have a similar situation in Al films, if we take the activation energy of lattice diffusion and grain-boundary diffusion in Al to be 1.3 eV and 0.7 eV, respectively, and the temperature of diffusion at 100 °C, which is roughly the operating temperature of Si devices. Then it is easy to see that electromigration in Al films is dominated by grain-boundary diffusion, because D_b is so much faster and the atomic flux along a large-angle grain boundary is significant.

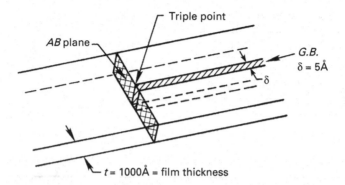

Fig. 9.3   Schematic diagram of a triple point at the intersection of two grain boundaries. There is a divergence of diffusional fluxes at the triple point.

To illustrate that grain-boundary flux is indeed significant and can lead to actual failure in a conducting line or through a diffusion barrier, we consider the case of mass transport of Al along a grain boundary starting from a triple point as shown in Fig. 9.3.

There is a divergence of the flux at the AB plane in Fig. 9.3. The number of M_b atoms transported along the grain boundary of a cross-sectional area A_b is given by

$$M_b = J_b A_b t \tag{9.4}$$

where $A_b = 0.5 \text{ nm} \times 100 \text{ nm} = 5 \times 10^{-13} \text{ cm}^2$, t is the diffusion time, and

$$J_b = -D_b \frac{\partial C_b}{\partial x} \tag{9.5}$$

If we assume that it is an Al line, the grain-boundary diffusivity at $127\,^\circ$C is

$$D_b \cong 0.1 \times e^{-0.7 \,\text{eV}/kT} \cong 10^{-10} \text{ cm}^2/\text{s}$$

We take

$$\frac{\partial C_b}{\partial x} = \frac{\Delta C_b}{\Delta x} = \frac{0.01 \times 10^{23} \text{ atom/cm}^3}{10^{-4} \text{ cm}} = 10^{25} \text{ atom/cm}^4$$

where we assume a 1% (0.01) change in concentration (ΔC_b) along the grain boundary length of 1 micron, 10^{-4} cm. We approximate the Al concentration to be $10^{23}/\text{cm}^3$ instead of $6 \times 10^{22}/\text{cm}^3$. We take the diffusion time, t, to be 100 days $= 10^7$ s, which is about the failure time of Al lines owing to electromigration during device operation.

Therefore,

$$M = JAt$$

$$= \left(-D_b \frac{\partial C_b}{\partial x}\right) A_b t$$

$$= (-10^{10} \text{ cm}^2/\text{s}) \left(-10^{25} \frac{\text{atoms}}{\text{cm}^4}\right) (5 \times 10^{-13} \text{ cm}^2)(10^7 \text{ s})$$

$$= 5 \times 10^9 \text{ atoms}$$

It means that we have transported 5×10^9 atoms. If all these atoms come from the grain behind the grain boundary, we can deplete a significant amount of Al from the line and create an opening or void. The line has a cross-section of $A = 1 \ \mu\text{m} \times 100 \ \text{nm} = 10^{-9} \ \text{cm}^2$; since there are 10^{15} atoms per cm^2, this means that there are only $10^{15} \times 10^{-9} = 10^6$ atoms on the cross-section. Thus, we have transported about 5×10^3 atomic layers, which is about $1 \ \mu\text{m}$ in length and is about the same as the line width. This is a big void, indicating an opening in the line.

The same calculation can be applied to grain-boundary penetration through a diffusion barrier having a grain size of 1 micron. The uncertainty in the above calculation is $\Delta C_b/\Delta x$. If this gradient changes by one to two orders of magnitude, the conclusion is still the same, i.e. grain-boundary diffusion is significant in thin films at moderate temperatures.

In the following, we shall consider how to measure D_b in grain-boundary diffusion analysis.

9.3 Fisher's analysis of grain-boundary diffusion

We consider the two-dimensional analysis of diffusion (y-direction) along the grain boundary ($D = D_b$) and the drain into the adjacent grains (x-direction) by lattice diffusion ($D = D_l$) in Fig. 9.4. Applying the continuity equation [6]

$$\frac{\partial C}{\partial t} = -(\nabla \cdot J) = -\left(\frac{\partial J_x}{\partial x} + \frac{\partial J_y}{\partial y}\right) \tag{9.6}$$

inside the grains, we have

$$\frac{\partial C}{\partial t} = D_l \left(\frac{\partial^2 C}{\partial x^2} + \frac{\partial^2 C}{\partial y^2}\right) \tag{9.7}$$

But in the grain-boundary slab, we first consider the continuity equation in a small area δdy and let

$$\Delta J_x = J_{x1} - J_{x2} = 2\left(-D_l \frac{\partial C}{\partial x}\bigg|_{x=\delta/2}\right) \tag{9.8}$$

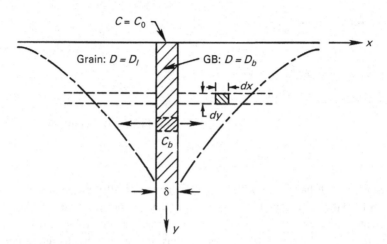

Fig. 9.4   Two-dimensional schematic diagram for Fisher's analysis of grain-boundary diffusion.

and $\Delta x = \delta$. Therefore, in the slab

$$\frac{\partial C_b}{\partial t} = D_b \frac{\partial^2 C_b}{\partial y^2} - \frac{\Delta J_x}{\Delta x} = D_b \frac{\partial^2 C_b}{\partial y^2} + \frac{2D_l}{\delta} \frac{\partial C}{\partial x}\Big|_{x=\frac{\delta}{2}} \tag{9.9}$$

For Fisher's analysis: $C = C(x, y, t)$
   Initial condition

$$C(x, 0, 0) = C_0$$

$$C(\pm \delta/2, y, 0) = 0$$

Boundary condition

$$C(x, 0, t) = C_0$$

$$C(\infty, \infty, t) = 0$$

(1) Assume steady state in the slab

$$0 = D_b \frac{\partial^2 C_b}{\partial y^2} + \frac{2D_l}{\delta} \frac{\partial C}{\partial x}\Big|_{x=\frac{\delta}{2}} \tag{9.10}$$

(2) Assume that $\partial C/\partial y = 0$ in grains, i.e. only diffusion in the x-direction

$$\frac{\partial C}{\partial t} = D_l \frac{\partial^2 C}{\partial x^2} \tag{9.11}$$

The physical picture of these assumptions is shown in Fig. 9.5 where we consider the
diffusion along a stack of layers parallel to the free surface. We then sum up diffusion

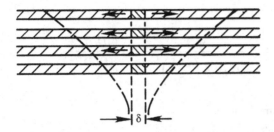

Fig. 9.5 Two-dimensional schematic diagram of a stack of layers parallel to the free surface. In Fisher's analysis, no communication between layers is assumed except at the grain boundary.

along each of the layers with no communication between layers except at the grain boundary.

The solution of Eq. (9.11) is

$$C(x, y, t) = C_b(y) \left\{ 1 - \text{erf}\left(\frac{x - \delta/2}{2\sqrt{D_l t}} \right) \right\} \qquad (9.12)$$

The partial derivative of Eq. (9.12) at $x = \delta/2$ gives

$$\left. \frac{\partial C(x, y, t)}{\partial x} \right|_{x=\frac{\delta}{2}} = \frac{C_b(y)}{(\pi D_l t)^{1/2}} \qquad (9.13)$$

Substituting Eq. (9.13) into Eq. (9.10)

$$D_b \frac{\partial^2 C_b}{\partial y^2} + \frac{2 D_l C_b}{\delta (\pi D_l t)^{1/2}} = 0 \qquad (9.14)$$

which has a solution of

$$C_b(y) = C_0 \exp \left\{ \frac{-\sqrt{2} y}{(\pi D_l t)^{1/4} (D_b \delta / D_l)^{1/2}} \right\} \qquad (9.15)$$

This can be substituted for $C_b(y)$ into Eq. (9.12)

$$C(x, y, t) = C_0 \exp \left\{ \frac{-\sqrt{2} y}{(\pi D_l t)^{1/4} \left(\frac{D_b \delta}{D_l} \right)^{1/2}} \right\} \left\{ 1 - \text{erf}\left(\frac{x - \delta/2}{2\sqrt{D_l t}} \right) \right\} \qquad (9.16)$$

Let

$$\eta = \frac{y}{(D_l t)^{1/2}}, \zeta = \frac{x - \delta/2}{2\sqrt{D_l t}}, \beta = \frac{D_b \delta}{2 D_l (D_l t)^{1/2}}$$

Then

$$\eta\beta^{-1/2} = \frac{\sqrt{2}y}{(D_l t)^{1/4}\left(\frac{D_b\delta}{D_l}\right)^{1/2}} = y\frac{(4D_l/t)^{1/4}}{(D_b\delta)^{1/2}} \tag{9.17}$$

Note that η, ζ, β are all dimensionless.
   Fisher's solution becomes

$$C(x,y,t) = C_0 \exp\left[-\pi^{1/4}\eta\beta^{-1/2}\right][1 - \mathrm{erf}\ \zeta] \tag{9.18}$$

This solution gives insight in the time-dependence of the grain-boundary penetration depth and the method of measuring D_b by sectioning radiotracer diffusion profiles to be shown below.

9.3.1   Penetration depth

We can evaluate the solution at the edge of the grain boundary. At $x = \delta/2$, $\mathrm{erf}(0) = 0$ and $\frac{C}{C_0} = \exp\left[-\pi^{-1/4}\eta\beta^{-1/2}\right]$,

$$\ln\frac{C}{C_0} = \ln\left(-\pi^{-1/4}\right) + \ln\eta - \frac{1}{2}\ln\beta$$

where β can be obtained from given C/C_0 and η (which depends on D_l only)

$$\frac{\partial \ln \beta}{\partial \ln \eta} = 2$$

Let C/C_0 be constant, then $\eta\beta^{-1/2}$ is constant, and from Eq. (9.17), we have

$$y \propto (D_b\delta)^{1/2} t^{1/4} \tag{9.19}$$

The penetration depth is proportional to $t^{1/4}$ rather than $t^{1/2}$. It is slower due to drain-out by the side diffusion into the grains adjacent to the boundary.

9.3.2   Sectioning

The average concentration $\bar{C}$ along the layer is measured by the difference in count rate of radioactive species between sequential layer removal steps (which remove a thickness Δy). The sectioning method gives the number of tracer atom/cm^2 in a layer of Δy thick.

$$\bar{C}(y,t)\Delta y = \Delta y \int_{-\infty}^{\infty} C(x,y,t)\,dx$$

$$= \Delta y C_0 \exp\left[-\pi^{-1/4}\eta\beta^{-1/2}\right]\int_{-\infty}^{\infty}\left[1 - \mathrm{erf}\left(\frac{x - \delta/2}{2\sqrt{D_l t}}\right)\right]dx \tag{9.20}$$

The integral is a constant. Therefore:

$$\ln \bar{C} = -\pi^{-1/4} \left(\beta^{-1/2} \eta \right) + \text{constant} \tag{9.21}$$

Consider differentiating Eq. (9.17)

$$\frac{\partial \beta^{-1/2} \eta}{\partial y} = \frac{(4D_l/t)^{1/4}}{(D_b \delta)^{1/2}}$$

$$D_b \delta = \frac{(4D_l/t)^{1/4}}{(\partial \beta^{-1/2} \eta / \partial y)^2} = \frac{(4D_l/t)^{1/4}}{(\partial \beta^{-1/2} \eta / \partial \ln \bar{C} / \partial y / \partial \ln \bar{C})^2} \tag{9.22}$$

Rearrange

$$D_b \delta = \left(\frac{\partial \ln \bar{C}}{\partial y} \right)^{-2} (4D_l/t)^{1/2} \left(\frac{\partial \ln \bar{C}}{\partial \beta^{-1/2} \eta} \right)^2$$

Note that the last term is equal to $\pi^{-1/2}$ from Eq. (9.21). Hence

$$D_b \delta = \left(\frac{\partial \ln \bar{C}}{\partial y} \right)^{-2} (4D_l/t)^{1/2} \pi^{-1/2} \tag{9.23}$$

Equation (9.23) is the key result in Fisher's solution for sectioning. It shows that by measuring $\ln \bar{C}$ as a function of y, and by knowing D_l, we can determine D_b.

An analysis of grain-boundary diffusion using Fisher's analysis can be made using radiotracer data of Ag in polycrystalline Ag shown in Fig. 9.6. If we take the data for 5 days ($t = 4.3 \times 10^5$ s) at 479 °C (725 K), we obtain graphically from the slope of the line

$$\frac{\partial \ln \bar{C}}{\partial y} = \frac{1}{2.5 \times 10^{-3} \text{cm}} = 400$$

and using Eq. (9.1) for the lattice diffusion of Ag,

$$D_l = 0.67e^{-30.1} = 5.7 \times 10^{-14}$$

We insert values in Eq. (9.23) and obtain

$$D_b \delta = \left(2.5 \times 10^{-3} \right)^2 \times \left(\frac{4 \times 5.7 \times 10^{-14}}{4.3 \times 10^5} \right)^{1/2} \pi^{-1/2} = 25.6 \times 10^{-16}$$

If the grain-boundary width is 0.5 nm, then

$$D_b = 5 \times 10^{-8} \text{ cm}^2/\text{s}$$

which is close to the value 8×10^{-8} cm^2/s obtained from Eq. (9.2) at 725 K.

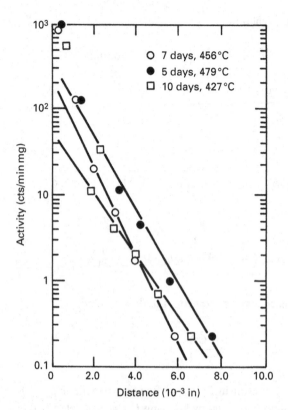

The activity data graph shows:
- O 7 days, 456°C
- ● 5 days, 479°C
- □ 10 days, 427°C

with Activity (cts/min mg) on y-axis and Distance (10^{-3} in) on x-axis.

Fig. 9.6   Radioactivity data of Ag in polycrystalline Ag. The grain-boundary diffusivity is obtained by using Fisher's analysis.

9.4   Whipple's analysis of grain-boundary diffusion

The grain boundary (GB) has a width and grain-boundary diffusivity D_b. Outside the grain boundary, we have D_l. The coordinates of the analysis are shown in Fig. 9.7 [7].
    The initial conditions are

$$C(x, 0, 0) = C_0$$

$$C(x, y, 0) = 0$$

The boundary conditions are

$$C(x, 0, t) = C_0$$

$$C(\infty, \infty, t) = 0$$

In the grains

$$\frac{\partial C}{\partial t} = D_l \left(\frac{\partial^2 C}{\partial x^2} + \frac{\partial^2 C}{\partial y^2} \right) \tag{9.24}$$

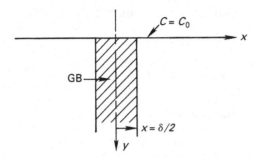

Fig. 9.7   Coordinates used in Whipple's analysis of grain-boundary diffusion.

Inside the grain boundary

$$\frac{\partial C_b}{\partial t} = D_b \left(\frac{\partial^2 C_b}{\partial x^2} + \frac{\partial^2 C_b}{\partial y^2} \right) \tag{9.25}$$

It is assumed that at the position $x = \pm \delta/2$, where the grain meets the grain-boundary slab,

$$C = C_b \tag{9.26}$$

$$D_b \frac{\partial C_b}{\partial x} = D_l \frac{\partial C}{\partial x} \tag{9.27}$$

i.e. the concentration and flux are continuous.

Whipple considered that within the grain-boundary slab, C_b is an even function of x. Note that the solution of a thin film diffusing into a rod shown below is an even function as given in Chapter 4:

$$C = \frac{C_0}{2 \, (\pi D t)^{1/2}} \exp \left(-\frac{x^2}{4Dt} \right)$$

It is then assumed that the concentration within the boundary is an even function,

$$C_b = C_b^0 + \frac{x^2}{2} C_b^1 (y, t) \tag{9.28}$$

where C_b^0 and C_b^1 are coefficients of the even function.

Substituting Eq. (9.28) into Eq. (9.25), we have

$$\frac{\partial C_b^0}{\partial t} = D_b \left(\frac{\partial^2 C_b^0}{\partial y^2} + C_b^1 \right) \tag{9.29}$$

At $x = \delta/2$, if δ is very small, we have from Eq. (9.26), (9.27) and (9.28) that

$$C = C_b^0$$

$$D_l \frac{\partial C}{\partial x} = D_b \frac{\delta}{2} C_b^1$$

Substituting C_b^0 and C_b^1 into Eq. (9.29), we obtain

$$\frac{\partial C}{\partial t} = D_b \left(\frac{\partial^2 C}{\partial y^2} + \frac{\delta}{2} \frac{D_l}{D_b} \frac{\partial C}{\partial x} \right) \tag{9.30}$$

(This is Fisher's Eq. (9.10) if $\partial C/\partial t = 0$.)

Now, if we substitute $\partial^2 C/\partial y^2$ from Eq. (9.24) into Eq. (9.30),

$$\frac{\partial C}{\partial t} = D_b \left(\frac{1}{D_l} \frac{\partial C}{\partial t} - \frac{\partial^2 C}{\partial x^2} + \frac{\delta}{2} \frac{D_l}{D_b} \frac{\partial C}{\partial x} \right) = \frac{D_b}{D_l} \frac{\partial C}{\partial t} - D_b \frac{\partial^2 C}{\partial x^2} + \frac{2 D_l}{\delta} \frac{\partial C}{\partial x}$$

Rearranging terms, we obtain

$$\left(\frac{D_b}{D_l} - 1 \right) \frac{\partial C}{\partial t} = D_b \frac{\partial^2 C}{\partial x^2} - \frac{2 D_l}{\delta} \frac{\partial C}{\partial x} \tag{9.31}$$

We have a pair of simultaneous equations, Eq. (9.24) and Eq. (9.31), for C. This is the starting point of Whipple's analysis.

Whipple's analysis (summary):

The initial conditions:　　　$C(x,0,0) = C_0$

$$C(x,y,0) = 0$$

The boundary condition:　　　$C(x,0,t) = C_0$

$$C(\infty,\infty,t) = 0$$

Eq. (9.24) $\Rightarrow$ $\dfrac{\partial C}{\partial t} = D_l \left(\dfrac{\partial^2 C}{\partial x^2} + \dfrac{\partial^2 C}{\partial y^2} \right)$ where $\dfrac{\partial^2 C}{\partial y^2} = 0$ in Fisher's analysis.

Eq. (9.31) $\Rightarrow$ $\left(\dfrac{D_b}{D_l} - 1 \right) \dfrac{\partial C}{\partial t} = D_b \dfrac{\partial^2 C}{\partial x^2} - \dfrac{2 D_l}{\delta} \dfrac{\partial C}{\partial x}$ where $\dfrac{\partial C}{\partial t} = 0$ in Fisher's analysis.

Equation (9.31) can be regarded as a boundary condition of C at $x = \delta/2$.

Solution:

$$C(x,y,t) = C_0 \left(1 - \text{erf} \frac{\eta}{2} \right) + \frac{C_0 \eta}{2\sqrt{\pi}}$$

$$\times \int_1^{\Delta(\infty)} \frac{d\sigma}{\sigma^{3/2}} \exp \left(\frac{-\eta^2}{4\sigma} \right) \text{erfc} \left[\frac{1}{2} \sqrt{\frac{\Delta - 1}{\Delta - \sigma}} \left(\xi + \frac{\sigma - 1}{\beta^1} \right) \right] \tag{9.32}$$

where σ is the variable of integration and

$$\eta = \frac{y}{(D_l t)^{1/2}}, \xi = \frac{x - \delta/2}{(D_l t)^{1/2}}, \Delta = \frac{D_b}{D_l}$$

$$\beta^1 = \left(\frac{D_b}{D_l} - 1\right) \frac{\delta/2}{(D_l t)^{1/2}} \sim \beta; \text{ see Eq. (9.17) when } \frac{D_b}{D_l} \gg 1;$$

Note that

$$\beta = \frac{D_b \delta}{D_l S} \cong \frac{\text{flux through grain boundary}}{\text{flux through grains}}$$

where $S = 2 (D_l t)^{1/2}$.

For application to the sectioning method, recall that:

$$\eta \beta^{-1/2} = y \frac{(4D_l/t)^{1/4}}{(D_l \delta)^{1/2}}$$

$$\left(\eta \beta^{-1/2}\right)^m = y^m \frac{(4D_l/t)^{m/4}}{(D_l \delta)^{m/2}} \tag{9.33}$$

$$\frac{\partial \left(\eta \beta^{-1/2}\right)^m}{\partial y^m} = \frac{(4D_l/t)^{m/4}}{(D_l \delta)^{m/2}}$$

$$\frac{\partial \left(\eta \beta^{-1/2}\right)^m / \partial \ln \bar{C}}{\partial y^m / \partial \ln \bar{C}} = \frac{(4D_l/t)^{m/4}}{(D_l \delta)^{m/2}}$$

$$D_b \delta = \left(\frac{\partial \ln \bar{C}}{\partial y^m}\right)^{-\frac{2}{m}} (4D_l/t)^{1/2} \left(\frac{\partial \ln \bar{C}}{\partial \left(\eta \beta^{-1/2}\right)^m}\right)^{\frac{2}{m}} \tag{9.34}$$

Note from Eq. (9.33) above that if

$$\frac{\partial \ln \bar{C}}{\partial y^m} = \text{const}$$

Then

$$\frac{\partial \ln \bar{C}}{\partial \left(\eta \beta^{-1/2}\right)^m} = \text{const}$$

Then, by evaluating the condition

$$\frac{\partial \ln \bar{C}}{\partial \left(\eta \beta^{-1/2}\right)^m} = \text{const}$$

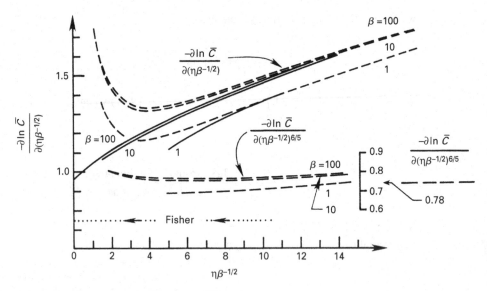

Fig. 9.8   Plot of the relation of $(\partial \ln \bar{C})/[\partial(\eta\beta^{-1/2})^m] = $ constant. The m and the constant are determined to be 6/5 and 0.78, respectively.

m is found to be 6/5; in fact, as shown in Fig. 9.8,

$$\frac{\partial \ln \bar{C}}{\partial \left(\eta\beta^{-1/2}\right)^{6/5}} = 0.78$$

So finally, we have

$$D_b\delta = \left(\frac{\partial \ln \bar{C}}{\partial y^{6/5}}\right)^{-5/3} \left(\frac{4D_l}{t}\right)^{1/2} (0.78)^{5/3} \tag{9.35}$$

We note that Eq. (9.35) is the key result in Whipple's solution. It has a form similar to that of Fisher's solution, Eq. (9.23). It shows that by plotting $\ln \bar{C}$ versus $y^{6/5}$ and by knowing D_l, we obtain $D_b\delta$. The slopes of $\ln \bar{C}$ versus $\eta\beta^{-1/2}$ are shown in Fig. 9.8.

9.5   Diffusion in small-angle grain boundaries

The effect of grain-boundary structure on diffusion has been shown in small-angle grain boundaries. Small-angle grain boundaries of the tilt type can be represented by a parallel array of edge dislocations. When these dislocations are far apart, the diffusion along each of them can be regarded as the diffusion in a pipe, "pipe diffusion." It is highly anisotropic, since diffusion along the pipe and normal to the pipe are very different, and the latter is much slower. Also, the activation energy of pipe diffusion is invariant as long as the dislocations are not too close to each other. It has been found that from $\theta = 9°$ to $16°$ of tilt angle, the activation energy is the same, except that at a higher angle of $\theta = 20°$, the diffusivity is faster because D_0 is greater.

If we assume that the distance between dislocation cores in a tilt-type grain boundary is given by

$$d = \frac{a}{2\sin\theta/2} \tag{9.36}$$

where a is the lattice constant, and θ is tilt angle, then we have

$$D_b\delta = \frac{D_p h^2}{d} = 2D_p h^2 \frac{\sin\theta/2}{a} \tag{9.37}$$

where D_p and h^2 are the diffusivity and the effective cross-sectional area of a dislocation pipe, respectively, and

$$D_p = A\exp\left(-\frac{Q}{kT}\right) \tag{9.38}$$

It can be seen that Q does not depend on θ. When the diffusion is conducted normal to the pipe, the measured diffusivity is much lower and varies with θ, too.

9.6 Diffusion-induced grain-boundary motion

In the above, we discussed the diffusion in a stationary grain boundary, i.e. the grain boundary does not move at all while atomic diffusion occurs in it. In a fine-grained thin film, the grain boundaries tend to migrate because of curvature and because of a reduction of grain-boundary energy. Therefore, we encounter the diffusion in a moving grain boundary. Diffusion in a moving grain boundary is a kinetic process of low activation energy of phase transformation in polycrystalline solids [8–10]. To consider such diffusion, there are two key issues: (a) the diffusion equation in a moving grain boundary, and (b) the driving force which moves the grain boundary.

We first discuss the diffusion equation used in Fig. 9.9 in which the grain boundary is moving to the right with a constant velocity, v. Following the derivation of the continuity equation as given in Chapter 4, we consider the fluxes flowing in and out of a tiny square

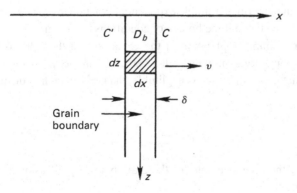

Fig. 9.9   Coordinates for a grain boundary in motion induced by diffusion along the grain boundary.

$dxdz$ in the grain boundary and we ignore the y-dimension because we assume that this is a two-dimensional problem. To simplify the problem, we further assume that we can ignore lattice diffusion for the moment. This assumption is acceptable for cases where the diffusion takes place at temperatures below half the melting point of the solid.

In the z-direction as shown in Fig. 9.9, we have

$$J_z = -D_b \frac{\partial C_b}{\partial z} \tag{9.39}$$

where D_b and C_b are diffusivity and concentration in the grain boundary, respectively. In the x-direction, we have

$$J_x = vC - vC'$$
$$\Delta x = \delta \tag{9.40}$$

where C and C' are lattice concentration before and behind the grain boundary. Then the divergence in the square $dxdz$ is given by

$$\nabla \cdot J = \frac{\partial J_x}{\partial x} + \frac{\partial J_z}{\partial z} = 0 \tag{9.41}$$

$$\frac{v(C - C')}{\delta} + \frac{\partial}{\partial z}\left(-D_b \frac{\partial C_b}{\partial z}\right) = 0 \tag{9.42}$$

$$\delta D_b \frac{\partial^2 C_b}{\partial z^2} - v(C - C') = 0$$

To solve this diffusion equation, we assume that

$$\frac{C}{C_b} = k \tag{9.43}$$

and the ratio k is called the segregation coefficient. We then have

$$\frac{\partial^2 C}{\partial z^2} - \frac{vk}{D_b \delta}(C - C') = 0 \tag{9.44}$$

This has the solution in the form of a simple exponential function if we let $p^2 = vk/D_b\delta$. The solution is readily available if the boundary conditions are given.

We shall consider a simple application of the equation in a thin film as shown in Fig. 9.10. The grain boundary is moving with a velocity v in the film with a thickness of Z. If we assume that lattice diffusion is negligible, the grain-boundary diffusivity can be estimated to be

$$D_b \cong \frac{Z^2 v}{\delta} \tag{9.45}$$

where δ/v is roughly the time available for the grain-boundary diffusion and Z is the distance of diffusion.

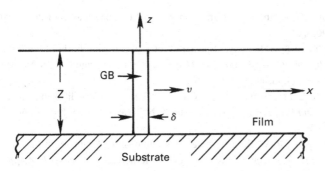

Fig. 9.10 Schematic diagram of a grain boundary in motion in a thin film.

The above kinetic process has been applied to the explanation of diffusion-induced grain-boundary motion (DIGM) phenomena in metals, dopant diffusion-induced grain growth in polycrystalline Si films, and oxygen diffusion-induced phase-boundary migration in copper oxide films. An example of DIGM is the interdiffusion between a polycrystalline Cu film and a polycrystalline Au film at around 160 °C. The temperature is so low that lattice diffusion is negligible in a reasonable period of time. Yet interdiffusion occurs by diffusion of Cu along the moving grain boundaries in Au, leading to the formation of CuAu solid solution in regions which had been swept by those moving grain boundaries.

Next, we discuss the driving force of grain-boundary motion. Note that a moving grain boundary leads to grain growth when the change across the grain boundary is only the crystallographic orientations of the two grains on either side of the grain boundary. In such grain-boundary motion, no long-range diffusion along the grain boundary is required, and atoms need to shuffle across the grain boundary of width δ, and to reorient themselves to the crystallographic axes of the growing grain. On the other hand, in DIGM a long-range grain-boundary diffusion is required. However, the major difference between the conventional normal grain growth and DIGM mode of transition is in the driving force.

In normal grain growth, the driving force behind the moving grain boundary comes from the reduction of grain-boundary energy (or area) and the grain boundary always moves against its curvature.

In the case of DIGM, the driving force is chemical in nature; in other words, it is due to the free energy change of a phase transition. Therefore, the grain-boundary motion can go with curvature, or it can move a straight grain boundary. This is because the chemical driving force is of the order of 1 eV which is much greater than the driving force of curvature change in grain growth. The latter is of the order of 0.01 eV.

References

[1] D. Turnbull and R. H. Hoffman, "The effect of relative crystal and boundary orientations on grain boundary diffusion rates," *Acta Met.* **2** (1954), 419.

[2] A. D. LeClaire, "The analysis of grain boundary diffusion measurements," *Brit. J. Appl. Phys.* **14** (1963), 351.

[3] R. W. Balluffi and J. M. Blakely, "Special aspects of diffusion in thin films," in *Low Temperature Diffusion and Applications to Thin Films*, eds A. Ganguler, P. S. Ho and K. N. Tu (Elsevier, Sequoia, Lausanne, 1975), 363.

[4] D. Gupta, D. R. Campbell and P. S. Ho, Ch. 7 of "Grain boundary diffusion" in *Thin Films—Interdiffusion and Reactions*, eds J. M. Poate, K. N. Tu and J. W. Mayer (Wiley-Interscience, New York, 1978).

[5] J. C. M. Hwang and R. W. Balluffi, "Measurement of grain boundary diffusion at low temperature by the surface accumulation method: I. Method and analysis," *J. of Appl. Phys.*, **50** (1979), 1339–48.

[6] J. C. Fisher, "Calculation of diffusion penetration curves for surface and grain boundary diffusion," *J. Appl. Phys.* **22** (1951), 74.

[7] R. T. P. Whipple, "Concentration contours in grain boundary diffusion," *Phil. Mag.* **45** (1954), 1225.

[8] K. N. Tu, "Kinetics of thin-film reactions between Pb and the AgPd alloy," *J. of Appl. Phys.*, **48** (1977), 3400.

[9] J. W. Cahn and R. W. Balluffi, "Diffusional mass-transport in polycrystals containing stationary or migrating grain boundaries," *Script. Met.* **13** (1979), 499–502.

[10] K. N. Tu, J. Tersoff, T. C. Chou and C. Y. Wong, "Chemically induced grain boundary migration in doped poly-crystalline Si films", *Solid State Communications* **66** (1988), 93–7.

Problems

9.1 In DIGM, is it a grain-growth process or is it a nucleation and growth process?

9.2 In DIGM, starting from a symmetrical tilt-type grain boundary, how can we decide the grain-boundary migration direction, to the right or to the left?

9.3 DIGM occurs in the diffusion of Ni into a Cu thin film. Grain-boundary velocities are recorded as a function of temperature for a sample at 350 °C; the velocity is 3.2×10^{-11} m/s, while at 900 °C, the velocity is 4.0×10^{-9} m/s. Estimate the activation energy for the grain-boundary diffusion of Ni in Cu.

9.4 At $T = 400$ °C and $t = 10$ min, diffusion length $= 150$ nm.
(a) Determine D (assuming that $D_0 = 1.0$ cm^2/s) and the activation energy.
(b) Decide whether the self-diffusion mechanism is due to lattice diffusion or to grain-boundary diffusion. Indicate the reason for your choice.

9.5 Determine the grain-boundary diffusion coefficient D_b (assume grain-boundary width $\delta = 0.5$ nm) from the radiotracer data in the figure below for a polycrystalline material. To obtain a lattice diffusion coefficient from a single crystal material, diffused at the same temperature (425 °C) and for the same time (10 days), the diffusion length was found to be $\lambda_D = 2 \times 10^{-5}$ cm.

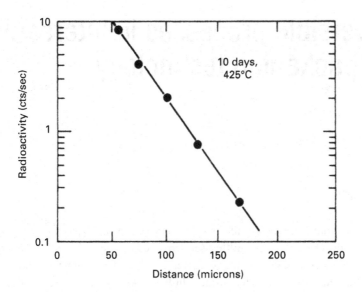

9.6 A 50 nm-thick polycrystalline Al sample has a square grain size of 100 nm, grain-boundary width of 1 nm, and grain-boundary diffusion coefficient $D_b = 10^{-10}$ cm^2/s. It was found using an isotope tracer that the solute penetrates 1.78 nm into adjacent grains in approximately 0.1 s. Using Fisher's analysis, estimate the amount of time it takes to penetrate the entire grain.

9.7 Solve Eq. (9.44) with the boundary condition that $dC/dz = 0$ at $z = 0$, and $C = C_e$ at $z = Z$.

10 Irreversible processes in interconnect and packaging technology

10.1 Introduction

Thin-film materials science is wafer-based and flux-driven. Up to now, most thin-film applications occur on devices built on semiconductor wafers. To process a microelectronic or opto-electronic device, the basic step consists of adding a monolayer of atoms on or subtracting it from a wafer surface. In these processes, we are not dealing with equilibrium states of materials; rather, we deal with kinetic states of a flux of atoms. Furthermore, for example, a *p-n* junction in a semiconductor is not at an equilibrium state. If we anneal the junction at a high temperature for a long time, it will disappear by interdiffusion of the *p*-type and *n*-type dopants. At device operation near room temperature, the dopants are supersaturated and frozen in place in the semiconductor to produce the electrical potentials, the built-in potentials, needed to guide the transport of charges. In doping a semiconductor, we need to diffuse or to implant a flux of atoms into the semiconductor to obtain the desired concentration profile of dopant. In device operation based on field effects, we pass an electric current or a flow of charge particles through the device to turn on or turn off the FETs. Thus, we consider flux-driven processes.

Generally speaking, we can have a flux or a flow of matter, a flow of energy (heat), or a flow of charge particles in a system. Indeed, in electronic devices, the operation can have all three kinds of flow coexist in the devices. Most importantly, from the point of view of device reliability, it is the interaction among these three kinds of flow or their cross-effects that can lead to device failure. In electromigration in Al and Cu interconnects, it may involve the interaction between atomic flow driven by stress gradient as well as by high current density. In thermomigration in eutectic flip-chip solder joints, a concentration gradient will not oppose phase separation in the eutectic structure under a temperature gradient.

In simple cases, the flows can be in a steady state, but the flow may not go from an equilibrium state to another. This is because the flow will stop in an equilibrium state, yet in many actual processes and systems, the flow keeps on going steadily under an applied driving force. Without equilibrium, we cannot use the condition of minimum Gibbs free energy to describe the states. Instead of states, we need to describe the flow process. We are in the domain of non-equilibrium or irreversible thermodynamics. In thermodynamics, the kinetic process going from one equilibrium state to another equilibrium state can be reversible or irreversible. Theoretically, it can be reversible. Practically, it tends to be irreversible. If the actual system is kept in homogeneous boundary conditions, i.e. at constant temperature and constant pressure, it goes irreversibly to the equilibrium end state, for example, in a phase transformation. On the other hand, if the system is kept in inhomogeneous boundary conditions, e.g. there is a temperature gradient or a pressure gradient, it will go irreversibly to a steady state instead of to an equilibrium state. Usually, we tend to call the steady-state processes "irreversible processes."

In bilayer thin-film reactions between Rd and Si or between Ni and Zr, for example, an amorphous alloy of RdSi or NiZr was formed upon slow heating rather than rapid quenching of the thin-film samples. Since crystalline intermetallic phases of RdSi or NiZr must have lower free energy than the amorphous alloy, it is clear that we cannot use maximum free-energy change to describe the formation of the amorphous alloys so that a minimum free-energy state is obtained. However, we may be able to use the maximum rate of free-energy change (or gain) in a short period of time to explain it. We have

$$\Delta G = \int_0^\tau \frac{\Delta G}{\Delta t} dt = \int_0^\tau \frac{\Delta G}{\Delta x}\frac{\Delta x}{\Delta t} dt = \int_0^\tau -Fv dt \qquad (10.1)$$

where $\Delta G/\Delta t$ is the rate of Gibbs free energy change, F is the driving force of the reaction or chemical potential gradient, and v is the reaction velocity or rate of formation of the phase under consideration. For the amorphous alloy, its driving force of formation will be lower than that of the competing crystalline phases, but if its rate of formation (in a short period of time, τ) is greater than that of the crystalline phases, it is possible that the product of Fv for the amorphous alloy is greater than that of the crystalline phase, so the amorphous phase forms. We can express Gibbs free energy change in a rate process as

$$\frac{dG}{dt} = -S\frac{dT}{dt} + V\frac{dp}{dt} + \sum_i^j \mu_i \frac{dn_i}{dt} \qquad (10.2)$$

Indeed, we can use rapid quenching (large change in dT/dt), rapid mechanical milling (large change in dp/dt), and ion implantation (large change in dn_i/dt) to produce amorphous or metastable materials. On the other hand, when time becomes infinitive, the equilibrium crystalline phase will win. In irreversible processes, however, it has been characterized by entropy change rather than free-energy change, to be discussed below by an example of heat transfer.

We shall use heat flow to illustrate entropy change in a flow process. In Fig. 10.1, we consider two heat chambers at temperatures of T_1 and T_2, where the temperature of T_1 is higher than T_2. If we connect these two chambers, heat will flow from T_1 to T_2. We assume a quantity of heat of δQ flows from T_1 to T_2; the change in entropy in the two

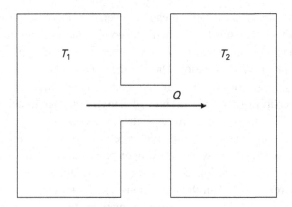

Fig. 10.1   Two heat chambers at temperature of T_1 and T_2, where the temperature of T_1 is higher than T_2.

chambers will be

$$dS_1 = -\frac{\delta Q}{T_1}$$

$$dS_2 = +\frac{\delta Q}{T_2}$$

The net change in entropy is equal to

$$dS_{net} = dS_1 + dS_2 = \delta Q \left(\frac{1}{T_2} - \frac{1}{T_1} \right) = \delta Q \left(\frac{T_1 - T_2}{T_1 T_2} \right) \tag{10.3}$$

which is positive. Entropy is not conserved, so the heat flow has generated a certain amount of entropy. Since the flow takes time to go from one chamber to the other, it is a rate process of entropy generation. We shall consider entropy generation in the flow of matter, heat, and electrical charges below. Before we do so, we need to define the flux and its driving force as well as the cross-effects.

10.2   Flux equations

There are three kinds of flux or flow, and they are governed by the three well-known phenomenological laws; the flow of matter (Fick's law), the flow of heat (Fourier's law), and flow of electric charge (Ohm's law). In one dimension, they are given below.

$$J = -D\frac{dC}{dx}$$

$$J_Q = -k\frac{dT}{dx} \tag{10.4}$$

$$j = -\sigma\frac{d\phi}{dx}$$

where the flux of matter, J, is equal to diffusivity times the gradient of concentration, the flux of heat, J_Q, is equal to thermal conductivity times the gradient of temperature, and the flux of charge carrier, j, is equal to conductivity times the gradient of electric potential (voltage) or the electric field. The last equation can be written as $E = j\rho$, where $E = -d\phi/dx$ is electric field, and ρ is resistivity. A direct comparison of the similarity in parameters between Fick's law and Ohm's law is given in Chapter 11.

We can consolidate the above three equations into a single equation

$$J = LX$$

For each force X, there is a corresponding conjugate primary flow of J, and L is the proportional constant or parameter. For example, the primary flow in a temperature gradient is heat and the parameter is heat conductivity. Typically, we take the negative gradient, $-dT/dx$, or $-dC/dx$, or $-d\phi/dx$, as the driving force in the flow process. However, note that none of them has the unit of mechanical force of "stress times area." Recall that kT is thermal energy and $e\phi$ is electric potential energy, since ϕ is defined as electric potential. Thus, we need to define conjugate force and flux in irreversible processes and this will be clear after we have defined rate of entropy production [1–5].

In heat conduction, it has been found that a temperature gradient can also drive a flow of electric charges as well as a flow of atoms. These are defined as cross-effects. When a temperature gradient induces an electric charge flux, it is called the thermal-electrical effect or Seebeck effect; the application of this effect in thermal couples for temperature measurement is well known. When an electric field induces a heat flow, it is called the Peltier effect. Whether these thermal-electrical effects have impact on device reliability is unclear. The Peltier effect may affect interfacial reaction at the hot end of the device. When a temperature gradient induces an atomic flux, it is called thermomigration, or the Soret effect when the thermomigration occurs in an alloy, and it is of interest here from the point of view of reliability. Another cross-effect which is of keen interest here is electromigration, which means atomic diffusion driven by an electric field under a high electric current density.

We can use a matrix to represent the relations:

$$J_i = \sum_i^j L_{ij} X_j \tag{10.5}$$

where L_{ii} are the coefficients between force and its primary flow, and L_{ij} are the coefficients of cross-effects when $i \neq j$.

It is worth mentioning that when we consider heat flow, temperature is not a constant and is a variable. Thus, in considering thermal-electrical effects or thermomigration, the temperature should be treated as a variable. However, atomic flow and charge carrier flow can occur at a constant temperature, so when we consider the cross-effect between them as in electromigration, we can assume a constant temperature process, so temperature is not a variable.

10.3   Entropy generation

To consider entropy generation in a flow process, we shall deal with inhomogeneous systems. The main thermodynamic supposition for such a system is a postulate of quasi-equilibrium for physically small volumes in the system. Each physically small volume or cell can be considered to be in quasi-equilibrium and its entropy can be determined from the relations of equilibrium thermodynamics by using thermodynamic variables and parameters. We can express a change of the internal energy of an ideal binary system, for example, the mixture of an element and its isotope, as

$$dU = TdS - pdV + \sum_{i=1}^{2} \mu_i dn_1 \qquad (10.6)$$

Note that when positive work is done by applying a pressure to the system, the volume decreases, so we have the $-pdV$ term which has the negative sign. Then

$$TdS = dU + pdV - \sum_{i=1}^{2} \mu_i dn_i = dH - Vdp - \sum_{i=1}^{2} \mu_i dn_i \qquad (10.7)$$

Here dS, dU, dH, dV and dn_i are respectively the changes of entropy, internal energy, enthalpy (thermal function), volume, and number of particles of i-species in the fixed physically small cell. The whole system's entropy (and its change) is defined here as simply the sum of entropies (and their changes) of all cells. The change of entropy in each cell can be represented as a sum of two terms; $dS = dS_e + dS_i$. The first term is due to divergence of entropy flux (describing the redistribution of already available entropy between cells); it corresponds to the difference between the entropy coming to a cell through its boundaries and the entropy going out from the cell through the boundaries. The second term means entropy production within the cell and is always positive (or zero when the system is in equilibrium and at zero fluxes). In a steady-state process, the sum dS is zero in every cell, but dS_e or dS_i may not be zero. This is because steady state means that the parameters of each cell are constant in time. Therefore, entropy which is a state function will remain constant, and dS will be zero. In a steady-state process, when the entropy production, dS_i, is not zero, it is compensated by the divergence of entropy flux, dS_e, in the cell. For example, in a steady-state process, when heat is generated in a cell, the heat must be transported away or dissipated, otherwise the temperature will increase and it cannot be kept in a steady state.

It is well known that the second term, the rate of entropy production, can always be represented as a product of "flux" and its corresponding or conjugate "force." We consider below the three basic processes of flow; heat conduction, atomic diffusion, and electrical conduction.

In heat conduction, temperature is not constant and should be treated as a variable. On the other hand, in atomic diffusion and electrical conduction, we can assume a constant temperature process. We should keep these in mind, especially when we consider cross-effects among them.

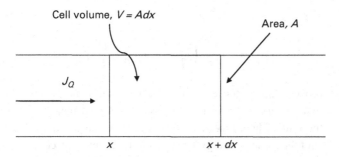

Fig. 10.2 Schematic diagram depicting on one-dimensional heat conduction. Let the cell be a thin layer having a cross-sectional area A and a length dx between x and $x + dx$.

10.3.1 Heat conduction

We consider one-dimensional heat flux caused by a temperature gradient along the x-axis. Let the cell be a thin layer having a cross-sectional area A and a width of dx between x and $x+dx$, shown in Fig. 10.2. The volume of the cell is $V = Adx$. Assume the process to be isobaric and with no atomic flow. Then the change of entropy in the cell is determined by the change of enthalpy, and in turn, this change of enthalpy will be determined by the difference of incoming and outgoing heat fluxes. From Eq. (10.7), we have

$$TdS = dH = J_Q(x) \cdot A \cdot dt - J_Q(x + \Delta x) \cdot A \cdot dt$$

$$= -V \frac{J_Q(x + \Delta x) - J_Q(x)}{\Delta x} \cdot dt = -V \frac{\partial J_Q}{\partial x} dt \qquad (10.8)$$

Then we divide both sides by $TVdt$ and use the well-known differentiation of a product; $[ydx = d(yx) - xdy]$. We have

$$\frac{\partial S}{Vdt} = -\frac{1}{T} \frac{\partial J_Q}{\partial x} = -\frac{\partial}{\partial x}\left(\frac{J_Q}{T}\right) + J_Q \frac{\partial}{\partial x}\left(\frac{1}{T}\right) \qquad (10.9)$$

In the above, temperature is treated as a variable. The ratio of J_Q/T in the first term in the right-hand side of the above equation is an entropy flux, and $\frac{\partial}{\partial x}(J_Q/T)$ is a divergence of this flux, meaning the rate of entropy change per cell volume caused by the difference of incoming and outgoing entropy fluxes. The second term in the right-hand side of the above equation is the rate of entropy production per unit volume per unit time within the cell. This second term is a product of heat flux and the gradient of inverse temperature which can be interpreted as the thermodynamic force (or the conjugate force to be discussed later) driving the heat flux. Since

$$\frac{\partial}{\partial x}\left(\frac{1}{T}\right) = -\frac{1}{T^2}\left(\frac{\partial T}{\partial x}\right) \qquad (10.10)$$

thus,

$$\frac{TdS}{Vdt} = -T\frac{\partial}{\partial x}\left(\frac{J_Q}{T}\right) - J_Q\frac{1}{T}\left(\frac{\partial T}{dx}\right) \tag{10.11}$$

In a steady-state process, the two terms on the right-hand side of the equation compensate each other. Note that the dimension of the above equation is correct. It is energy/cm^3 s, and, recall that the unit of heat flux J_Q is energy/cm^2 s, but the unit of the force of $(1/T)(dT/dx)$ is not the same as energy/length as in mechanical forces.

10.3.2   Atomic diffusion

We consider a binary system in isobaric condition,

$$TdS = dH - \sum_{i=1}^{2}\mu_i dn_i$$

If we take the binary system to be an element and its isotope or an ideal solution, then $dH = 0$, so

$$\frac{TdS}{dt} = -\sum_{i=1}^{2}\mu_i\left(\frac{\partial n_i}{\partial t}\right)$$

If we divide the above equation by volume V and take $n_i/V = C_i$ and for simplicity recall the continuity equation in one dimension, we have

$$\frac{\partial C_i}{\partial t} = -\nabla\cdot J_i = -\frac{\partial J_i}{\partial x}$$

Again, we use the relation of $ydx = d(xy) - xdy$, obtaining

$$\frac{TdS}{Vdt} = \sum_{i=1}^{2}\mu_i\frac{\partial J_i}{dx} = -\frac{\partial}{\partial x}\left(-\sum_{i=1}^{2}\mu_i J_i\right) + \sum_{i=1}^{2}J_i\left(-\frac{\partial\mu_i}{\partial x}\right) \tag{10.12}$$

In a steady-state process, the sum of the two terms on the right-hand side of the equation will be zero and they compensate each other. We see that the last term, the rate of entropy production, is equal to the product of flux times its driving force.

10.3.3   Electrical conduction

We consider the one-dimensional conduction of a pure metal at constant temperature. In this case, we have $dn_i = 0$, and $dV = 0$ when we assume a constant volume:

$$TdS = dU$$

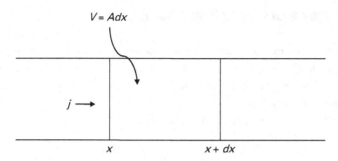

$V = Adx$

$j \longrightarrow$

x

$x + dx$

Fig. 10.3   Schematic diagram depicting electric conduction in a conductor of constant cross-section of A between x and $x + dx$, and the voltage drop is $\Delta\phi$.

To consider the internal energy change in electric conduction, recall a relation in physical unit conversion that "1 newton-meter = 1 joule = 1 coulomb-volt." It means that the internal energy change can be expressed as a product of charge and its potential. Recall that the unit of electric energy is eV, which is coulomb-volt. In electrical conduction, we define j as current density as A/cm^2 as C/cm^2 s, and also we define ϕ as voltage as electric potential. Figure 10.3 depicts the electric conduction in a conductor of constant cross-section of A between x and $x + dx$, and the voltage drop is $\Delta\phi$. Thus, jAt will give us the charge or coulomb. If we consider a conduction of constant current density, $j =$ constant, the conduction across a cell of volume $V = Adx$ will cause a voltage drop of $\Delta\phi$. Then we write

$$TdS = dU = jAdt\Delta\phi = jAdt[\phi(x) - \phi(x + \Delta x)]$$

$$= -jVdt\left[\frac{\phi(x + \Delta x) - \phi(x)}{\Delta x}\right] = jVdt\left[-\frac{d\phi}{dx}\right]$$

where $V = Adx$. So we obtain

$$\frac{TdS}{Vdt} = j\left[-\frac{d\phi}{dx}\right] = jE = j^2\rho \tag{10.13}$$

where $E = -d\phi/dx = \rho j$ is electric field, ρ is resistivity, and $j^2\rho$ is called "joule heating" per unit volume. Thus, in this simple case the rate of entropy production is a product of electric current density or flux (C/cm^2 s) and the force generating this flux (which is a negative gradient of electric potential or the electric field). Entropy production is joule heating.

In Eq. (10.13), note that joule heating alone cannot be in a steady state. While the applied current can be steady, the temperature without heat dissipation will raise. So, to reach a steady state, the system needs outgoing heat fluxes of $J_Q = j^2\rho$, which simultaneously will be accompanied by the outgoing entropy fluxes of J_Q/T.

10.4 Conjugate forces with varying temperature

What has been shown by the three examples above is that in irreversible processes, TdS/Vdt is equal to the product of flux J times its driving force X. Onsager (see [1]) defined them as the conjugate forces and fluxes. The unit of the conjugate force may not be the same as mechanical force.

From Eq. (10.11) on thermal conduction, we have

$$\frac{TdS}{Vdt} = J_Q X_Q = -J_Q \left(\frac{1}{T} \frac{dT}{dx} \right) \tag{10.14}$$

So the conjugate force to J_Q is $X_Q = -1/T(dT/dx)$. Thus, we write

$$J_Q = L_{QQ} X_Q = L_{QQ} \left(-\frac{1}{T} \frac{dT}{dx} \right) = -k \frac{dT}{dx} \tag{10.15}$$

We obtain $L_{QQ} = Tk$, or the coefficient L_{QQ} is equal to temperature times thermal conductivity.

For the conjugate forces in atomic diffusion and in electrical conduction, note that we have already obtained them in the last section. However, we have assumed implicitly that they occur under constant temperature. When we have temperature as a variable, we shall have the following conjugate forces for the three kinds of flow:

$$X_Q = T \frac{d}{dx} \left(\frac{1}{T} \right) \tag{10.16a}$$

$$X_{Mi} = -T \frac{d}{dx} \left(\frac{\mu_i}{T} \right) \tag{10.16b}$$

$$X_E = -T \frac{d}{dx} \left(\frac{\phi}{T} \right) \tag{10.16c}$$

The unit of force in Eq. (10.16a) and Eq. (10.16c) is different from that in Eq. (10.16b). The latter has the unit as mechanical force, which is a gradient of potential energy, but the former is not. In non-equilibrium thermodynamics, since a conjugate force is coupled to its corresponding flux, their product is defined to be the heat of entropy production per unit time per unit volume, TdS/Vdt, which has the unit of energy/cm^3 s. Thus, the dimension of the conjugate force is determined by the dimension of the corresponding flux. In heat conduction, the flux has the unit of energy/cm^2 s, so the conjugate force which accompanies the heat flux is $Td(1/T)/dx) = -(1/T)(dT/dx)$, which has the unit of cm^{-1}. In atomic diffusion, the flux has the unit of number of particle/cm^2 s, so the conjugate force has the unit of energy/cm. In electrical conduction, the flux has the unit of number of charge particle/cm^2 s, or e/cm^2 s, so the conjugate force has the unit of V/cm and their product is eV/cm^3 s.

When we consider thermomigration, since the atomic flux is in number of atom/cm^2 s, we need to add Q^*, heat of transport, to the conjugate force. On the other hand, atomic diffusion and electric conduction can take place at a constant temperature. Then, when we consider the interaction between atomic flow and electric flow, e.g. in electromigration, we can assume a constant temperature, so we take $X_{Mi} = -d\mu_i/dx$ and $X_E = -d\phi/dx$, in Eq. (10.16b) and (10.16c), respectively.

Knowing the fluxes and their conjugate forces, we can apply them to Eq. (10.5), obtain the equations in irreversible processes and study the cross-effects in the interaction among heat flow, mass flow, and electric flow. Because heat conduction occurs under a temperature gradient, so temperature is not constant and it becomes a variable in the conjugate forces given above. When we consider thermal-electrical effects and thermomigration, we have to take temperature as a variable to be shown below.

10.4.1 Atomic diffusion

To consider the conjugate force in atomic diffusion, for simplicity, we shall consider a constant pressure process. We rewrite Eq. (10.7) as

$$dS = \frac{1}{T}dH - \frac{1}{T}\sum_{i=1}^{2}\mu_i dn_i$$

$$\frac{\partial S}{V\partial t} = \frac{1}{T}\frac{\partial h}{\partial t} - \sum_{i=1}^{2}\frac{\mu_i}{T}\frac{\partial C_i}{\partial t} \tag{10.17}$$

where $h = H/V$ is enthalpy per unit volume, and $C_i = n_i/V$ is concentration of i-species. We recall that in one-dimensional analysis

$$\frac{\partial h}{\partial t} = -\nabla \cdot J_Q = -\frac{\partial J_{Qx}}{\partial x}$$

$$\frac{\partial C_i}{\partial t} = -\nabla \cdot J_i = -\frac{\partial J_{ix}}{\partial x}$$

Again for simplicity, we assume there is no heat source and sink and we use the one-dimensional equation to substitute the above continuity equations into Eq. (10.17), and then apply the relation of $ydx = d(xy) - xdy$:

$$\frac{dS}{Vdt} = -\frac{\partial}{\partial x}\left(\frac{J_{Qx}}{T}\right) + J_{Qx}\frac{\partial}{\partial x}\left(\frac{1}{T}\right) + \frac{\partial}{\partial x}\left(\sum_{i=1}^{2}\frac{\mu_i}{T}J_{ix}\right) - \sum_{i=1}^{2}J_{ix}\left(\frac{\partial}{\partial x}\frac{\mu_i}{T}\right)$$

$$= -\frac{\partial}{\partial x}\frac{1}{T}\left(J_{Qx} - \sum_{i=1}^{2}\mu_i J_{ix}\right) + J_{Qx}\frac{\partial}{\partial x}\left(\frac{1}{T}\right) + \sum_{i=1}^{2}J_{ix}\left(-\frac{\partial}{\partial x}\frac{\mu_i}{T}\right)$$

If we multiply the last equation by T, we obtain the conjugate force from the last term as given in Eq. (10.16b). It is worth noting that under a temperature gradient, the energy flux, $J_{Qx} - \mu J$, contains not only the heat flux, but also the flux of chemical potential energy transported by the diffusing atoms.

10.4.2   Electrical conduction

To consider electric conduction, we shall assume one-dimensional case under constant current and constant pressure process in a pure element, and both ϕ and T are functions of x. From Section 10.3.3 and taking $V = Adx$ and $H/V = h$, we have

$$dS = \frac{dH}{T} + \frac{jAdt}{T}[\phi(x + \Delta x) - \phi(x)]$$

$$\frac{dS}{Vdt} = \frac{1}{T}\frac{dh}{dt} - j\left(\frac{1}{T}\frac{\partial\phi}{\partial x}\right)$$

$$\frac{dS}{Vdt} = -\frac{\partial}{\partial x}\left(\frac{J_{Qx}}{T}\right) + J_{Qx}\frac{\partial}{\partial x}\left(\frac{1}{T}\right) - j\left(\frac{\partial}{\partial x}\left(\frac{\phi}{T}\right) - \phi\frac{\partial}{\partial x}\left(\frac{1}{T}\right)\right)$$

$$= -\frac{\partial}{\partial x}\left(\frac{J_{Qx}}{T}\right) + (J_{Qx} + j\phi)\frac{\partial}{\partial x}\left(\frac{1}{T}\right) + j\left(-\frac{\partial}{\partial x}\left(\frac{\phi}{T}\right)\right)$$

When we multiply the last equation by T, we obtain from the last term the conjugate force presented in Eq. (10.16c). It is worth noting that under a temperature gradient, the energy flux, $J_{Qx} + j\phi$, contains not only the heat flux, but also the flux of electric energy transported by the charge going with the current and the charge equal to jAt.

10.5   Joule heating

We have shown in the above that entropy production in electric conduction is joule heating. Usually, the power of joule heating is written as

$$P = I^2R = j^2\rho V \tag{10.18}$$

where I is applied current and $I/A = j$, and A is the cross-sectional area of the sample, R is the resistance of the sample and $R = \rho l/A$, and l is the length of the sample, so the volume $V = Al$. Thus, I^2R is joule heating per unit time (power = energy/time) of the entire sample, and $j^2\rho$ is joule heating per unit volume per unit time of the sample.

Since Si devices operate electrically, joule heating is a built-in cause of reliability. To calculate joule heating, we consider flip-chip technology, in which typically every two solder joints connect by an Al interconnect on the chip side, shown in Fig. 10.4. The current density in the Al is about 10^6 A/cm^2 and the current density in the solder joint is about 10^4 A/cm^2. This is because the cross-section of the Al interconnect line is about two orders of magnitude smaller than that of the solder joint. To calculate the joule heating in them, if we take $j = 10^6$ A/cm^2 and $\rho = 10^{-6}$ Ω-cm for Al, joule heating of $\rho j^2 = 10^6$ J/cm^3 s. For the solder joint, if we take $j = 10^4$ A/cm^2 and $\rho = 10^{-5}$ Ω-cm for SnAgCu solder, joule heating of $\rho j^2 = 10^3$ J/cm^3 s. Thus, the Al interconnect will be hotter than the solder joint; in turn the chip side will be hotter than the substrate side, and consequently there will be a temperature gradient across the solder joint.

While joule heating is a kind of waste heat and cannot be used to do work, however, it generates heat and will increase the temperature of the conductor. Consequently, it will

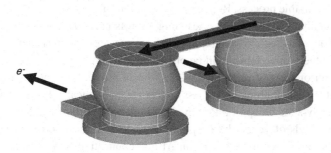

Fig. 10.4   Schematic diagram of flip-chip technology, in which typically every two solder joints are
connected by an Al interconnect on the chip side.

increase the resistance of the conductor, and in turn will cause more joule heating. An
increase of temperature of the conductor will lead to thermal expansion of the conductor,
and this is the basic reason for thermal stress in devices when various materials of different
thermal expansion coefficients are used. Stress and stress gradient may lead to fracture
and creep. When the thermal stress is cyclic because semiconductor devices are being
turned on and off frequently, fatigue can happen because of the accumulation of plastic
strain energy in the system. Furthermore, the higher temperature caused by joule heating
will increase atomic diffusion, rate of phase change, and interdiffusion and interfacial
reactions in the devices. In addition, joule heating can lead to temperature gradient in
the device, for example, across a flip-chip solder joint as discussed before, it will cause
thermomigration. Thus, joule heating is a major concern of device reliability.

It is worth mentioning the application of joule heating in fuses to prevent overheating
in household appliances. Thin solder strips are used as fuses. When the appliance draws
a current close to 10 A, the fuse will melt and the circuit will have an open. If we assume
that the fuse has a cross-section of 2 mm $\times$ 0.1 mm, the current density in the fuse will
be about 5×10^4 A/cm^2. It will lead to melting of the solder fuse. Indeed when such a
high current density was applied to flip-chip solder joints, melting of the solder bump
was observed. Knowing the heat capacity of the solder, we can calculate the temperature
increase due to joule heating, provided that we know the rate of heat dissipation. Since Si
chip itself is a very good heat conductor, this is the reason why Al and Cu interconnects
on Si can take a current density close to 10^6 A/cm^2, which is much higher than that in a
household wire or an extension cord.

10.6   Electromigration, thermomigration, and stress-migration

Electromigration and thermomigration are fluxes of atoms driven by electron flow and
heat flow, respectively, under an electrical potential gradient and temperature gradient.

Stress-migration is creep and it is a flux of atoms driven by a stress potential gradient. In other words, creep does not occur under a uniform pressure or hydrostatic pressure, hence it is an irreversible process. Both electromigration and thermomigration are cross-effects in irreversible processes, but stress-migration is not. Stress potential is part of the chemical potential that drives atomic diffusion, so creep is a primary flow. The fundamentals of these three topics will be presented in later chapters.

From the point of view of reliability, a steady-state process may not lead to device failure. When there is no failure, it is of less concern technologically. For example, if we consider the steady-state creep of a Pb pipe hanging on the wall of an old house under gravity over hundreds of years, the Pb pipe crept and bent, but there was no failure. Another example is electromigration of an Al line connecting two infinitely large Al bond pads. If the electromigration occurs at a steady state, there is no failure when a uniform flux of Al atoms is being transported steadily from the cathode to the anode. Only when there is a divergence in the transport and also when the divergence leads to excess vacancies or atoms, can the excess vacancies or atoms condense to form voids and hillocks, respectively, so failure occurs due to electrical open or short in the circuit, and then it is of reliability concern.

Electromigration in Al and Cu interconnects has been the most persistent reliability problem in very large-scale integration (VLSI) of circuits in Si technology. Under high current density, above 10^5 A/cm^2, at the device operation temperature around $100\,^\circ$C, voids and hillocks have been found to form at the cathode and the anode of Al interconnects, respectively. The study of electromigration in Al short strips showed that there is a critical length, below which there is no electromigration, as shown in Fig. 10.5. Blech and Herring (see Chapter 11) proposed that this is due to the effect of back stress in the Al strips. The gradient of back stress potential has induced an atomic flux to counter the atomic flux of electromigration. When these fluxes balance each other, there is no electromigration. Below, we shall consider the interaction between electromigration and stress-migration by using irreversible processes. It is an interaction between atomic flow and charge flow.

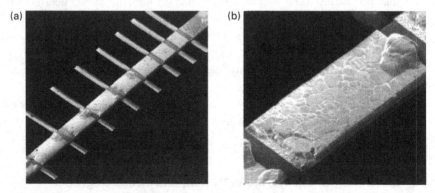

Fig. 10.5　(a) SEM images of a set of Al short strips on a TiN baseline after electromigration. There is a critical length below which there is no electromigration. (b) A high magnification image of one of the Al strips (courtesy of Dr. Alexander Straub, MPI Stuttgart, Germany).

10.7 Irreversible processes in electromigration

Figure 10.6 depicts a schematic diagram of a short Al strip patterned on a baseline of TiN. Since TiN is a poor conductor, electrons will make a detour from the TiN to the Al to reduce the resistance. Under a high current density of electrons going from left to right, electromigration transports Al atoms from the cathode at the left to the anode at the right, leading to depletion or void formation at the cathode and pile-up or hillock formation at the anode. Hence, the damage of electromigration can be recognized directly using the short strip test samples. The depletion rate at the cathode can be measured and the drift velocity of electromigration can be deduced. Furthermore, it was found that the longer the strip, the longer the depletion at the cathode side in electromigration. However, below a critical length, there was no observable depletion as shown in Fig. 10.5.

The dependence of depletion on strip length was explained by the effect of back stress by Blech and Herring. In essence, when electromigration transports Al atoms in a strip from the cathode to the anode, the latter will be in compression and the former in tension. On the basis of the Nabarro–Herring (see Section 14.3) model of equilibrium vacancy concentration in a stressed solid, to be discussed in Chapter 13, the tensile region has more vacancy and the compressive region has less vacancy than the equilibrium vacancy in the unstressed region, so there is a vacancy concentration gradient decreasing from the cathode to the anode in the Al strip, as shown in Fig. 10.7. The vacancy gradient induces a flux of Al atoms diffusing from the anode to the cathode, and it opposes the Al flux driven by electromigration from the cathode to the anode. The vacancy concentration gradient depends on the length of the strip; the shorter the strip, the greater the slope or the gradient. At a certain short length defined as the critical length, the vacancy gradient

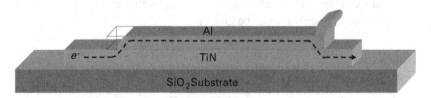

Fig. 10.6   Schematic diagram of a short Al strip patterned on a baseline of TiN.

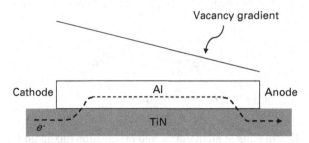

Fig. 10.7   Schematic diagram to depict a vacancy concentration gradient decreasing from the cathode to the anode in the Al strip, due to the back stress in the Al stripe.

generated by the back stress is large enough to balance electromigration, so no depletion at the cathode and no extrusion at the anode occur [10–13].

10.7.1   Electromigration and creep in Al strips

In analyzing the back-stress effect, we shall use irreversible processes by combining electrical and mechanical forces on atomic diffusion. We assume that these processes occur at constant temperature, so temperature is not a variable. The electrical force proposed by Huntington and Grone (see Chapter 11) is taken to be

$$F_{em} = Z^* eE$$

The mechanical force is taken as the gradient of chemical potential in a stressed solid; see Chapter 13. Since we consider pure Al interconnects, there is no concentration gradient in Al, so the chemical potential due to concentration change is zero. Instead, we have a stress potential gradient to drive atomic flux:

$$C_v = C_V^{eq} \exp\left(\frac{\pm\sigma\Omega}{kT}\right) = C \exp\left(\frac{-\Delta G_f \pm \sigma\Omega}{kT}\right)$$

where C_v, C_V^{eq}, and C are the equilibrium concentration of vacancy in the stressed region, the equilibrium concentration of vacancies in the unstressed region, and concentration of atoms, respectively. ΔG_f is the formation free energy of a vacancy. The positive sign of the stress indicates that there are more vacancies in the tensile region, and the negative sign indicates fewer vacancies in the compressive region. Thus, there is a vacancy concentration gradient going down from the tensile region to the compressive region. The driving force can be given as

$$F_{me} = -\nabla\mu = -\frac{d\sigma\Omega}{dx} \tag{10.19}$$

where σ is hydrostatic stress in the metal, Ω is atomic volume, and $\sigma\Omega$ is defined as stress potential energy. In essence, it is a creep process, in which pressure is not constant but temperature is constant.

Thus, we have a pair of phenomenological equations for atomic flux and electron flux [6],

$$J_{em} = -C\frac{D}{kT}\frac{d\sigma\Omega}{dx} + C\frac{D}{kT}Z^* eE \tag{10.20a}$$

$$j = -L_{21}\frac{d\sigma\Omega}{dx} + n\mu_e eE \tag{10.20b}$$

where J_{em} is atomic flux in units of atom/cm^2 s, and j is electron flux in units of C/cm^2 s. C is the concentration of atoms per unit volume, and n is the concentration of conduction electrons per unit volume. D/kT is atomic mobility, E is electric field, and $E = \rho j$, ρ is resistivity, j is electric current density, and μ_e is electron mobility. The unit of D/kT and the unit of μ_e are basically the same, and they are cm^2/eV s and cm^2/V s, respectively, where V is electric potential. L_{21} is the phenomenological coefficient of irreversible processes and contains the deformation potential.

In Eq. (10.20a), the first term is creep and is the primary flux, and the second term is the cross-effect of electrical force on atomic diffusion due to electromigration. In Eq. (10.20b), the last term is electron flux in electric conduction and the first term is the cross-effect of mechanical force on electric conduction.

Note that in the flux equation of Eq. (10.20a), neither divergence nor lattice shift is considered. In other words, the equation does not indicate whether or not failure should occur. In Chapter 15, we shall discuss further that even if flux divergence exists, it is a necessary but insufficient condition of failure. For failure to occur due to the formation of a void, for example, we require excess vacancies and so the distribution of vacancy is non-equilibrium. Recall that in Darken's analysis of interdiffusion as presented in Chapter 5, flux divergence occurs in the samples, yet there is no Kirkendall void formation and also no stress since the condition of vacancy equilibrium is assumed, and lattice shift occurs to accommodate the creation and annihilation of vacancies for equilibrium.

It is worth mentioning that in the Blech and Herring model on back stress, vacancy is assumed to be in equilibrium with the stress everywhere in the sample. The change of equilibrium vacancy concentration under stress is small, hence we may assume that the source and sink of vacancies in the sample can do so. However, at equilibrium, neither void nor hillock should form. When the electromigration flux is larger than the stress-migration flux in those Al stripes longer than the critical length as shown in Fig. 10.5, void and hillock were found in the cathode and anode, respectively. To interpret the failure, we will need to assume that in the longer strips, there is no vacancy equilibrium and no lattice shift. Thus, the extra flux due to electromigration will require extra lattice sites to be created in the cathode to form the void as well as in the anode to form the hillock. In hillock growth, stress-migration from the interior to the free surface of the Al strip is required too.

In Eq. (10.20), if we take $J_{em} = 0$, for instance, there is no net electromigration flux or no damage. In other words, it reaches steady-state equilibrium in an irreversible process. The expression for the critical length is obtained as

$$\Delta x = \frac{\Delta \sigma \Omega}{Z^* e E} \tag{10.21}$$

Since the resistance of the conductor can be taken to be constant at a constant temperature, we have instead the critical product or threshold product of $j\Delta x$ by moving the current density from the right-hand side to the left-hand side of the above equation:

$$j\Delta x = \frac{\Delta \sigma \Omega}{Z^* e \rho} \tag{10.22}$$

In the above equation, all the parameters on the right-hand side are given for a conductor. Under a constant applied current density, a bigger value of critical product in Eq. (10.22) means a longer critical length, in turn a larger back stress in Eq. (10.21). For Al and Cu interconnects, we take $j = 10^6$ A/cm^2 and $\Delta x = 10\,\mu$m; we have a typical value of critical product to be about 10^3 A/cm.

Note that while the back stress in the above is induced by electromigration, the interaction between an applied stress and electromigration is the same. For example, an applied compressive stress at the anode will retard electromigration. Or if we can increase the yield stress in the conductor, meaning if we can increase $\Delta\sigma$ in Eq. (10.22), we can increase the critical product as well as the critical length.

The critical length can be measured experimentally by extending the time of electromigration to a sufficiently long period until the mass transport in the strip ceases, as shown in Fig. 10.5. We can use Eq. (10.21) to calculate the critical length Δx of Al strips. If we assume that the extrusion at the anode is accompanied by a certain amount of plastic deformation, the change in elastic stress can be taken to be the value corresponding to the elastic limit. For Al, we have $\sigma_{Al} = -1.2 \times 10^9$ dyn/cm^2, $\Omega_{Al} = 16 \times 10^{-24}$ cm^3, $e = 1.6 \times 10^{-19}$ C, and $E = j\rho$, where $j = 3.7 \times 10^5$ A/cm^2 and $\rho_{Al} = 4.15 \times 10^{-6}$ Ω-cm at 350 °C. By substituting these values into Eq. (10.18), we obtain

$$\Delta x_{Al} = -\frac{78\,\mu\text{m}}{Z^*}$$

By taking $Z^* = -26$ for bulk Al, we find the critical length to be 3 μm, which is of the right order of magnitude but shorter than the experimental finding of 10 to 20 μm. Since the Al strips are polycrystalline thin films, grain-boundary diffusion has played a dominant role in electromigration, so the Z^* for atoms diffusing in grain boundaries should be different from that in the bulk or lattice diffusion. It seems that for polycrystalline Al thin films, Z^* is below 10.

In Eq. (10.20b), if we take the coefficient $L_{21} = ne\mu_e N^*$ where N^* is a parameter to be discussed below, and if we consider a short strip deposited on an insulating substrate and take $j = 0$, we have

$$N^* = -\frac{1}{\Omega}\left|\frac{d\phi}{d\sigma}\right| \quad \text{at } j = 0 \tag{10.23}$$

where $d\phi/d\sigma$ at $j = 0$ is deformation potential defined as the electrical potential per unit stress difference at zero current. By using Onsager's reciprocity relation, $L_{12} = L_{21}$, we obtain the expression of

$$\frac{d\phi}{d\sigma} = -\frac{Z^* D\rho e}{kT} \tag{10.24}$$

The dimensions of $d\phi/d\sigma$ and N^* are cm^3/C and C^{-1}, respectively.

To calculate the deformation potential as shown in Eq. (10.24), we take $Z^* = -26$ for Al at $T = 500\,°C$, $kT = 0.067$ eV, lattice diffusivity of Al at $500\,°C$ to be about 2×10^{-10} cm^2/s, and resistivity to be about 4.83×10^{-6} Ω-cm. We obtain

$$\frac{d\phi}{d\rho}_{j=0} = 3.7 \times 10^{-13} \text{ cm}^3/\text{C}$$

However, we expect the order of magnitude of deformation potential to be close to Ω/e, which is about 10^{-4} cm^3/C or 10^{-10} V/Nm^{-2}. It seems that what we have calculated on the basis of Eq. (10.24) is too small. This is because the deformation process involved in Eq. (10.24) is by creep, which depends on long-range atomic diffusion and it is a very slow deformation process. On the other hand, the mechanical deformation in the usual sense does not involve thermally activated atomic diffusion; rather, the atomic displacement under an applied mechanical stress proceeds by the acoustic mode of atomic motion, i.e. at the speed of sound which is about 10^5 cm/s or equal to $a_0\nu$, where $a_0 = 3 \times 10^{-8}$ cm is the interatomic distance and $\nu = 10^{13}$ s^{-1} is atomic vibration frequency. Thus, taking $D = (a_0)^2\nu$ in Eq. (10.24), we have $\frac{d\phi}{d\rho}_{j=0} = 0.2 \times 10^{-4}$ cm^3/C, which is of the expected order of magnitude. Hence, the deformation potential due to creep is indeed very small.

10.8 Irreversible processes in thermomigration

When a temperature gradient is applied to a homogeneous alloy, the alloy becomes inhomogeneous. This is known as the Soret effect. One of the components in the alloy will diffuse against its concentration gradient, leading to segregation. Finally, it establishes a concentration gradient and reaches a steady state. Thus, in thermomigration, we shall consider the interaction between heat flow and atomic flow. The conjugate forces are X_Q and X_M as shown in Eq. (10.25) and (10.26), respectively.

$$J_M = L_{MM}X_M + L_{MQ}X_Q = C\frac{D}{kT}\left[-T\frac{d}{dx}\left(\frac{\mu}{T}\right)\right] - C\frac{D}{kT}\frac{Q^*}{T}\frac{dT}{dx} \tag{10.25}$$

$$J_Q = L_{QM}X_M + L_{QQ}X_Q = L_{QM}\left[-T\frac{d}{dx}\left(\frac{\mu}{T}\right)\right] + \kappa\frac{dT}{dx} \tag{10.26}$$

where Q^* is the heat of transport which is negative when the fluxes of atoms diffuses from the hot to the cold end since they lose heat, and Q^* is positive when atoms are driven from the cold end to the hot end and gain heat. The last term in the equation of J_Q is Fourier's law and κ is thermal conductivity.

10.8.1 Thermomigration in unpowered composite solder joints

Figure 10.8(a), (b), and (c) show respectively the schematic diagram of a Flip-chip on a substrate, the cross-section of a composite flip-chip solder joint between the chip

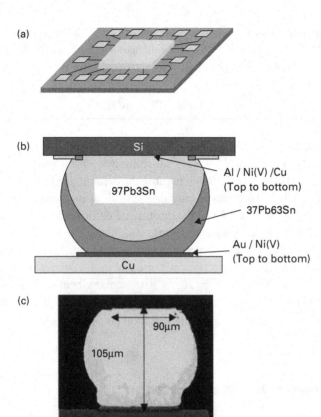

Fig. 10.8   (a) Schematic diagram of a flip-chip on a substrate. (b) The cross-section of a composite flip-chip solder joint between the chip and the substrate. (c) SEM image of a flip-chip solder joint.

and the substrate, and an SEM image of the cross-section of a flip-chip solder joint. In Fig. 10.8(a), the small squares on the substrate are electrical contact pads. The composite solder joint was composed of 97Pb3Sn on the chip side and eutectic 37Pb63Sn on the substrate side. The solder bump has a height of 105 μm. The contact opening on the chip side has a diameter of 90 μm. The tri-layer thin films of UBM on the chip side were Al (∼0.3 μm)/Ni(V) (∼0.3 μm) /Cu (∼0.7 μm). On the substrate side, the bond-pad metal layers were Ni (5 μm) coated with a thin film of Au (0.05 μm).

As a control experiment, a constant temperature annealing of the flip-chip composite samples was performed in an oven at 150 °C and atmospheric pressure for a period of one month as shown in Fig. 10.9. The microstructures of the cross-section of the composite solder joint were examined under optical microscope and SEM. Composition was analyzed using energy dispersive X-ray (EDX) and electron probe microanalysis (EPMA). No mixing between the high-Pb and the eutectic was observed, and the image was nearly the same as that shown in Fig. 10.8(c). According to the eutectic phase

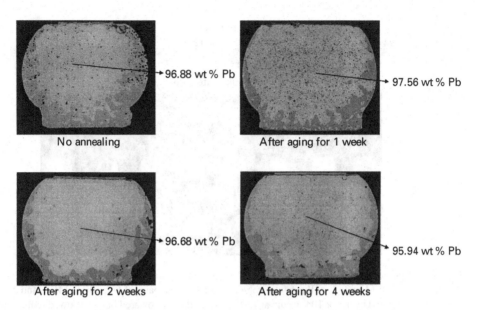

96.88 wt % Pb

No annealing

97.56 wt % Pb

After aging for 1 week

96.68 wt % Pb

After aging for 2 weeks

95.94 wt % Pb

After aging for 4 weeks

Fig. 10.9   SEM cross-sectional images of a composite solder bump under aging at a constant temperature of 150 °C.

diagram of SnPb, at 150 °C, the chemical potential of the high-Pb phase of 97Pb3Sn and the eutectic phase of 37Pb63Sn is about the same. Thus, there is no driving force for intermixing.

To conduct thermomigration by using the temperature gradient induced from joule heating, a set of 24 bumps on the peripheral of the Si chip in a flip-chip sample was tested. Figure 10.10(a) depicts a row of 24 solder bumps from right to left at the peripheral of a chip and each bump has the original microstructure as shown in Figure 10.8(c) before electromigration stressing. Recall that the darker region in the bottom area of each bump is the eutectic SnPb and the brighter region in the top part is 97Pb3Sn.

Electromigration was conducted through only four pairs of bumps in the row of 24 bumps on the peripheral of the chip. They were the pairs of numbers 6/7, 10/11, 14/15, and 18/19 as numbered in Fig.10.10(b). The arrows indicated the electron path during electromigration. The electron current went from one of the contact pads to the bottom of one of the bumps, up the bump to the Al thin-film interconnect on the Si chip, then to the top of the next bump, down the bump, and to the other contact pad on the substrate. It is worth noting that we can pass current through just one pair of bumps or several pairs in the row to conduct electromigration. The joule heating from the Al thin-film line on the chip is the heat source. Due to the excellent thermal conduction of Si, the neighboring unpowered solder joints would have experienced a thermal gradient similar to the pairs stressed by current.

Cross-sectioning was performed after the bump-pair 10 and 11 failed after 5 h of current stressing at 1.6×10^4 A/cm^2 at 150 °C. To study thermomigration, the unpowered

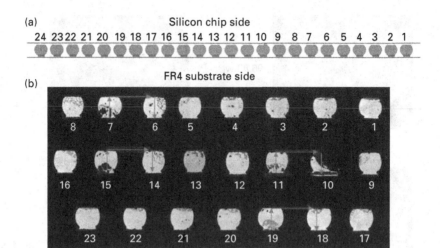

Fig. 10.10   (a) Schematic diagram of a row of 24 solder bumps from right to left at the peripheral of a chip. (b) SEM images of the row of solder bump after the pairs of 6/7, 10/11, 14/15, and 18/19 were stressed by 1.6×10^4 A/cm^2 at 150 °C. The other unpowered bumps showed redistribution of Sn and Pb due to thermomigration.

neighboring bumps were examined. The effect of thermomigration is clearly visible across the entire row of the unpowered solder joints, as shown in Fig. 10.10(b). In all of them Sn has migrated to the Si side, the hot end, and Pb has migrated to the substrate side, the cold end. The redistribution of Sn and Pb or the redistribution of the eutectic phase and the high-Pb phase was caused by the temperature gradient across the solder joints since no current was applied to them. For the unpowered bumps which were the nearest neighbors of the powered bumps, the Sn redistribution is also tilted toward the powered bumps. For example, the powered bump 10 is to the left of the unpowered bump 9, the Sn-rich region in bump 9 is tilted to the left and a void is observed. Then the powered bump 15 is to the right of the unpowered bump 16, the Sn-rich region in bump 16 is tilted to the right. In those bumps further away from the powered bumps, for example from bump 1 to bump 4 and bump 21 to 23, Sn accumulated rather uniformly on the Si side.

The coupling between thermomigration and creep will be presented in Chapter 13.

10.9   Irreversible processes in thermo-electric effects

These are interactions between heat flow and electric flow. The thermal-electrical effects are well known, especially the Seebeck effect and the Peltier effect. They are cross-effects. The Seebeck effect is the generation of an electric flow or electric potential by a temperature gradient. It is the basis of application of thermocouples to measure the temperature of materials. The Peltier effect is the generation of a heat flow by an electric potential gradient. It is the basis of solid-state cooling devices.

The interaction between heat conduction and electric conduction can be represented by a pair of equations of irreversible processes. Because temperature is not constant, it becomes a variable in the forces in these two equations:

$$J_Q = L_{QQ}X_Q + L_{QE}X_E = L_{QQ}T\frac{d}{dx}\left(\frac{1}{T}\right) - L_{QE}T\frac{d}{dx}\left(\frac{\phi}{T}\right)$$

$$J_E = L_{EQ}X_Q + L_{EE}X_E = L_{EQ}T\frac{d}{dx}\left(\frac{1}{T}\right) - L_{EE}T\frac{d}{dx}\left(\frac{\phi}{T}\right)$$

(10.27)

where $L_{QQ} = T\kappa$, $L_{EE} = ne\mu_e$, and L_{QE} and L_{EQ} are coefficients of cross-effects. L_{QE} is for heat flow induced by an electric field, and L_{EQ} is for electric flow induced by a temperature gradient. On the basis of Onsager's reciprocity relationship, $L_{QE} = L_{EQ}$. We shall analyze the two equations for an understanding of thermal-electrical effects. We can rewrite the pair of equations in Eq. (10.27) in terms of dT/dx and $d\phi/dx$ as below:

$$J_Q = (-L_{QQ} + \phi L_{QE})\frac{1}{T}\left(\frac{dT}{dx}\right) - L_{QE}\frac{d\phi}{dx}$$

$$J_E = (-L_{EQ} + \phi L_{EE})\frac{1}{T}\left(\frac{dT}{dx}\right) - L_{EE}\frac{d\phi}{dx}$$

(10.28)

10.9.1 Thomson effect and Seebeck effect

Figure 10.11(a) depicts a single metal wire of a given length, with its two ends kept at two temperatures, T_1 and T_2, where $T_1 > T_2$. Thus, there is a temperature gradient and we expect it to drive an electric flow of charge carriers. Since the wire is open-ended, no electric flow will occur. Thus, in the second equation in Eq. (10.28), we have $J_E = 0$ and

$$0 = (-L_{EQ} + \phi L_{EE})\frac{1}{T}\left(\frac{dT}{dx}\right) - L_{EE}\frac{d\phi}{dx}$$

Thus, we have

$$\frac{\Delta\phi}{\Delta T} = \frac{(-L_{EQ} + \phi L_{EE})}{TL_{EE}}$$

(10.29)

This is called the Thomson effect. It means that there will be an electric potential difference between the ends due to the temperature gradient. In the above, we obtain $\Delta\phi/\Delta T$ by assuming that $J_E = 0$.

However, if we apply an electric current through the wire in Fig. 10.11(a), the applied potential will be increased in one direction but decreased if it is applied in the opposite direction. In addition to the conventional joule heating when electric conduction occurs at constant temperature, there will be an addition or reduction in joule heating due to the Thomson effect.

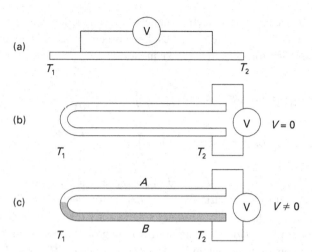

Fig. 10.11   (a) Schematic diagram of a single metal wire of a given length, with its two ends kept at different temperatures, T_1 and T_2, where $T_1 > T_2$. (b) Schematic diagram of a bent wire having two branches. The bent end is placed at T_1 and the other two ends at T_2. The potential difference across the two ends at T_2 is zero, $V = 0$. (c) Schematic diagram of two kinds of metallic wire, A and B, joined at one end. The joined end is kept at T_1, and the unjoined ends at T_2. There is a potential difference between the two open ends at T_2. The joined end is the probe in a thermal couple.

Now we double the length of the wire and bend it at the middle into two branches and place the bent end at T_1 and the other two ends at T_2, as shown in Fig. 10.11(b). While there will be an electric potential difference between the end at T_1 and the two ends at T_2, there will be no net potential difference between the two ends kept at T$_2$ since the potential changes in both branches are the same. If we use a potentiometer to measure the potential between the two ends at T_2, it is zero.

Then, if we replace one branch of the bent wire by a different metal, or if we take two kinds of metallic wire, A and B, and join them at the end kept at T_1, as shown in Fig. 10.11(c), we have a thermal couple. If we place the joined end at a high temperature and the unjoined ends of the couple at a reference temperature of $0\,^\circ$C or room temperature, we have a thermal couple and we can use a potentiometer to measure a potential difference $\Delta\phi$ between the two ends kept at the reference temperature. This is because the potential changes in the two wires are not the same, so a potential difference occurs. If we have calibrated the couple at different temperatures, we obtain the Seebeck coefficient,

$$\frac{\Delta\phi}{\Delta T} = \varepsilon_{AB} = \varepsilon_A - \varepsilon_B$$

where ε_{AB} is the combination of thermal electric properties of wire A and wire B in the couple as shown in Fig. 10.11(c). For convenience, thermal electric properties of individual material of ε_A, ε_B, and others have been measured, so we can choose a pair to serve as a thermal couple to measure temperature in various temperature ranges.

10.9.2   Peltier effect

The Peltier effect is opposite to the Seebeck effect. If we keep the two ends of the sample as shown in Fig. 10.11(c) at constant temperature, and if we apply an electric field, we can transfer heat from one end to the other end. If we let $J_Q = 0$ in the first equation in Eq. (10.28), we have

$$0 = (-L_{QQ} + \phi L_{QE})\frac{1}{T}\frac{dT}{dx} - L_{QE}\frac{d\phi}{dx}$$

Thus, we have

$$\frac{\Delta T}{\Delta \phi} = \frac{TL_{QE}}{(-L_{QQ} + \phi L_{QE})} \tag{10.30}$$

In the above equation, ΔT is proportional to $\Delta \phi$ and L_{QE}. The larger the ΔT, the better the cooling effect. However, due to the existence of ΔT, heat will be transferred from the hot end to the cold end and will reduce the cooling effect. In order to reduce the heat transfer, we will need to reduce heat conductivity in the sample. Yet, for most conductors, heat conductivity is linearly proportional to electric.

References

[1] I. Prigogine, *Introduction to Thermodynamics of Irreversible Processes*, 3rd edn (Wiley-Interscience, New York, 1967).

[2] David V. Ragone, "Nonequilibrium thermodynamics", Ch. 8 of *Thermodynamic of Materials*, Vol. II (Wiley, New York, 1995).

[3] Paul Shewmon, *Diffusion in Solids*, 2nd edn (TMS, Warrendale, PA, 1989).

[4] R. W. Balluffi, S. M. Allen and W. C. Carter, "Irreversible thermodynamics: coupled forces and fluxes", Ch. 2 of *Kinetics of Materials* (Wiley-Interscience, New York, 2005).

[5] J. C. M. Li, "Caratheodory's principle and the thermodynamic potential in irreversible thermodynamics", *J. Phys. Chem.*, **66** (1962), 1414–20.

[6] K. N. Tu, "Electromigration in stressed thin films", *Phys. Rev.* **B45** (1992), 1409–13.

Problems

10.1   What is conjugate force and flux? In heat conduction, what is the conjugate force of heat and the corresponding heat flux?

10.2   Can we apply the heat from joule heating to do work? If not, why not?

10.3   In electromigration of an Al interconnect under a current density of 10^5 A/cm^2, what are the joule heating due to electrical conduction and the joule heating due to electromigration?

10.4   Calculate the critical length of electromigration of Cu when the back stress can be taken to be its elastic limit.

10.5 In thermomigration under a temperature gradient of 3000 °C/cm, what is the driving force of atomic motion?

10.6 In a flip-chip solder joint, the Al interconnect on the chip side has a cross-section of 0.5 μm × 80 μm and a length of 300 μm. When we apply a current density of 10^5 A/cm^2 through the Al interconnect, what is the joule heating? When the current flows through the solder joint which has a cylindrical cross-section of 100 μm in diameter and a height of 100 μm, what will the joule heating in the solder joint be if we assume a uniform current density in the solder joint? Assume the resistivity of Al and solder to be 10^{-6} Ω-cm and 10^{-5} Ω-cm, respectively.

11 Electromigration in metals

11.1 Introduction

In Chapter 1, Section 1.2, we discussed the operation of a metal-oxide-semiconductor FET. There are hundreds of millions or even billions of such transistors on a Si chip the size of a fingernail. To interconnect all these transistors by VLSI circuit technology, multilayers of thin-film interconnect wires made of Al or Cu were used. Electromigration is the most serious and persistent reliability problem in the interconnect structure on a Si chip in microelectronic technology. This is because typically a current density of 10^5 to 10^6 A/cm^2 is conducted by the thin-film wires. Under such high current density, atomic diffusion and rearrangement are enhanced, leading to void formation in the cathode and extrusion in the anode of an interconnect. The void can become an open and the extrusion a short in the circuit. On the other hand, it is worth mentioning that there is no electromigration in an ordinary extension cord used at home and in laboratories. The electric current density in the cord is low, about 10^2 A/cm^2, and also the ambient temperature or room temperature is too low for atomic diffusion to occur in the copper wire in a cord.

The free electron model of conductivity of metals assumes that the conduction electrons are free to move in the metal, unconstrained by the perfect lattice of atoms except for scattering due to phonon vibration and structural defects such as grain boundaries, dislocations, and vacancies in the lattice. The scattering is the cause of electrical resistance and joule heating. When an atom is out of its equilibrium position, for example, a diffusing atom at the activated state, it possesses a very large scattering cross-section and in turn a very large resistance. Nevertheless, when the electric current density is low, the scattering or the momentum exchange between the electrons and the diffusing atom does not enhance the displacement of the latter, and it has no net effect on atomic diffusion. However, the scattering by electrons in a high current density, above 10^4 A/cm^2, enhances atomic displacement of the diffusing atom in the direction of electron flow. The enhanced atomic displacement and the accumulated effect of mass transport under the influence of an electric field (mainly due to a high-density electric current rather than a high electric voltage) is called electromigration [1–8].

It is worth noting that a household cord is allowed to carry only a very low current density, otherwise joule heating will burn the fuse. Yet a thin-film interconnect on a Si chip can carry a much higher current density, which facilitates electromigration. This is because the Si chip is a very good heat conductor and can remove most of the joule heating effectively, hence the interconnect circuit on Si can carry a very high current density without overheating. On the other hand, in a device having a very dense integration of circuits, the heat management or heat removal is a very serious issue, and it is becoming the limiting factor in ultra-large-scale circuit integration in the near-future Si devices. Typically, a device is cooled by a fan or other heat sink in order to maintain the working temperature at around 100 °C.

In VLSI of circuits on a Si device, assuming an Al or Cu thin-film line of 0.5 μm wide and 0.2 μm thick carrying a current of 1 mA, for example, the current density will be 10^6 A/cm^2. Such current density can cause electromigration in the line at the device working temperature of 100 °C and can lead to void formation at the cathode and extrusion at the anode. As device miniaturization demands smaller and smaller interconnects, the current density goes up, as does the probability of circuit failure induced by electromigration. This is the reason why electromigration has been the most persistent and most serious reliability failure in thin-film integrated circuits. It is a subject which has demanded and attracted much attention.

Figs. 11.1(a) and (b) are cross-sectional SEM images of void formation due to electromigration in the upper and the lower ends, respectively, of an interconnecting via between two levels of wires at the cathode end in a dual damascene structure of Cu interconnects. They were detected by a very large resistance increase and had caused open circuit. The kinetic process of such void formation will be discussed later and it was due to the propagation and accumulation of an array of small voids on the upper surface of the Cu interconnect. Electromigration in Cu interconnects is dominated by surface diffusion.

Note that Fig. 11.1 was obtained from a failed device. The failed site was located and then a cross-section of the location was prepared for examination. However, the phenomenon of electromigration can be observed directly from the morphological changes of a set of short Cu strips on a baseline of Ta as shown in Fig. 11.2. The short-strip structure for direct observation of electromigration was developed by Blech [3]. This kind of test structure is called the Blech structure for electromigration, as it has been shown in Fig. 10.5 in Chapter 10. Fig. 11.2(a) shows the morphology of a Cu strip in electromigration with a current density of 5×10^5 A/cm^2 at 350 °C for 99 h. At the cathode end of the strip, a depleted region can be seen, but at the anode end, an extrusion is seen. By conservation of mass, the volume of depletion (void) equals the extrusion in the same strip. The rate of depletion at the cathode can be measured so the drift velocity can be calculated for determining the driving force in electromigration. The drift velocity or the velocity of growth of the depletion was approximately 2 μm/h. Fig. 11.2(b) is an SEM image of the depletion at the cathode of a Cu 2 wt. % Sn strip taken at the same test conditions as the Cu strip shown in Fig. 11.2(a), indicating that the drift velocity or electromigration of the Cu(Sn) strip was much slower.

The phenomenon of electromigration from the response of a set of short Al strips on a baseline of TiN is shown in Fig. 11.3. The Al strips have a line width of 20 μm, a thickness of 100 nm, and lengths of 10, 20, 30, and 85 μm. The applied electric current in the Mo baseline took a detour to go along the Al strips because the latter are paths of low resistance. When the current density and temperature are high enough, atomic

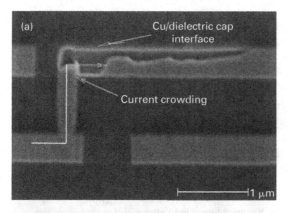

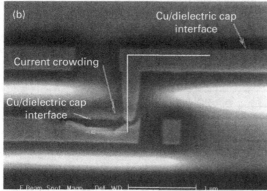

Fig. 11.1   (a) and (b) Cross-sectional SEM images of void formation due to electromigration in the upper end and the lower end, respectively, of an interconnecting via between two levels of wires at the cathode end in a dual damascene structure of Cu interconnects. (Courtesy of Prof. S. G. Mhaisalkar, Nanyang Technological University, Singapore.)

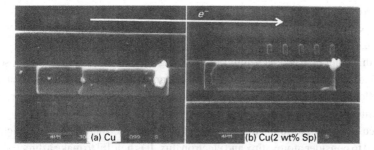

Fig. 11.2   (a) SEM image of a Cu strip on W baseline after electromigration with a current density of 5×10^5 A/cm^2 at 350 °C for 99 h. The drift velocity is approximately 2 μm/h. (b) SEM images of the depletion at the cathode of a Cu 2 wt.% Sn strip taken at the same test conditions as the Cu strip shown in Fig. 11.2(a).

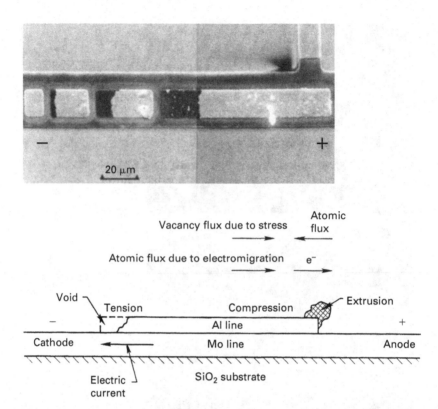

Fig. 11.3 SEM images of electromigration in a set of short Al strips on a baseline of TiN under applied current density of 3.7×10^5 A/cm^2 at 350 °C for 15 h. The Al strips have a line width of 20 µm, a thickness of 100 nm, and lengths of 10, 20, 30, and 85 µm. (b) Schematic diagram depicting void and hillock formation in the strip. The atomic displacement and mass transport are in the same direction as the electron flow: from left to right.

transport occurs and void and extrusion formation can be observed directly. Under an applied current density of 3.7×10^5 A/cm^2 at 350 °C for 15 h, depletion at the cathode end and extrusion at the anode end of the longer strips can be seen in Fig. 11.3(a). It is worth noting that no electromigration damage can be seen in the shortest strip in the left-hand side. Fig. 11.3(b) is a schematic diagram of one of the strips, depicting void and hillock formation; note that the atomic displacement and mass transport are in the same direction as the electron flow from left to right.

Since electromigration is the interaction between electron flow and atomic diffusion, we need to consider atomic flux and electron flux. It is helpful to make a direct comparison of the definition and the variables used to define these two kinds of flux. In Table 11.1, we list them side-by-side for comparison.

To compare electrical conduction and atomic diffusion, we have "force" defined as negative potential energy gradient in atomic diffusion, so the chemical force acting on the diffusing atom is given as $F = -d\mu/dx$, where μ is chemical potential energy. In electrical conduction, $\varepsilon = -d\phi/dx$ is called electric field, rather than electric force.

Table 11.1. Comparison between atomic flux and electron flux

Atomic flux	Electron flux
Chemical potential: μ	Electric potential: Φ
	Applied voltage: V
Chemical force: $F = -\frac{\partial \mu}{\partial x}$	Electric field: $\varepsilon = -\frac{\partial \phi}{\partial x} = -\frac{\partial V}{\partial x}$
Mobility: $M = \frac{D}{kT}$	Electron mobility: $\mu = \frac{e\tau}{m}$
Drift velocity: $v = MF$	Drift velocity: $v = \mu\varepsilon$
Atomic flux: $J = Cv = CMF$	Electron flux: $j = nev = ne\mu\epsilon = \frac{\epsilon}{\rho} = \frac{ne^2\tau\epsilon}{m}$
	(electron current density)
Viscosity (friction coefficient): $1/M$	Resistivity: $\rho = \frac{1}{ne\mu} = \frac{m}{ne^2\tau}$
Divergency: $\nabla \cdot J = \frac{\partial J}{\partial x} + \frac{\partial J}{\partial y} + \frac{\partial J}{\partial z} = -\frac{\partial C}{\partial t}$	Divergency (Gauss' theorem): $\nabla \cdot j = -\frac{\partial (ne)}{\partial t}$

Electric force is given as $e\epsilon$ (or eE). It means that if we place a charge of "e" in the electric field, the charge will feel a force of $F = e\epsilon = -d(e\phi)/dx$ acting on it. Recall that ϕ or V is called electric potential, which is not electrical potential energy. Electrical potential energy is defined as "$e\phi$" or "eV." Again, recall that we express thermal energy as kT and electrical energy as eV. In diffusion, the activation energy can be given as kcal/mole or eV/atom.

To compare mobility of a charge and mobility of an atom, their unit should be the same. Charge carrier mobility μ_e is given by v/ε, where v is velocity and has the unit of cm/s, and ε is electric field and has the unit of V/cm, so the unit of charge mobility is cm^2/V s. Also, charge carrier mobility $\mu_e = e\tau/m^*$, where τ is scattering time and m^* is electron mass. Using Newton's law that $F = ma$, we have the unit of mass as force/acceleration, which is equal to energy s^2/cm^2 or eV s^2/cm^2. Again, we obtain the unit of charge carrier mobility ($= e\tau/m^*$) to be cm^2/V s. On the other hand, atomic mobility is given as D/kT, where D is diffusivity which has the unit of cm^2/s and kT is thermal energy, so the unit of atomic mobility is cm^2/J s or cm^2/eV s. In Chapter 4, we discussed the difference between electron mobility and atomic mobility. This is because electrical force is given as $F = e\varepsilon = -d(e\phi)/dx$, so the charge "$e$" has been cancelled in the carrier mobility.

On electric current and electric current density, note that electric current is defined as the total number of charges passing per unit time through the cross-section of area A of a conductor. Thus, $I/A = j$, where j is defined as electric current density. The unit of I is A or coulomb/s, and the unit of j is A/cm^2 or C/cm^2 s.

For comparison, we see that atomic flux J is defined as $J =$ number of atom/cm^2 s.

The electron flux j or current density is defined as $j =$ number of charge/cm^2 s $=$ C/cm^2 s $=$ A/cm^2. In electromigration, the electrical force acting on a diffusing atom (ion) is expressed in terms of electric field, where the electric field or force is defined as the gradient of electric potential, or is equal to electric current density times resistivity to be discussed later; $Z^* eE = Z^* e\rho j$, where Z^* is the effective charge number of an ion.

11.2   Ohm's law

In Chapter 10, we presented Ohm's law as

$$j = -\sigma \frac{\partial \phi}{\partial x} = -\frac{1}{\rho}\frac{\partial \phi}{\partial x}$$

where σ is conductivity and ρ is resistivity. The above equation can be written as $E = -d\phi/dx = j\rho$. More often, Ohm's law relates the electric current I to the applied voltage drop ΔV in a conductor. It is given as

$$\Delta V = IR$$

where R is the resistance of the conductor. The units of V, I, and R are volt, ampere, and ohm, respectively. Ohm's law indicates that due to the resistance in the conductor, there will be a voltage drop or electric potential drop as an electric current passes through the conductor. Fig. 11.4 depicts the simple circuit of measurement of the resistance of a piece of metal having a length l and cross-sectional area A. We can express the resistance R in term of resistivity ρ of the conductor,

$$R = \rho \frac{l}{A}$$

Now, we rearrange the voltage drop equation of $\Delta V = IR$ as

$$\frac{\Delta V}{l} = \frac{I}{A}\frac{RA}{l} = j\rho$$

Note that $\Delta V/l$ is negative. This is because $l = x_2 - x_1 > 0$ is positive, but $\Delta V = V_2 - V_1 < 0$ is negative due to voltage drop. Since $\Delta V/l = \varepsilon = -d\phi/dx$ is the electrical

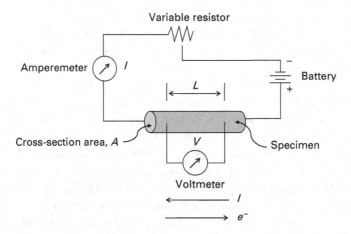

Fig. 11.4   Schematic diagram depicting the circuit of measurement of the resistance of a piece of metal having a length l and cross-sectional area A.

field, we can express Ohm's law as ε (or E) $= j\rho$, where j is electric flux or current density in units of A/cm^2 or C/cm^2 s and ρ is resistivity in units of Ω cm. Since the physical meaning of electric field is that a charge of e in the field, the charge will feel a force of "eE" acting on it; thus, if an ion has an effective charge of Z^*e, it will feel a force of Z^*eE when we place the ion in the field, or we have a force of $Z^*e\rho j$. In electromigration, the meaning of Z^* has been given by Huntington and Grone [1].

11.3 Electromigration in metallic interconnects

Electromigration is the result of a combination of thermal and electrical effects on mass transport. If the conducting line is kept at a very low temperature (e.g. liquid nitrogen temperature), electromigration cannot occur because there is no atomic diffusion because of the very low mobility of atoms, even though there is an electrical driving force. The contribution of thermal effect can be recognized by the fact that electromigration in a bulk eutectic solder bump occurs at about three-quarters of its melting point in absolute temperature, electromigration in a polycrystalline Al thin-film line occurs at less than one-half of its melting point in absolute temperature, and electromigration in a Cu thin-film line having bamboo-type grain structure occurs at about one-quarter of its melting point in absolute temperature. At these homologous temperatures, there are atoms which undergo random walk in the bulk of the solder bump, in the grain boundaries of the Al thin-film line, and on the free surface of the Cu damascene interconnect, respectively, and these are the atoms which take part in electromigration under the applied high current density.

Indeed, we assume the Si device working temperature to be 100 °C, which is about three-quarters of the melting point of solders, slightly less than half of the melting point of Al, and about one-quarter of the melting point of Cu. In this homologous temperature scale, lattice diffusion, grain-boundary diffusion, and surface diffusion occur predominantly at three-quarters, half, and a quarter of the absolute temperature of solder, Al interconnect, and Cu interconnect, respectively.

Table 11.2 lists the melting points and diffusivities which are relevant to the electromigration behaviors in Cu, Al, and eutectic SnPb. The diffusivities for Cu and Al were calculated on the basis of the following equations from the master plot of $\log D$ versus T_m/T for fcc metals [9].

$$D_l = 0.5 \exp\left(-34T_m/RT\right)$$
$$D_{gb} = 0.3 \exp\left(-17.8T_m/RT\right) \qquad (11.1)$$
$$D_s = 0.014 \exp\left(-13T_m/RT\right)$$

where D_l, D_{gb}, and D_s are lattice diffusivity, grain-boundary diffusivity, and surface diffusivity, respectively, T_m is the melting point, and the units of 34 T_m, 17.8 T_m, and 13 T_m are in cal/mole. As shown in Table 11.2, at 100 °C the lattice diffusivity of Cu and Al is insignificantly small, and the grain-boundary diffusivity of Cu is three orders of

Table 11.2. Diffusivity of Al, Cu, and SnPb

	Melting point (K)	Temperature ratio 373K/T_m	Diffusivity at 100 °C (cm²/s)	Diffusivity at 350 °C (cm²/s)
Cu	1356	0.275	Lattice $D_l = 7 \times 10^{-28}$ Grain boundary $D_{gb} = 3 \times 10^{-15}$	$D_l = 5 \times 10^{-17}$ $D_{gb} = 1.2 \times 10^{-9}$
Al	933	0.4	Surface $D_s = 10^{-12}$ Lattice $D_l = 1.5 \times 10^{-19}$ Grain boundary $D_{gb} = 6 \times 10^{-11}$	$D_s = 10^{-8}$ $D_l = 10^{-11}$ $D_{gb} = 5 \times 10^{-7}$
Eutectic SnPb	456	0.82	Lattice $D_l = 2 \times 10^{-9}$ to 2×10^{-10}	Molten state $D_l > 10^{-5}$

magnitude smaller than the surface diffusivity of Cu. At 350 °C the difference between surface diffusivity and grain-boundary diffusivity of Cu is much less, indicating that we cannot ignore the latter. The lattice diffusivity of eutectic SnPb (not a fcc metal) at 100 °C given in Table 11.2 is an average value of tracer diffusivity of Pb and Sn in the alloy. It depends strongly on the lamellar microstructure of the eutectic sample. Since a solder joint of 100 μm in diameter typically has several large grains, the smaller diffusivity is better for our consideration. The surface diffusivity of Cu, grain-boundary diffusivity of Al, and lattice diffusivity of the solder are actually rather close at 100 °C. To compare atomic fluxes transported by these three kinds of diffusion in a metal, we should have multiplied the diffusivity by their corresponding cross-sectional area of path of diffusion, but the outcome is the same.

Table 11.2 also shows the homologous temperature of Cu, Al, and solder at the device operation temperature of 100 °C; they are 0.25, 0.5, and 0.82, respectively. The homologous temperature of solder is very high. It means that the application of a solder joint in devices will be affected by the high-temperature properties of the solder, or affected by thermally activated processes such as diffusion. For example, the mechanical properties of solder joints at the device working temperature will be influenced greatly by creep. This is a very important point to remember when we study the mechanical and other physical properties of solder joints.

In fcc metals such as Al and Cu, atomic diffusion is mediated by vacancies. When a flux of Al atoms is driven by electromigration to go to the anode, it requires a flux of vacancies to go to the cathode in the opposite direction. If we can stop the vacancy flux, we will be able to stop electromigration. To maintain a vacancy flux, we must supply vacancies from vacancy sources continuously. Hence, we can stop a vacancy flux by removing the sources or supplies of vacancies. Within a metal interconnect, dislocations and grain boundaries are sources of vacancies, but the free surface is generally the most important and most effective source of vacancies. For Al, its native oxide is protective, which means that the interface between the metal and its oxide is not a good source or sink of vacancies.

This is also true for Sn. Thus, Al and Sn do not have free surfaces as sources and sinks of vacancies. When vacancies are to be removed without replenishment or to be added without effective sinks, equilibrium vacancy concentration cannot be maintained, as a consequence, back stress will be generated. This is a topic to be discussed in Section 11.7.

If the atomic flux or the opposing vacancy flux is uniform and continuous in the interconnect, i.e. the anode can supply vacancies and the cathode can accept them continuously, and if there is no flux divergence in between the cathode and the anode, vacancy concentration is in equilibrium everywhere, then there will be no electromigration-induced damage such as void and extrusion formation. In other words, without mass flux divergence, no electromigration damage will occur in an interconnect when fluxes of atoms and vacancies can pass through it uniformly. Hence, atomic or mass flux divergence is a necessary condition concerning electromigration failure in real devices. The most common mass flux divergences are the triple points of grain boundaries and also the interfaces between dissimilar materials. Since a flip-chip solder joint has two interfaces, one at the cathode and the other at the anode, they are the common failure sites, especially the cathode interface where accumulation of vacancies occurs leading to the formation of voids. However, in Chapter 12 on electromigration-induced failure, we shall emphasize that mass flux divergence is a necessary but insufficient condition of structure failure; for structure failure to occur, we must consider whether or not there is lattice plane shift.

In summary, electromigration involves both atomic and electron fluxes. Their distribution in interconnects is the most important consideration when electromigration damage is of concern. In a region where both distributions are uniform, there will be electromigration, but there will be no electromigration-induced damage because of lack of flux divergence.

Concerning atomic or vacancy flux, the first most important factor is the temperature scale shown in Table 11.2. Atomic diffusion must be thermally activated. The second most important factor is the design and processing of the interconnect structure. Non-uniform flux distribution or divergence in interconnects occurs at microstructure irregularities such as grain-boundary triple points and interphase interfaces, and they are the sites of failure initiation. The third most important factor is the sources and sinks of vacancies and lattice plane motion. The physical analysis of the effects of microstructure, solute, and stress on electromigration in interconnects of Al and Cu will be discussed in Chapter 12. The statistical MTTF analysis on the basis of void formation in flip-chip solder joints due to electromigration will be presented in Chapter 15.

Concerning the electron flux, the current density must be high enough for electromigration to occur. Because FETs in devices are turned on by pulsed direct current, we consider only electromigration under direct current and pulsed current for transistor-based devices such as computers. While a uniform current distribution is expected in straight lines, non-uniform current distribution or current crowding occurs at corners where a conducting line turns, at interfaces where conductivity changes, and also around voids or precipitates in a metallic matrix.

It has been widely observed that electromigration-induced damages tend to occur at low current density regions, rather at high current density regions. This is against our intuition on the basis of Huntington's electron wind force model. To explain the discrepancy, we need to consider a new driving force of electromigration, the current density gradient force, to be discussed in Section 11.10.

One of the most important factors to affect electromigration in a flip-chip solder joint is the unique line-to-bump geometry, where a very large change of current density takes place from the line to the bump, in turn leading to a large current crowding at the contact between the line and the bump. The effect of current crowding on electromigration-induced damage in flip-chip solder joints will be discussed in Chapter 15.

11.4   Electron wind force of electromigration

The electrical force acting on a diffusing atom (ion) is taken to be [1]

$$F_{em} = Z^* eE = (Z_{el}^* + Z_{wd}^*)eE \qquad (11.2)$$

where e is the charge of an electron and E is the electric field, and Z^* is the effective charge number of electromigration and it consists of Z_{el}^* and Z_{wd}^*. Z_{el}^* can be regarded as the nominal valence of the diffusing ion in the metal when the dynamic screening effect is ignored; it is responsible for the electric field effect and $Z_{el}^* eE$ is called the direct force, and it is acting in the direction opposing electron flow. The magnitude of Z_{el}^* is taken to be the nominal number of valence electrons of the metal atom. Z_{wd}^* is an assumed charge number representing the effect of momentum exchange between electrons and the diffusing ion, and $Z_{wd}^* eE$ is called the electron wind force, and it is acting in the same direction as electron flow. Generally, Z_{wd}^* is found to be of the order of 10 for a good conductor, so the electron wind force is much greater than the direct force for electromigration in metals. Hence, in electromigration, the enhanced flux of atomic diffusion is moving in the same direction as electron flow.

To appreciate the electron wind force, we depict in Fig. 11.5(a) the configuration of a shaded Al atom and a neighboring vacancy in a fcc lattice structure before they exchange positions along a <110> direction. They have four nearest neighbors in common, including the two shown by the broken circles, the one on top and the other one on the bottom of the close-packed atomic plane. When the shaded atom is diffusing halfway towards the vacancy as depicted in Fig. 11.5(b), it is at the activated state, sitting at a saddle point while displacing the four nearest-neighbor atoms. Since the saddle point is not part of the lattice periodicity, the atom at the saddle position is out of its equilibrium position and will experience a greater electron scattering. Because the neighboring atoms are displaced too, the cluster of atoms, including the diffusing atoms and the neighbors, will make a much larger contribution to the resistance to electrical current than a normal lattice atom. Hence, the diffusing atom experiences a greater electron scattering and a greater electron wind force which will push it to an equilibrium position, the vacant site,

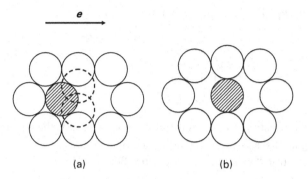

Fig. 11.5   Schematic diagram of electromigration of a diffusing atom (a) before, and (b) at the activated state, when it possesses a very large scattering cross-section.

in order to reduce the resistance. The diffusion of the atom is enhanced in the direction of the electron flow. We emphasize that the diffusing atom will experience the electron wind force, not just at the saddle point, but all the way from the beginning to the end in the entire jumping path of the diffusion.

To estimate the electron wind force, the ballistic approach to the scattering process was developed by Huntington and Grone [1]. The model postulates a transition probability of free electrons per unit time from one free electron state to another free electron state due to the scattering by the diffusing atom. The force, i.e. the momentum transfer per unit time, is calculated by summing over the initial and final states of the scattered electrons. The step-by-step derivation of the model is presented in *Appendix C*. A simple derivation is given below.

During elastic scattering of electrons by a diffusing atom, the system momentum is conserved. The average change in electron momentum in the transport direction is equal to $m_e <v>$, rather than $2m_e <v>$, where m_e is the electron mass and $<v>$ is the mean velocity of electrons in the direction of current flow. This is because the atom moves. The force on the moving ion induced by the scattering is

$$\overrightarrow{F_{wd}} = \frac{m_e <v>}{\tau_{col}} \tag{11.3}$$

where τ_{col} is the mean time interval between two successive collisions. The net momentum lost per second per unit volume of electrons to the diffusing ions is then $nm_e <v>/\tau_{col}$, and the force on a single diffusing ion is

$$\overrightarrow{F_{wd}} = \frac{nm_e <v>}{\tau_{col} N_d} \tag{11.4}$$

where n is the electron density and N_d is the density of diffusing ions. The electron current density can be written as

$$j = -ne <v> \tag{11.5}$$

Substituting $<v>$ in Eq. (11.5) into Eq. (11.4), we obtain

$$\overrightarrow{F_{wd}} = -\frac{m_e j}{e\tau_{col} N_d} = -\frac{m_e}{ne^2\tau_{col}} \frac{neE}{\rho N_d}$$

$$= -\left[\frac{\rho_d}{N_d}\right]\left[\frac{n}{\rho}\right] eE \qquad (11.6)$$

where $\rho = E/j$ is the resistivity of a conductor, $\rho_d = m/ne^2\tau_{col}$ is the metal resistivity due to the diffusing atoms, and E is the applied electrical field.

Aside from the electron "wind" force, the electric field E will produce a direct force on the diffusing ion to be given by

$$\overrightarrow{F_d} = Z^*_{el_z} eE \qquad (11.7)$$

where Z^*_{el} can be regarded as the nominal valance of the metal ion when the dynamical scattering effect around the ion is ignored. So the total force will be

$$\overrightarrow{F_{EM}} = \left[Z^*_{el} - Z\left[\frac{\rho_d}{N_d}\right]\left[\frac{N}{\rho}\right]\right] eE \qquad (11.8)$$

where N is the atomic density of the conductor and $n = NZ$ is used. Equation (11.8) can be written as

$$\overrightarrow{F_{EM}} = Z^* eE \qquad (11.9)$$

and

$$Z^* = \left[Z^*_{el} - Z\left[\frac{\rho_d}{N_d}\right]\left[\frac{N}{\rho}\right]\right] \qquad (11.10)$$

Z^* is called the effective charge number of the ion in electromigration.

The electromigration model based on the ballistic scattering of electrons is the first and simplest model of the phenomenon of electromigration. Theoretical understanding is further developed by the contributions of numerous researchers to date. In spite of these theoretical developments, the model developed by Huntington and Grone, especially the drift velocity of electromigration, is employed as the theoretical basis in nearly all experimental studies of electromigration. For example, the drift velocity is taken as

$$v_d = MF = \frac{D}{kT} Z^* ej\rho \qquad (11.11)$$

It indicates that if we measure the drift velocity using short strips and know the diffusivity, D, we will be able to calculate Z^*.

The above model shows that the effective charge number can be given in terms of specific resistivities of a diffusing atom and a normal lattice atom,

$$Z^*_{wd} = -\frac{\rho_d/N_d}{\rho/N} \frac{m_0}{m^*} \qquad (11.12)$$

where $\rho = m_0/ne^2\tau$ and $\rho_d = m^*/ne^2\tau_d$ are the resistivity of the equilibrium lattice atoms and the diffusing atoms, respectively; m_0 and m^* are the free electron mass and effective electron mass, respectively, and we can assume that they are equal; and τ and τ_d are the relaxation times of a lattice atom and a diffusing atom, respectively. In a fcc lattice, there are 12 equivalent jump paths along the <110> directions. For a given current direction, the average specific resistivity of a diffusing atom must be corrected by a factor of one-half. By rewriting Eq. (11.10), we have

$$Z^* = -Z\left[\frac{1}{2}\frac{\dfrac{\rho_d}{N_d}}{\dfrac{\rho}{N}}\frac{m_0}{m^*} - 1\right] \tag{11.13}$$

where Z_{el} has been taken as Z, the nominal valence of the metal atom. This is the Huntington and Grone equation for the effective charge number of electromigration. To calculate Z^*, we need to know the specific resistivity of a diffusing atom, or its ratio to that of a lattice atom.

11.5   Calculation of the effective charge number

If the specific resistivity of an atom in a metal is assumed to be proportional to the elastic cross-section of scattering, which in turn is assumed to be proportional to the average square displacement from equilibrium, or $<x^2>$, the cross-section of a normal lattice atom can be estimated from the Einstein model of atomic vibration in which the energy of each mode is

$$\frac{1}{2}m\omega^2 <x^2> = \frac{1}{2}kT \tag{11.14}$$

where the product $m\omega^2$ is the force constant of the vibration, and m and ω are atomic mass and angular vibrational frequency, respectively.

To obtain the cross-section of scattering of a diffusing atom, $<x_d^2>$, we assume that the atom and its surrounding atoms as shown in Fig. 11.5(b) have acquired the motion energy of diffusion, ΔH_m, which is independent of temperature,

$$\frac{1}{2}m\omega^2 <x_d^2> = \Delta H_m \tag{11.15}$$

Then, the ratio of the last two equations gives the ratio of cross-section of scattering,

$$\frac{<x_d^2>}{<x^2>} = \frac{2\Delta H_m}{kT} \tag{11.16}$$

It shows that the ratio varies inversely with temperature. This dependence comes from the well-known fact that the resistivity of normal metals varies linearly with temperature

Table 11.3. Comparison of the measured and calculated values of Z^*

Metal	Measured Z^*	Temp. (°C)	ΔH_m (eV)	Calculated Z^*
Monovalent				
Au	-9.5 to -7.5	850 to 1000	0.83	-7.6 to -6.6
Ag	-8.3 ± 1.8	795 to 900	0.66	-6.2 to -5.5
Cu	-4.8 ± 1.5	870 to 1005	0.71	-6.3 to -5.4
Trivalent				
Al	-30 to -12	480 to 640	0.62	-25.6 to -20.6
Quadrivalent				
Pb	-47	250	0.54	-44

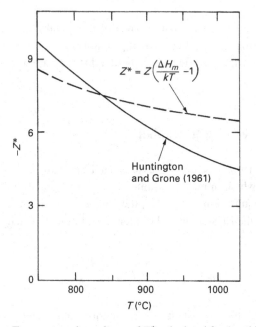

Fig. 11.6   Temperature dependence of Z^* calculated for Au; this agrees well with the measured values.

above the Debye temperature. Substituting the last equation into the equation of Z^*, we obtain

$$Z^* = -Z\left[\frac{\Delta H_m}{kT}\frac{m_0}{m^*} - 1\right] \qquad (11.17)$$

In the above equation, the numerical factor of $1/2$ has been canceled when the probability of averaging jumps in a given direction (i.e. the direction of electron flow) from among the 12<110> paths in a fcc metal has been measured. Now the value of Z^* can be calculated at a given temperature by using the last equation. The calculated values of Z^* agree quite well with those measured for Au, Ag, Cu, Al, and Pb; see Table 11.3. For example, at 480 °C, the measured and calculated Z^* for Al (taking $\Delta H_m = 0.62$ eV/atom) are about -30 and -26, respectively. The temperature dependence of Z^* calculated for Au is also found to agree well with the measured values; see Fig. 11.6.

Roughly speaking, we can see from Fig. 11.5(b) that the diffusing atom at the activated state processes a scattering cross-section of about 10 atoms, therefore its effective charge number will be approximately equal to $10Z$, where Z is its nominal valence number, so we have the order of magnitude of Z^* of -10, -30, and -40 for Cu (noble metal), Al, and Pb (or Sn), respectively.

11.6   Effect of back stress and measurement of critical length, critical product, and effective charge number

In Chapter 10, Fig. 10.6 depicts a schematic diagram of a short Al strip patterned on a baseline of TiN. In electromigration of the short strip, a high current density of electrons goes from left to right and it transports Al atoms from the cathode to the anode, leading to depletion or void formation at the cathode and pile-up or hillock formation at the anode. Furthermore, it was found that the longer the strip, the more the depletion at the cathode side in electromigration. But below a critical length, there was no observable depletion as shown in Fig. 11.3(a) [1, 3].

The dependence of depletion on strip length was explained by the effect of back stress. In essence, when electromigration transports Al atoms in a strip from the cathode to the anode, the latter will be in compression and the former in tension. On the basis of the Nabarro–Herring model of equilibrium vacancy concentration in a stressed solid to be discussed in Chapter 14, Section 14.3, the tensile region has more vacancies and the compressive region has fewer vacancies than the unstressed region, so there is a vacancy concentration gradient decreasing from the cathode to the anode, as depicted in Fig. 10.7. The gradient induces an atomic flux of Al diffusing from the anode to the cathode, and it opposes the Al flux driven by electromigration from the cathode to the anode. The vacancy concentration gradient depends on the length of the strip; the shorter the strip, the greater the gradient. At a certain short length defined as the critical length, the gradient is large enough to balance electromigration so that no depletion at the cathode and no extrusion at the anode occur.

In analyzing this stress effect on electromigration, irreversible processes have been presented by combining electrical and mechanical forces on atomic diffusion, as discussed in Chapter 10, Section 10.7.1. The critical length in Eq. (10.21) for Al strips was calculated to be about 3 μm, which is of the right order of magnitude, but shorter than the experimental value found in between 10 to 20 μm due to grain-boundary diffusion.

The temperature dependence of the critical length can be examined by substituting Z^* into Eq. (10.21), and we have

$$\Delta x = \frac{\Delta \sigma \Omega}{-Z\left[(\Delta H_m/kT)(m_0/m^*) - 1\right]ej\rho} \tag{11.18}$$

For normal metals whose electrical resistivity increases linearly with temperature above the Debye temperature, the last equation shows that the critical length is rather insensitive to temperature, provided that $\Delta H_m \gg kT$ so that the unity in the denominator can be dropped.

To calculate the effective charge number, we can use Eq. (10.21) provided that we have measured Δx and $\Delta \sigma$. On the other hand, if we use a very long strip and ignore the back stress effect and measured the drift velocity, we have

$$v_d = MF = \frac{D}{kT} Z^* e j \rho = \frac{D_0}{kT} \exp\left(-\frac{Q}{kT}\right) Z^* e j \rho \qquad (11.19)$$

It indicates that if we know the diffusivity, D, we will be able to calculate Z^*. Furthermore, if we measure the drift velocity at several temperatures, we can take ln of the last equation and obtain

$$\ln(v_d T) = \ln\left(\frac{D_0}{k} e Z^* j \rho\right) - \frac{Q}{kT} \qquad (11.20)$$

Thus, by plotting ln $(v_d T)$ vs. $1/kT$, we can determine the activation energy of the diffusion process in the electromigration.

11.7   Why is there back stress in an Al interconnect?

While the Blech structure has been used very often in experimental studies of electro-migration in Al strips, there has been a question about the origin of the back stress. If we confine a short strip by rigid walls as shown in Fig. 11.7, we can easily envisage the compressive stress at the anode induced by electromigration. In a fixed or constant volume of V at the anode, the stress change in the volume by adding atoms or adding ΔV into it by electromigration is

$$\Delta \sigma = -B \frac{\Delta V}{V} = -B \frac{\Delta V / \Omega}{V / \Omega} - B \frac{\Delta C}{C} \qquad (11.21)$$

where B is bulk modulus and Ω is atomic volume. The negative sign indicates that the stress is compressive. In other words, we are adding atomic volume into the fixed

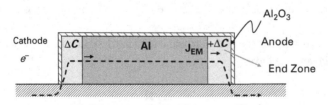

Fig. 11.7   Short strip confined by rigid walls.

volume. If the lattice plane in the fixed volume cannot shift and if the fixed volume cannot expand, a compressive stress will occur. When more and more Al atoms, say n Al atoms, diffuse into the volume, V, the stress in Eq. (11.21) increases. Theoretically, a fixed volume means the constraint of a constant volume. Thus, the implicit assumption about the origin of back stress is the assumption of a constant volume constraint. Why we have such a constraint in Al short strips will be explained below.

In a fixed volume confined within rigid walls, the compressive stress increases with the addition of atoms if no lattice shift is allowed. However, in short-strip experiments, there are no rigid walls to cover the Al strips, except native oxide. How can the back stress build up at the anode if the native oxide is not a rigid wall?

In Chapter 5, Section 5.3.2.1, we mentioned that in the classic Kirkendall effect of interdiffusion in a bulk diffusion couple of A and B, while more A atoms diffuse into B, no stress was assumed in the analysis of Darken's model of interdiffusion. Since more A atoms are diffusing into B, we might expect that there will be a compressive stress in B, on the basis of the constant volume argument given in the above. However, Darken has made a key assumption that vacancy concentration is in equilibrium everywhere. Vacancies (or vacant lattice sites) can be created and/or annihilated as needed in the sample, leading to lattice shift; in other words, lattice sites and lattice planes can migrate out of the fixed volume, so there is no stress and no void formation. Thus, if we assume a constraint of constant volume, we must allow the excess lattice sites and planes to migrate out of the volume so that no stress is generated. Otherwise, stress will build up.

A plausible explanation of the back stress in Al short strips is that since the Al thin film has a native and protective oxide on the surface, the native oxide has removed the sources and sinks of vacancies from the surface. Therefore, when electromigration drives atoms into the anode region, the out-diffusion of vacancies will reduce the equilibrium vacancy concentration in the anode region quickly if there is no vacancy source to replenish it. Furthermore, if the oxide ties down the end of lattice planes and prevents them from moving, a compressive stress will be generated. The oxide is effective because the Al film is thin. This is the basic mechanism of back-stress generation in Al interconnects.

On the basis of the above discussion, it is clear that the origin of back stress, in turn the existence of a critical product or critical length of electromigration in Al short strips, depends on the effectiveness of sources and sinks of vacancies of the surface oxide in the samples and lattice shift. If the sources and sinks are effective and lattice shift can occur, as in the assumption of Darken's model of interdiffusion, there will be no back stress and no critical product of electromigration.

Electron wind force can be regarded as a driving force of atomic diffusion, and the latter is a thermally activated process. Theoretically, even at 1 K, atomic diffusion can take place except that the probability or the frequency of exchange jumps will be infinitely small, so does electromigration. However, in real devices, it is not electromigration itself but rather the structural damage induced by electromigration that is of concern, and the damage should not occur within the lifetime of the device. The Blech short-strip structure has enabled us to see electromigration-induced damage of void formation at the cathode

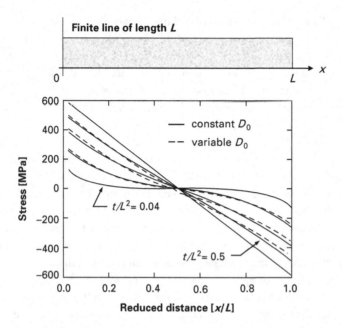

Fig. 11.8   Solution for a finite line and the manner of stress build-up as a function of time.

and hillock formation at the anode very conveniently. Indeed, there is a threshold current density, a back stress, and a critical length of Al short strip, but they are unique because of the surface oxide on the thin-film Al short strips. For a Cu interconnect, the situation is different, since it has no protective oxide.

The time-dependence of stress build-up in a short strip by electromigration can be obtained by solving the continuity equation, since stress is energy density and a density function obeys the continuity equation, and we can convert ΔC to $\Delta \sigma$ from the last equation,

$$\frac{C}{B}\frac{\partial \sigma}{\partial t} = -\frac{D}{kT}\frac{\partial^2 \sigma}{\partial x^2} - \frac{D}{BkT}\left(\frac{\partial \sigma}{\partial x}\right)^2 - \frac{CDZ^* eE}{BkT}\frac{\partial \sigma}{\partial x} \tag{11.22}$$

The solution for a finite line and the manner of stress build-up as a function of time is shown in Fig. 11.8 [10, 11]. Clearly, in the beginning of electromigration the back stress is non-linear along the length of the strip, represented by the curved lines. In reality, the build-up is asymmetrical since the hydrostatic tensile stress at the cathode can hardly be developed.

11.8   Measurement of back stress induced by electromigration

Serious efforts have been dedicated to the measurement of back stress in Al strips during electromigration. It is not an easy task since the strip is thin and narrow; typically, it is only a few hundred nanometers thick and a few microns wide, so a very high intensity and

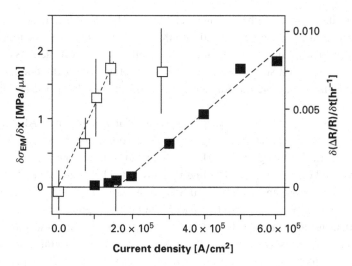

Fig. 11.9 Steady-state rate of resistance increase, $\delta(\Delta R/R)/\delta t$, and the electromigration-induced steady-state compressive stress gradient, $\delta\sigma EM/\delta x$, are plotted against current density.

focused X-ray beam is needed in order to determine the strain in Al grains by precision lattice parameter measurement. Microbeam X-ray diffraction using synchrotron radiation has been employed to study back stress. White X-rays of $10\,\mu m \times 10\,\mu m$ beam from the National Synchrotron Light Source (NSLS) at National Brookhaven Laboratory were used to study electromigration-induced stress distribution in pure Al lines by Cargill *et al.* [12]. The line was 200 μm long, 10 μm wide, and 0.5 μm thick with 1.5 μm SiO$_2$ passivation layer on top, 10 nm Ti/60 nm TiN shunt layer at the bottom, and 0.2 μm thick W pads at both ends which connected the line to contact pads. The electromigration tests were performed at 260 °C. The results of steady-state rate of resistance increase, $\delta(\Delta R/R)/\delta t$, and the electromigration-induced steady-state compressive stress gradient, $\delta\sigma_{EM}/\delta x$, versus current density are shown in Fig. 11.9. No electromigration occurred below the threshold current density j_{th} of 1.6×10^5 A/cm^2. Below the threshold current density, the electromigration-induced steady-state stress gradient increased linearly with current density, wherein the electron wind force was counter-balanced by the mechanical force, so no electromigration drift was observed.

The microbeam X-ray diffraction apparatus at the Advanced Light Source (ALS) in Lawrence Berkeley National Laboratory is capable of delivering white X-ray beams (6–15 keV) focused to 0.8 to 1 μm by a pair of elliptically bent Kirkpatrick–Baez mirrors. In the apparatus, the beam can be scanned over an area of 100 $\mu m \times 100\mu m$ in steps of 1 μm [13, 14]. If the diameter of the grains in the strip is about 1 μm, each grain can be treated as a single crystal with respect to the microbeam. Structural information such as stress/strain and orientation can be obtained by using white-beam Laue diffraction. Laue patterns were collected with a large area (9×9 cm^2) charge-coupled device (CCD) detector with an exposure time of 1 s or longer, from which the orientation and strain tensor of each illuminated grain can be deduced and displayed by software. The resolution of the white microbeam Laue technique is 0.001 % strain. In addition, a four-crystal monochromator can be inserted into the beam to produce monochromatic light for diffraction. The combined white and monochromatic beam diffractions are capable of determining the total strain–stress tensor in each grain. The technique and applications of scanning X-ray microdiffraction (μSXRD) has been described by MacDowell et al. [13].

The transient stress state in the early stage of electromigration in an Al–0.5 wt.% Cu interconnect has been studied by using in-situ synchrotron radiation microbeam X-ray diffraction at ALS by Wang et al. [12]. The dimensions of the line were 4.1 μm wide, 30 μm long, and 0.75 μm thick. The line is sandwiched between 10 nm Ti on top and 45 nm Ti at the bottom and with 0.7 μm SiO_2 passivation layer. Both ends of the line were connected by W vias to contact pads. The electromigration was carried out at 224 °C with a current density of 1×10^6 A/cm^2 up to 25 h. The polygonization by active dislocation slip systems as a function of stress and the rotation of grains in the cathode and anode regions were characterized by X-ray microdiffraction Laue patterns using the scan of a micron-size white beam. Electromigration-induced plastic deformation in grains resulted in broadening of the Laue spots corresponding to the bending due to polygonization of the geometrically necessary dislocations to form subgrain boundaries. Analysis of the Laue spot broadening patterns allows for the identification of the active dislocation slip systems that produced the bending. The dislocations are edge-type and are oriented within 5° with respect to the current flow direction. It was proposed that the dislocations are aligned in this way in order to minimize the electrical resistance of the dislocations.

It is worth mentioning that while dislocation slip was observed, not much dislocation climb was found under electromigration. Yet stress relaxation due to vacancy creation and annihilation and lattice shift requires dislocation climb. The lack of dislocation climb indicates that the back stress induced by electromigration will build up with time in the Al(Cu) interconnects. After 25 h stressing, the strain was found to be higher than the elastic limit and plastic deformation occurred.

Whether back stress exists in electromigration in a Cu damascene structure at the device working temperature has not been confirmed. If electromigration occurs by surface diffusion in a Cu damascene structure, we need a mechanism of back-stress generation in the bulk of the structure induced by surface diffusion. Without a protective surface oxide, surface diffusion occurs in Cu, so it means that the surface is a good source and sink of vacancies.

11.9   Current crowding

Interconnect in VLSI technology is a three-dimensional multilevel structure. When electrical current turns, e.g. at vias in the three-dimensional structure, where the current passes from one level of interconnect to another, current crowding occurs and it affects electromigration significantly. We postulate that defects such as vacancies and solute atoms will have a higher potential in the high current density region than in the low

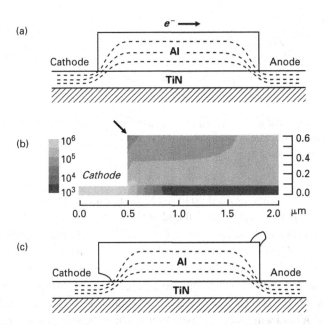

Fig. 11.10 (a) Sketch of the cross-section of the well-known Blech–Herring short-strip test structure of electromigration. (b) Simulation of the current crowding picture. (c) If a void nucleates and grows in the high current density region, it will not be able to deplete the cathode completely.

current density region. The equilibrium concentration of vacancies and solute atoms in the high current density region is lower than that in the low-current density region. The potential gradient in the current crowding region provides a driving force to push these excess defects from the high current density region to the low current density region. As a consequence, the voids tend to form in the low current density region rather than in the high current density region. In other words, failure tends to occur in a low current density region in three-dimensional interconnects.

Figure 11.10(a) is a sketch of the cross-section of the well-known Blech short-strip test structure of electromigration. We assume that electrons go from the left (cathode) to the right (anode). The TiN baseline has a higher resistance than the short Al stripe, so the electrons will make a detour from the TiN into the Al because the latter is a better conductor. In the region where the current enters or leaves the Al, current crowding occurs. The current crowding on the left half of the strip has been simulated and shown in Fig. 11.10(b), where the arrow indicates that the upper left corner and its neighborhood are the low current density regions. Clearly, the lower left corner of the Al–TiN interface, where current crowding occurs, is the high current density region.

Many SEM images have shown that void formation occurs at the cathode of the strip as a result of electromigration. If a void nucleates and grows in the lower left corner of the high current density region, as depicted in Fig. 11.10(c), it will not be able to extend to the low current density region in the upper left corner because the void is an open, and the current entering the Al will be pushed back toward the anode. In order to deplete

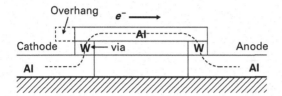

Fig. 11.11   Sketch of the cross-section of a two-level Al interconnect structure connected by W vias. One way to delay the wear-out failure is to add an overhang of the Al interconnect above the W via, as shown by the broken line.

the entire cathode, the vacancies must go to the low current density region, so the void must start from the upper corner of the left end of the strip.

Figure 11.11 is a sketch of the cross-section of a two-level Al interconnect structure connected by W vias. Again it is assumed that electrons go from left to right and current crowding occurs in passing through the vias. Since atomic diffusion in W is much slower than that in Al, the W–Al interface is a flux divergence plane of diffusion, where more Al atoms are leaving. The reverse flux of vacancies would lead to vacancy condensation near the interface. A void will form above the W via. When the void size is large enough to cover the entire via, a circuit open occurs. In the microelectronic industry, this has been called the wear-out mechanism of failure. What is very interesting is that the way to delay the wear-out failure adopted by the industry is to add an overhang of the Al interconnect above the W via, as shown by the arrow and the broken line in Fig. 11.11. The overhang provides an additional volume or reservoir for void growth, so it can lengthen the MTTF. However, it is implicitly assumed in this remedy that vacancies will go to the low current density region of the overhang. Actually, there is no electromigration in the overhang. Why do vacancies go there?

Often it is assumed that there is a stress gradient to drive the vacancies to the low current density region in Fig. 11.10. We should examine whether or not the void nucleation at the upper left corner can be explained by stress-migration. If we follow the electron wind force, vacancies are being driven to the high-current density region (we could assume it to be in tension), a concentration gradient of vacancy is created between the high and low current density regions. The latter can be assumed to have no stress. The stress gradient will drive vacancies from the high to the low current density region. Then the question is: will it lead to void formation in the low current density region? Since a vacancy concentration gradient is assumed, the vacancy concentration in the high current density region is always higher than that in the low current density region. Since nucleation of a void requires supersaturation of vacancies and since the vacancy concentration in the high current density region is higher, it is unreasonable to assume that void will be nucleated in the low current density region rather than in the higher current density region.

Following the electron wind force, we expect vacancies to go to the high current density region at the Al–W interface and reach supersaturation, and void nucleation occurs there, but this cannot be true since void formation has to occur in the low current density region, otherwise the overhang cannot be depleted.

11.10   Current density gradient force of electromigration

Besides the electron wind force of electromigration, we need another force, the current density gradient force, to explain the void formation in the low current density region. The force will divert vacancies to the low current density region before they reach the high current density region. Since vacancies go to the low current density region, the concentration of vacancy can reach supersaturation and void can nucleate there.

The existence of the current density gradient force is due to the existence of a gradient of electric potential in current crowding [15, 16]. In field theory, the gradient of a potential energy is force, as we have defined the chemical force on diffusion by the gradient of chemical potential energy or we have defined the mechanical force in creep by the gradient of a stress potential energy. However, the electric potential gradient across an atom is quite small and the gradient force is negligible, but since the electrical resistance of a vacancy is about 100 times larger than that of an atom, the electric potential gradient force on a vacancy is significant.

We postulate that defects such as vacancies and solute atoms have a higher potential in the high current density region than in the low current density region. The potential gradient in the current crowding area provides a driving force normal to the current flow direction (or normal to the electronic wind force), which pushes the excess defects from the high current density region to the low current density region. As a consequence, the voids tend to form in the low current density region rather than in the high current density region. In other words, electromigration failure indeed tends to occur in low current density regions in a three-dimensional interconnect.

When an electric potential is applied to the conductor, the chemical potential energy of every atom and vacancy is increased. Because the resistance of a defect is much higher than a lattice atom, the increase in potential energy of vacancies or solute atoms is much larger. Furthermore, the increase is proportional to current density. In the high current density region, the vacancies will have a higher potential energy, so the concentration of vacancies in the high current density region will be reduced. In other words, the formation energy of a vacancy in the high current density region is higher, so the concentration will be lower, similar to what we have discussed in the compressive stress region in Chapter 6.

We consider the electromigration in a short strip as shown in Fig. 11.10. In the middle of the strip, the current density is uniform and there is a uniform vacancy flux moving from the anode to the cathode side. Approaching the cathode end, the electric current will turn to enter the TiN baseline and current crowding occurs. As vacancies follow the high current density, its potential energy increases and its equilibrium concentration decreases with respect to those in the uniform region. The electric potential gradient will push the excess vacancies out of the high current density region to the low current density region. The low current density region has a higher equilibrium vacancy concentration than that in the uniform region. Quantitatively, an analysis is given below.

To envisage the driving force that enables vacancies to go from the high current density region to the low current density region, we consider a single crystal Al strip and assume that a vacancy in the Al crystalline lattice has a specific resistivity of ρ_v. The specific resistivity may depend on current density due to joule heating since resistivity depends on temperature. However, for simplicity, we ignore the temperature effect here and assume that the specific resistivity is independent of current density for the simple analysis to be given below. Since vacancy is a lattice defect, we can regard its specific resistivity as the excess resistance over that of a lattice atom. Under electromigration in a current density of j_e, a voltage drop of $j_e A R_v$ occurs across the vacancy, where A is the cross-sectional area of a vacancy and R_v is the resistance of a vacancy. From the energy viewpoint, we can regard the vacancy as having a potential of $j_e A R_v$ above the surrounding lattice atoms. Knowing the charge of the vacancy, we have the potential energy of the vacancy in the current density of j_e. Let the charge of the vacancy be $Z^{**}e$, where Z^{**} is the effective charge number of the vacancy and e is the charge of an electron; we have the potential energy of $P_v = Z^{**}ej_e A R_v$ for a vacancy stressed by the current density j_e.

If we assume the equilibrium vacancy concentration in the crystal without any electrical current ($j_e = 0$) to be C_v,

$$C_v = C_0 \exp\left(-\Delta G_f / kT\right) \tag{11.23}$$

where C_0 is the atomic concentration of the crystal and ΔG_f is the formation energy of a vacancy. When we stress the crystal by a current density of j_e, the vacancy concentration will be reduced to

$$C_{ve} = C_0 \exp\left[-(\Delta G_f + Z^{**}ej_e A R_v)/kT\right] \tag{11.24}$$

Under a uniform high current density, the equilibrium vacancy concentration in the crystal decreases. In other words, the electric current dislikes any excess high resistive obstacles (or defects) and prefers to get rid of them until the equilibrium is reached. When there exists a current density gradient as in current crowding, a driving force exists to do so,

$$F = -dP_v/dx \tag{11.25}$$

This force drives the excess vacancies to diffuse in the direction normal to that of current flow. Now if we go back to Fig. 11.10(a) and consider the electromigration flux in the short strip, in the middle section of the strip, the current density is constant so there is a constant flux of Al atoms moving from left to right and a balancing flux of vacancies moving from right to left. A few of the vacancies near the surface or the substrate may escape to the surface or to the substrate interface, but the concentration as given by Eq. (11.24) is maintained. When the vacancy flux approaches the cathode and enters the current crowding region, some of the vacancies become excess and a force to divert them to the low current density region comes into play. Consequently, a component of the vacancy flux is moving in the direction normal to the current flow,

$$J_{cc} = C_{ve}(D_v/kT)(-dP_v/dx) \tag{11.26}$$

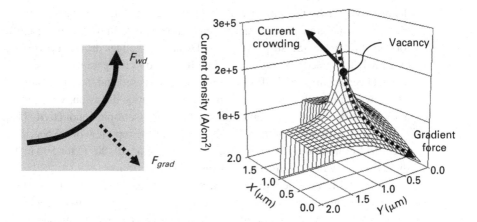

Fig. 11.12 Vacancies driven by two forces in the current crowding region.

where D_v/kT is the mobility and D_v is the diffusivity of vacancies in the crystal. Since a constant flux of vacancies keeps coming from the anode to the cathode due to electromigration, the total flux of vacancies moving towards the cathode is given by the sum of two components,

$$J_{sum} = J_{em} + J_{cc} = C_{ve}(D_v/kT)(-Z^*eE - dP_v/dx) \quad (11.27)$$

where the first term is due to electromigration driven by the current density (electron wind force) and the second term is due to current crowding driven by the current density gradient force. In the first term, Z^* is the effective charge number of the diffusing Al atom, and $E = j_e\rho$ (where ρ is the resistivity of the Al). Note that here we assume the vacancy flux to be opposite but equal to the Al flux. Also, it is important to note that the sum in Eq. (11.27) is a vector sum; the first term is directed along the current and the second term is directed normal to the current. In other words, the vacancies are driven by two forces in the current crowding region. They are depicted in Fig. 11.12. Since the current turns in the current crowding region, the direction of J_{sum} changes with position. Clearly, a detailed simulation is needed in order to unravel the magnitude and distribution of the forces in the current-crowding region.

The size of the gradient force is of interest. If we apply a current density of 10^5 A/cm^2 through the strip and assume that the current density will drop to zero across the thickness of the strip of 1 μm, the gradient can be as high as 10^9 A/cm^3. The gradient force is of the same magnitude as the electron wind force. Under such a large gradient, a high-order effect might exist, but we will ignore it at this moment.

11.11 Electromigration in an anisotropic conductor of beta-Sn

White tin (or beta-Sn) has a body-center tetragonal crystal structure, and its lattice parameters are $a = b = 0.583$ nm and $c = 0.318$ nm. Its electrical conductivity is anisotropic;

the resistivity along the a and b axes is 13.25 $\mu\Omega$-cm and along the c-axis is 20.27 $\mu\Omega$-cm. In electromigration under an applied current density of 6.25×10^3 A/cm^2 at 150 °C for one day, the beta-Sn strips were found to show a voltage drop of up to 10%, reported by Lloyd. The electromigration in white tin (melting temperature $T_m = 232$ °C) is of interest because most of the Pb-free solders are Sn-based and the electromigration at the device working temperature occurs primarily by lattice diffusion, which may lead to a noticeable microstructural change affected by its anisotropic conductivity [17].

Figure 11.13(a) and (b) shows SEM images of the top view of a Sn strip before and after electromigration, respectively, at 2×10^4 A/cm^2 at 100 °C for 500 h. The images show that a few of the grains rotated after electromigration.

In Fig. 11.14, the schematic cross-section of a beta-Sn grain is shown. It has a body-centered-tetragonal structure and we assume that its c-axis is making an angle θ with the x-axis, its a-axis is on the plane of the figure and is making an angle of $(90° - \theta)$ with the x-axis, and its b-axis is normal to the plane of the figure. The applied current density of electrons, j, is from left to right, along the x-axis. The resistivity along the a- and b-axes is the same and smaller than that along the c-axis. Due to the anisotropy of resistivity, the electrical field, E, can be written in two components, E_a and E_c, along the

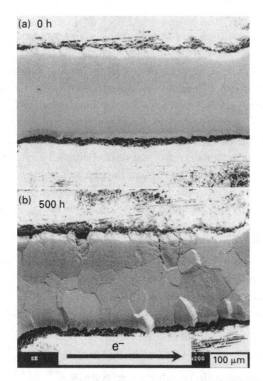

Fig. 11.13   (a) and (b) SEM images of the top view of a Sn strip before and after electromigration, respectively, at 2×10^4 A/cm^2 at 100 °C for 500 h. The lower image shows that a few of the grains rotated after electromigration. (Courtesy of Prof. C. R. Kao, National Central University, Jhongli, Taiwan, ROC.)

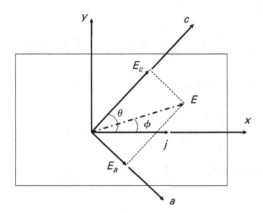

Fig. 11.14   Cross-section of a beta-Sn grain. E versus j.

a- and c-axes, respectively. The electrical field along the a-axis is $E_a = \rho_a j_a$; the field along the c-axis is $E_c = \rho_c j_c$, where j_a and j_c are, respectively, the two components of the electrical (electron) current density, j, along the a- and c-axes.

In isotropic materials, such as Cu or Al, they have the same resistivity along all the axes; therefore the magnitude of $\rho_c j_c$ is the same as $\rho_a j_a$. It also means that $E_a = E_c$, the overall electric field within the grain will coincide with the current flow direction, j. However, due to the difference of the magnitude between E_a and E_c in anisotropic materials, such as beta-Sn, there will be an angle ϕ, as shown in Fig. 11.14, between the combined electric field E and the direction of j inside the grain. This is a unique property of anisotropic conducting material and it is important to examine analytically how the angle ϕ would affect the interaction between the electrical current and the electrical force exerting on the grain.

In Fig. 11.14, the two components of current density j along the a-axis and c-axis would be

$$j_a = j \sin\theta, \quad j_c = j \cos\theta \tag{11.28}$$

Therefore the resulting electrical fields along these two axes are

$$E_a = \rho_a j \sin\theta, \quad E_c = \rho_c j \cos\theta \tag{11.29}$$

The combined electric field, E, would be

$$E = \sqrt{E_a^2 + E_c^2} = \sqrt{(\rho_a j \sin\theta)^2 + (\rho_c j \cos\theta)^2} = j\sqrt{(\rho_a \sin\theta)^2 + (\rho_c \cos\theta)^2} \tag{11.30}$$

To evaluate the magnitude of the angle ϕ, we consider the components of E, E_a and E_c, on the y-axis. From Fig. 11.14,

$$E \sin\varphi = E_c \sin\theta - E_a \cos\theta \tag{11.31}$$

By rearrangement, we have

$$\sin \phi = \frac{E_c \sin \theta - E_a \cos \theta}{E} = \frac{(\rho_c j \cos \theta) \sin \theta - (\rho_a j \sin \theta) \cos \theta}{E} \tag{11.32}$$

By substituting E from Eq. (11.30), we have

$$\sin \phi = \frac{j[(\rho_c \cos \theta) \sin \theta - (\rho_a \sin \theta) \cos \theta]}{j \sqrt{(\rho_a \sin \theta)^2 + (\rho_c \cos \theta)^2}} \tag{11.33}$$

The current term, j, can be cancelled out and by substituting $2 \sin \theta \cos \theta = \sin 2\theta$, the last equation becomes

$$\sin \varphi = \frac{(\rho_c - \rho_a) \sin 2\theta}{2 \sqrt{(\rho_a \sin \theta)^2 + (\rho_c \cos \theta)^2}} \tag{11.34}$$

Similarly, if we consider the components of E, E_a and E_c, on the x-axis, we obtain

$$\cos \phi = \frac{\rho_c \cos^2 \theta + \rho_a \sin^2 \theta}{\sqrt{(\rho_a \sin \theta)^2 + (\rho_c \cos \theta)^2}} \tag{11.35}$$

Either Eq. (11.34) or (11.35) defines the magnitude of the angle ϕ between the electric field and the applied current density from the data of resistivity and the orientation of the grain. Eq. (11.34) shows that if $\theta = 0°$ and $\theta = 90°$, then $\phi = 0°$ or in these cases E will be parallel to j.

Since the direction of the electric field deviates from that of the current density, it follows that the force originating from this field will also deviate from the current flow direction. The effect on force can have two significant consequences. The first is a torque and the second is that the force has a component parallel to the grain-boundary plane or the y-axis as shown in Fig. 11.14. It can be seen that the force generated from momentum exchange between electrons and atoms exerted on the boundaries of the grain is also making an angle to the current density. The existence of a pair of totally opposite forces provides a torque. Note that the boundaries are not necessary to be grain boundaries; they could also be the interface between the sample and the substrate.

11.12   Electromigration of a grain boundary in anisotropic conductor

We consider the electromigration of a grain boundary. The electron flux is normal to the plane of the grain boundary. In the case of grain boundaries in Al thin films, electromigration-induced migration of the grain boundaries has been observed. We shall consider a grain boundary in beta-Sn, which is an anisotropic conductor, and we shall demonstrate that electromigration will lead to an atomic flux along the plane of the grain boundary. In other words, the induced atomic flux along the grain boundary is moving in a direction normal to the electron flux or the electron wind force.

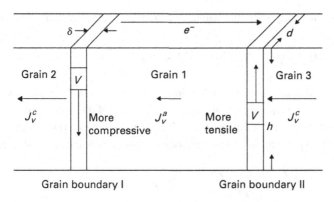

Fig. 11.15   Schematic diagram of a simple and geometrically ideal situation of a sandwiched grain structure.

In Fig. 11.15, a simple and geometrically ideal situation of a grain boundary between two beta-Sn grains is depicted; grain 1 on the right and grain 2 on the left of the grain boundary 1. We assume that grain 2, on the left, has its crystallographic c-axis directed along the flow direction of electron current, j, which is directed from left to right as indicated by a long arrow. We further assume that grain 1, on the right, has its crystallographic a-axis directed also along the current flow direction. Since both resistivity and diffusivity along the a- and c-axes in beta-Sn are different, the electron wind force and the corresponding vacancy fluxes in the c-axis (grain 2) and a-axis (grain 1) grains will be different:

$$J_v^c = \frac{C_v^{bulk} D_v^{c,bulk}}{kT} Z^* e\rho^c j \text{ and } J_v^a = \frac{C_v^{bulk} D_v^{a,bulk}}{kT} Z^* e\rho^a j \qquad (11.36)$$

where J_v^c and J_v^a are the vacancy fluxes in grain 2 along the c-axis and in grain 1 along the a-axis, respectively. We have the reference data below for the diffusivity and resistivity of Sn atoms along these two directions:

$$D^c = 5.0 \times 10^{-13} \text{ cm}^2/\text{s}, \rho^c = 20.3 \times 10^{-6} \Omega\text{-cm}$$

$$D^a = 1.3 \times 10^{-12} \text{ cm}^2/\text{s}, \rho^a = 13.3 \times 10^{-6} \Omega\text{-cm}$$

The atomic flux under electromigration should be in the same direction of electron flow. Therefore, a counter-flux of vacancy flows from right to left, as indicated by the two short arrows in Fig. 11.15. The effective charge, Z^*, is considered to be the same in both directions. Since $D^c\rho^c < D^a\rho^a$, from Eq. (11.36), we find that a larger vacancy flux reaches the grain boundary from grain 1 to grain 2, yet a smaller vacancy flux leaves the grain boundary going into grain 2. In grain 2, supersaturation of vacancies occurs at the grain boundary and corresponding tensile stresses occur near the grain boundary.

If we consider a small volume within the grain boundary of $d \times h \times \delta$, where d is the width of the grain, h is the height of the grain, and δ is the effective grain-boundary width, we have in steady state, the balance of the flux or conservation of mass,

$$\frac{C_v^{bulk} Z^* ej}{kT} (D_v^{a,bulk} \rho^a - D_v^{c,bulk} \rho^c) \times d \times h \cong D_v^{GB} \frac{C_v^\infty - C_v^L}{h} \times d \times \delta \qquad (11.37)$$

The first term of this equation states the difference of the flux through bulk diffusion; the second term of the equation mean that the difference of the vacancy flux goes to the surface via grain-boundary diffusion, since the surface is a good sink/source of vacancy.

From Eq. (11.37), the difference in vacancy concentrations can be evaluated as

$$\frac{\Delta C_v}{h} \cong \frac{C_v^{bulk} Z^* ej}{kT \delta D_v^{GB}} (D_v^{a,bulk} \rho^a - D_v^{c,bulk} \rho^c) \times h \qquad (11.38)$$

where $\Delta C_v = C_v^\infty - C_v^L$, where C_v^∞ is the equilibrium vacancy concentration of the free surface and C_v^L is the vacancy concentration in the grain boundary. Thus, we have obtained a flux of vacancies (or atoms) along the grain boundary, provided that a sink for vacancies exists at the end of the grain boundary, which can be a free surface or a void. Again, note that this flux is moving in a direction normal to the electron flux. We recall that this is the second case where atomic or vacancy flux is moving normal to electron flux, and the first case is due to current density gradient presented in Section 11.10.

If we extend the above analysis to a three-grain structure of one c-axis grain sandwiched between two c-axis grains, it will lead to grain rotation of the c-axis grain in the sandwiched structure. Furthermore, the analysis presented in the above can also be applied to interphase interfaces. For example, if we consider the interface between solder and Cu_6Sn_5 in a flip-chip solder joint, since the resistivity and diffusivity in these two phases are different, there will be a vacancy flux along the interface under electromigration with electron flow normal to the plane of the interface. This interfacial flux could lead to void formation and morphological change of the interface, to be discussed in Chapter 15, Section 15.4.3.

11.13   AC electromigration

Electromigration in interconnects is commonly a DC behavior. In computer devices based on FETs, such as DRAM devices, the gate of the transistor is turned on and off by pulsed DC current. On the other hand, in most communication devices, AC current is used. Especially in power switching devices, and radio frequency and audio power amplifiers, large AC swings occur during operation. The question whether AC can induce electromigration is often asked. Typically, it is believed that AC has no effect on electromigration.

We follow Huntington and Grone's model that the driving force of electromigration is due to momentum exchange in the scattering of electrons by diffusing atoms. A diffusing atom will be out of equilibrium state and will have a large scattering cross-section. If we consider an AC of frequency of 60 Hz or 60 cycles/s, it means that the scattering will be reversed once in a period of 1/120 s. For lattice diffusion in Pb assuming a vacancy mechanism of diffusion at $100\,^\circ\mathrm{C}$, the concentration of equilibrium vacancy in $1\,\mathrm{cm}^3$ of Pb is given by

$$\frac{n_V}{n} = \exp\left(-\frac{\Delta G_f}{kT}\right)$$

where ΔG_f is the formation free energy of a lattice vacancy. Taking $\Delta G_f = 0.55\,\mathrm{eV/atom}$, we have $n_V/n = 10^{-7}$. If we take $n = 10^{22}\,\mathrm{atom/cm}^3$, we have $n_V = 10^{15}$ vacancy/cm^3, meaning that these many vacancies are attempting to jump. The successive jumps in diffusion are limited by the frequency factor of

$$\nu = \nu_0 \exp\left(-\frac{\Delta G_m}{kT}\right)$$

where ν_0 is the Debye frequency or attempt frequency of jumping of the diffusing atom and is about $10^{13}\,\mathrm{Hz}$ for metals above their Debye temperature. Taking $\Delta G_m = 0.55\,\mathrm{eV/atom}$, which is assumed to be the motion-free energy of a vacancy in Pb, we obtain $\nu = 10^6\,\mathrm{jump/s}$. Then, in $1/120$ s, there are about 10^4 successive jumps of each of the 10^{15} vacancy/cm^3. Implicitly, we have assumed that the lifetime of the transition state or the activated state is very short. In other words, in each cycle of the 60 Hz AC, there are a large number of vacancies (or atoms) jumping in one direction in the first half of the cycle driven by electromigration and then an equal number of vacancies will jump in the opposite direction in the second half cycle. They cancel out each other statistically, so there is no net atomic flux driven by AC current. Nevertheless, AC will generate joule heating, and joule heating may develop a temperature gradient that induces atomic diffusion.

In the above analysis, an implicit assumption is that the electric field or electric current is uniform. However, when the current distribution is non-uniform, it is unclear whether or not AC electromigration can occur. Non-uniform current distribution occurs when the electric current turns, as in a flip-chip solder joint, or at an interface between Al and W via, or in a two-phase alloy where a precipitate and its matrix have different resistivities, or at a reactive interphase interface such as Sn–Cu. In the last case, if the interface is not at equilibrium, the atomic jumps across the interface in one direction are not the same as the jumps in the reverse direction. Since it is irreversible, the AC effect of electromigration may enhance the jumping in one direction. At a Schottky barrier across a metal–n-type semiconductor interface, the carrier flow is one-way from the semiconductor to the metal, so a high current density of AC may enhance the diffusion of the semiconductor into the metal. Thus, because of current crowding and rectified interfaces, it is possible for AC electromigration to occur.

References

[1] H. B. Huntington and A. R. Grone, "Current-induced marker motion in gold wires," *J. Phys. Chem. Solids* **20** (1961), 76.

[2] I. Ames, F. M. d'Heurle and R. Horstman, "Reduction of electromigration in aluminum films by copper doping," *IBM J. Res. Develop.* **4** (1970), 461.

[3] I. A. Blech, "Electromigration in thin aluminum films on titanium Nitride," *J. Appl. Phys.* **47** (1976), 1203–8.

[4] I. A. Blech and C. Herring, "Stress generation by electromigration," *Appl. Phys. Lett.* **29** (1976), 131–3.

[5] P. S. Ho and T. Kwok, "Electromigration in metals," *Rep. Prog. Phys.* **52** (1989), 301.

[6] K. N. Tu, "Electromigration in stressed thin films," *Phys. Rev. B* **45** (1992), 1409–13.

[7] R. S. Sorbello, in *Solid State Physics*, eds H. Ehrenreich and F. Spaepen (Academic Press, New York, 1997), Vol. 51, pp. 159–231.

[8] R. Kirchheim, "Stress and electromigration in Al-lines of integrated circuits," *Acta Metall. Mater.* **40** (1992), 309–23.

[9] N. A. Gjostein, in *Diffusion*, ed. H. I. Aaronson (American Society for Metals, Metals Park, OH, 1973), Ch. 9, p. 241.

[10] M. A. Korhonen, P. Borgesen, K. N. Tu and Che-Yu Li, "Stress evolution due to electromigration in confined metal lines", *J. Appl. Phys.* **73** (1993), 3790–9.

[11] J. J. Clement and C. V. Thompson, "Modeling electromigration-induced stress evolution in confined metal lines," *J. Appl. Phys.* **78** (1995), 900.

[12] P. C. Wang, G. S. Cargill III, I. C. Noyan and C. K. Hu, "Electromigration-induced stress in aluminum conductor lines measured by X-ray microdiffration," *Appl. Phys. Lett.* **72** (1998), 1296.

[13] A. A. MacDowell, R. S. Celestre, N. Tamura, R. Spolenak, B. Valek, W. L. Brown, J. C. Bravman, H. A. Padmore, B. W. Batterman and J. R. Patel, "Submicron X-ray Diffraction," *Nuclear Inst. and Meth.* **A 467** (2001), 936.

[14] K. Chen, N. Tamura, B. C. Valek and K. N. Tu, "Plastic deformation in Al (Cu) interconnects stressed by electromigration and studied by synchrotron polychromatic X-ray microdiffraction," *J. Appl. Phys.* **104** (2008), 013513.

[15] K. N. Tu, C. C. Yeh, C. Y. Liu and Chih Chen, "Effect of current crowding on vacancy diffusion and void formation in electromigration," *Appl. Phys. Lett.* **76** (2000), 988–90.

[16] C. C. Yeh and K. N. Tu, "Numerical simulation of current crowding phenomena and their effects on electromigration in VLSI interconnects," *J. Appl. Phys.* **88** (2000), 5680–86.

[17] Albert T. Wu, A. M. Gusak, K. N. Tu and C. R. Kao, "Electromigration induced grain rotation in anisotropic conduction beta-Sn," *Appl. Phys. Lett.* **86** (2005), 241902.

Problems

11.1   In a fcc metal, show that the displacement of an atom from its equilibrium lattice position to the activated position is $\sqrt{2}a/4$ whare a is the lattice constant. Take $<x_d^2> = (\sqrt{2}a/4)^2$ and show that $1/2(m\omega^2)/8$ is a good appromixation of the activation energy of motion, ΔG_m.

11.2   Plot Z^* versus temperature for Al from $350\,°C$ to $650\,°C$ by using Eq. (11.12).

11.3   Given drift displacement x as a function of time, calculate the drift velocity $<v>$, and also calculate Z^* by knowing D.

11.4 Calculate the electrical force Z^*eE at a current density of 10^5 A/cm^2 and the chemical potential $\sigma\Omega$ at the elastic limit for Au, and calculate the critical length for Au using Eq. (11.16).

11.5 Two Al short strips with length of 20 μm and 30 μm undergo electromigration. Calculate the stress at their anode ends when they carry a current density of 10^5 A/cm^2.

12 Electromigration-induced failure in Al and Cu interconnects

12.1 Introduction

While electromigration is the most persistent reliability problem in interconnects of microelectronic devices, it does not necessarily lead to microstructure failure. In interconnect regions where electromigration is uniform and steady, there may not be microstructure damage. While atoms are being driven from the cathode to the anode and vacancies are being driven in the opposite direction at the same time, no void or hillock occurs as long as the vacancy distribution is in equilibrium and the source and sink of vacancies are effective at the anode and the cathode, respectively. Only if vacancy distribution is not at equilibrium in electromigration will it lead to microstructure failure such as void and extrusion formation. The occurrence of these failures must involve atomic flux divergence and change in total number of lattice sites. It is a time-dependent reliability problem [1–5].

We have discussed in Chapter 11 that the Blech structure of short strips can show electromigration failure directly. This is because the short strips have the built-in divergence of atomic flux at the two ends, where atoms of Al cannot diffuse out of Al into the TiN-based line, nor from the TiN into the Al. Thus, atomic flux divergence is a necessary condition of electromigration-induced damage. Sometimes voids or hillocks are found in the middle of a strip where the electric current seems uniform, but they are due to flux divergence in the middle of the line caused by triple points of grain boundaries or other kinds of structural defect. Therefore, to understand electromigration failure, we need to discuss the causes of mass flux divergence in interconnects.

In addition, interconnect in VLSI technology is a three-dimensional multilevel structure. When electrical current turns or converges, e.g. at vias in the three-dimensional structure, current crowding occurs and it affects electromigration significantly. The current crowding can have a very great effect on atomic flux divergence and, in turn, structure damage formation. A simulation of the current distribution will be very useful in knowing whether or not the design of the structure has a serious problem electromigration.

We shall discuss atomic flux divergence and electric current crowding before we discuss the damages in Al and Cu interconnects caused by electromigration.

12.2   Electromigration-induced failure due to atomic flux divergence

In electric conduction, Kirchhoff's law states that the sum of all currents in and out of a point must be zero, thus the current in-flux equals the current out-flux, so there is no electric flux divergence at the point, except if the point is a source or sink of charges. But atomic flux can have flux divergence at a physical point, for example, at a triple point in two dimensions where three grain boundaries meet in a crystalline microstructure. If we assume that the diffusivity and the effective grain-boundary width of diffusion are the same, there will be atomic flux divergence at the triple point when the in-flux comes from one grain boundary, but the out-flux goes out along the other two grain boundaries, or vice versa. The net effect of flux divergence after a long time event of electromigration is the accumulation or depletion of vacancies (or atoms). The accumulation of vacancies can lead to void formation at the site of divergence. The question is: can we have a systematic classification of sites of atomic flux divergence in interconnects? In general, we can classify them according to the mechanism of vacancies absorption and emission in a microstructure or we can use the source and sink of vacancies to do so.

The most common site of vacancy absorption and emission is the free surface of a thin film. However, when the surface is oxidized and the oxide is protective, as with the Al oxide, the Al–oxide interface is no longer an effective source and sink of vacancies, so flux divergence will occur. Next, the interphase interfaces, such as the interface between the W via and Al line, is another common site of flux divergence. This is because solubility and diffusivity change dramatically across the interface. Then, grain boundaries, especially the triple points, are sites of flux divergence. Other planar defects such as twin boundary and stacking fault are not effective flux divergence sites. Yet we must emphasize that when a twin boundary or a stacking fault meets a grain boundary, a free surface or an interface, the triple point can become a flux divergence point and has been shown to have a very great effect on electromigration. We shall give a detailed discussion on the effect of nanotwin on electromigration in Cu in Section 12.5.4.

No doubt, we must consider dislocations, especially the kink site on an edge dislocation line. As in dislocation climb, the kink sites can absorb and emit atoms and vacancies. Consequently, lattice sites and even lattice planes can be created or destroyed via dislocation climb. The lattice plane migration is manifest by marker motion in Darken's analysis of interdiffusion as discussed in Chapter 5. The event of lattice plane migration has a very great effect on microstructure damage. When lattice plane migration (lattice shift) is absent in interdiffusion, it leads to Kirkendall void formation. Void formation in electromigration also requires the absence of lattice shift.

12.3   Electromigration-induced failure due to electric current crowding

Current crowding occurs typically when the flow of electrons makes a turn, where all electrons take the shortest path in order to reduce resistance. This is shown in Fig. 12.1(a) and (b), where a Ni silicide line is formed under a high current density in the heavily doped Si channels [6]. In Fig. 12.1(a), the silicide line is formed in a straight manner

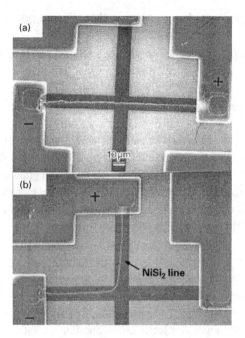

Fig. 12.1   (a) A Ni silicide line is formed under a high current density in heavily doped Si channels. The silicide line is formed in a straight manner connecting the cathode and the anode. (b) The line turns as the two electrodes and the Si channel were making an angle of 90°. The effect of current crowding on silicide formation is shown.

connecting the cathode and the anode. In Fig. 12.1(b), the line turns as the two electrodes and the Si channel were making an angle of 90°. What is significant is that the silicide line was formed around the inner corner of the channel, rather than in the middle of the channel. Since its formation was induced by current, the formation shows the current crowding behavior.

Using simulation of current distribution in a conductor, we can easily show the current crowding phenomenon as a function of geometry and resistance of an interconnect structure such as a multilevel interconnect structure of Al line and W via or a flip-chip solder joint. Generally speaking, in the three-dimensional multilayered interconnect structure, there are vias connecting two levels of interconnect, so there are many turns in the conduction path, and current crowding is common. Besides turns, when the conductor changes its thickness or width, current crowding also occurs. At interfaces, for example between Al and TiN, current crowding depends not only on the resistance of the Al and TiN, but also on the contact resistance of the Al–TiN interface. The larger the contact resistance, the lesser the current crowding.

12.3.1   Void formation in the low-current density region

What is unique in current crowding-induced failure due to void formation is that the void does not occur in the high current density region as expected; rather, it occurs in the nearby region of low current density. In Chapter 11, we have shown in Fig. 11.10 and Fig. 11.11 the phenomenon of void formation in regions of low current density in Al short strips. We shall use several more experimental observations to illustrate this phenomenon below.

Fig. 12.2 (a) is a schematic diagram of an Al strip on a TiN base line patterned for the in-situ TEM study of void formation in electromigration. The width of the Al line was designed to be thin enough to be imaged laterally from the side as indicated by the long and straight arrow which shows the direction of the viewing electron beam in the TEM. The applied current of 50 mA for electromigration, from right to left along the Al strip, is indicated by two short arrows in the shape of "Z". In electromigration, a void was seen at the upper-right corner of the strip, i.e. the upper corner of the cathode end, by Okabayashi *et al.*, as shown in Fig. 12.2(b)–(c) [7]. Note that the upper-right corner has a very low current density, as given by the current distribution simulation in the short strip. Furthermore, after a long electromigration time, the polarity of the applied current was reversed, and again void formation was observed at the tip of the cathode. Fig. 12.3 shows that a void was again formed at the tip of the cathode (the cathode was the anode before the change of polarity). Again, the current density at the tip of the cathode should

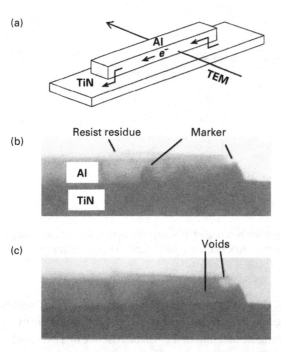

Fig. 12.2   (a) Schematic diagram of an Al strip on a TiN baseline patterned for an in-situ TEM study of void formation in electromigration. The width of the Al line is designed so that it can be imaged laterally from the side as indicated by the long and straight arrow which shows the direction of the viewing electron beam in the TEM. (b) TEM image of the cathode of Al strip before elecromigration. (c) TEM image of void formation at the upper corner of the cathode after electromigration at 50 mA for 14 s.

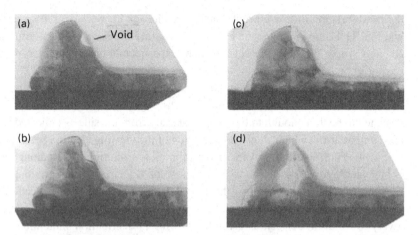

Fig. 12.3   (a)–(d) In electromigration under reverse current, TEM images show that a void was formed at the tip of the cathode (which was the anode before the change of polarity). Again, the current density at the tip of the cathode should be very low, yet the void kept growing with electromigration, as shown here. These results show that electromigration-induced void formation occurs in the low current density region.

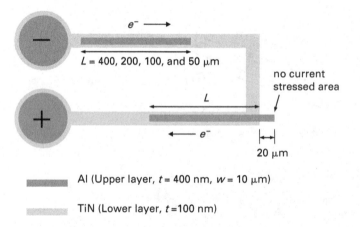

Fig. 12.4   Schematic diagram of a U-shape baseline of TiN, on which Al short strips were deposited. Some of the short strips were made to stick out from the U-turn, so the Al short strip has an overhang out of the baseline.

be very low, yet the void kept on growing with electromigration. These results show that electromigration-induced void formation occurs in the low current density region.

Fig. 12.4 shows a schematic diagram of a U-shape baseline of TiN, on which Al short strips were deposited [8]. Some of the short strips were made to stick out from the U-turn, so the Al short strip has an overhang out of the baseline. The overhang was about 20 μm long. It is clear that there will be no current in the overhang during electromigration when current is applied to the baseline. However, voids were observed to form in the overhang in electromgiration.

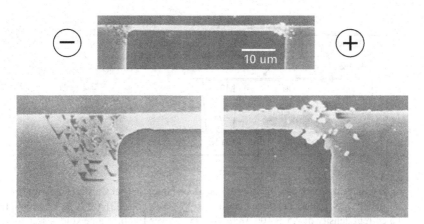

Fig. 12.5   SEM images of a single crystal line and pads of Al before and after electromigration. In order to remove the effect of grain-boundary diffusion, test samples of a single-crystal Al line and pads were produced by cluster beam deposition. Upon stressing by a current density at 3.5×106 A/cm^2 at 240 °C for 19 h, void and hillocks were observed at the cathode and anode, respectively.

Fig. 12.5 shows SEM images of a single crystal line and pads of Al before and after electromigration [9]. In order to remove the effect of grain-boundary diffusion, the test sample of single crystal Al line and pads was produced by cluster beam deposition. Upon stressing by a current density at 3.5×10^6 A/cm^2 at 240 °C for 19 h, voids and hillocks were observed at the cathode and the anode, respectively. What is significant is that the damage of void and hillock formation is asymmetrical. In the enlarged images of the two ends of the line, we see that the hillocks were formed in the high current density region in the line, but the voids were formed in the low current density region in the pad, rather than in the high current density region in the line. The pad has a much larger cross-section than the line, so the current density in the pad is much lower than that in the line. The voids have a triangular shape, which indicates that the single-crystal Al thin film is [111] oriented. If we calculate the diffusivity of vacancies in Al, we find that in the accelerated test at 240 °C for 19 h the vacancies can diffuse a distance more than 100 μm, which is the range of void formation in the pad as shown in Fig. 12.5.

Fig. 12.6(a) is a schematic diagram of a three-level damascene interconnect structure of Cu with two Cu vias (V1 and V2) connecting the three levels of lines (M1, M2, and M3) [10]. In an electromigration test at current density of 2.5×10^6 A/cm^2 at 295 °C over 100 h, a large void is seen to form to the left of V1 via as shown in the image in Fig. 12.6(b). However, the curved and thick arrow in Fig. 12.6(b) indicates the flow direction of electrons in the test, meaning that the current density in the void formation region is low. Also, voids were found to form on the surface of the line above the V1 via, which is in agreement with the fact we know now that electromigration in Cu occurs by surface diffusion, as discussed early in Chapter 11.

The observations presented above have a common feature that electromigration-induced void formation occurs in the low current density region in interconnects. This

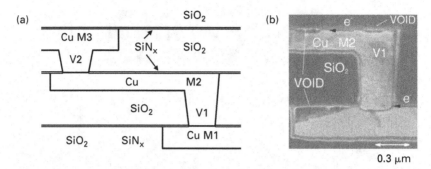

Fig. 12.6   (a) Schematic diagram of a three-level damascene interconnect Cu structure with two Cu vias (V1 and V2) to connect the three lines levels (M1, M2, and M3). (b) In an electromigration test at a current density of 2.5×106 A/cm^2 at 295 °C over 100 h, a large void is seen to form to the left of the V1 via as shown.

is quite unexpected on the basis of our understanding of the electron wind force of electromigration. The force means that the higher the current density, the greater the driving force; in turn more atoms (or vacancies) will be driven into the region of high current density. Actually, in the low current density region, the electron wind force is so weak that there should be no electromigration.

12.4   Electromigration-induced failure in Al interconnects

The microelectronic industry has used Al as the interconnect conductor for 40 years even though Cu is a better electrical conductor. The choice of Al is because Al has certain advantages in lithographic processing. It has good adhesion to a SiO$_2$ surface. It can be deposited by e-beam evaporation or sputtering, and it can be patterned by dry or reactive ion etching. It does not poison Si as Cu does, so Al can be deposited directly on Si to serve as contact metallization on Si devices.

Electromigration damage in Al lines was discovered in the 1960s and was soon recognized to occur by grain-boundary diffusion. In the early stage of study of electromigration in Al, the effect of microstructure and the effect of solute were emphasized before the Blech structure was invented. In order to reduce electromigration, grain-boundary structure and diffusion must be studied. Actually, a bamboo-type microstructure and even single-crystal Al thin-film lines were investigated. On the solute effect, the finding that the addition of 1 atomic % of Cu to Al can greatly reduce grain-boundary diffusion in Al was probably one of the most important inventions in interconnect technology.

12.4.1   Effect of microstructure in Al on electromigration

The triple points of grain boundaries can serve as centers of atomic flux divergence. The mechanism of void formation at a triple point and growth into an opening along a grain boundary were studied. Consequently, the processing of Al interconnects with a bamboo-type microstructure, containing no grain-boundary triple points and no continuous grain-boundary path, attracted much attention. Interestingly, it was recognized that when the line-width becomes smaller than grain size, the microstructure naturally becomes bamboo-like. But it was also found to be very hard to fabricate a line having a perfect or 100% bamboo-type microstructure.

12.4.2 Wear-out failure mode in multilayered Al lines and W vias

When a single level of Al interconnect was advanced to a multilevel Al interconnect, atomic flux divergence shifted from grain-boundary triple points to the interface between an Al line and a W via. The interfacial divergence is due to the fact that atomic diffusivities in Al and W are very different. The wear-out failure mechanism as depicted in Fig. 11.11 was established. In actual devices having the multilevel structure of Al lines and W vias, the lifetime has been found to be reduced by a factor of 50 as compared to that of a single-level Al interconnect due to flux divergence at the interface. Then, the technology of overhang was invented to improve the lifetime of wear-out.

12.4.3 Solute effect of Cu on electromigration in Al

In bulk alloys, certain solutes have the effect of retarding or enhancing solvent diffusion. For example, solute atoms of Cu in bulk Al are known to enhance the lattice diffusion of Al solvent atoms. This effect can be calculated on the basis of atomic jump frequencies around the Cu vacancy pair in a Al lattice. On this reasoning of the use of alloying effect to retard grain-boundary diffusion of Al, we would not choose Cu. However, when a small amount of Cu was co-deposited with Al, the co-deposited thin-film sample actually showed much less electromigration. Now it is a general practice in industry to add 1 atomic % or so of Cu to Al, and the lifetime improvement against electromigration can be orders of magnitude better than that of pure Al. The excess Cu forms Al_2Cu precipitates in Al grain boundaries. These precipitates dissolve and serve as sources of Cu to replenish the loss of Cu in Al grain boundaries when electromigration depletes the Cu atoms by driving them to the anode.

Why Cu is capable in retarding grain-boundary electromigration in Al has been a question of keen interest. Because of the difficulty in knowing grain-boundary structure precisely, no definitive answer of the effect of alloying has been given. Most likely the answer is either a reduction of driving force or a reduction of kinetics or both. Kinetically, Cu may either reduce the concentration of vacancy in the grain boundary or increase the activation energy of grain-boundary diffusion of Al. What is important to industrial manufacturing is the finding that the process of adding Cu to Al is forgiving, meaning that it tends to work well.

12.4.4 Mean-time-to-failure in Al interconnects

Black's equation of MTTF has been given as

$$\text{MTTF} = B\frac{1}{j^n} \exp\left(\frac{E}{kT}\right) \tag{12.1}$$

where B is a pre-factor constant, j is current density, $n = 2$ is a power factor of j, E is activation energy, and kT has the usual meaning of thermal energy. Note that there is no negative sign in front of E/kT. For most experimental data, the power factor of n has been found to be close to 2, yet why n is equal to 2 has been the most common question on Black's equation.

The most plausible explanation to $n = 2$ was given by Shatzkes and Lloyd [11]. They proposed a model by solving the time-dependence diffusion equation and obtained a solution for MTTF in which the square-power dependence on current density was obtained. It is due to the nucleation of a void which needs supersaturation of vacancies in the cathode side. The incubation time of nucleation is much longer than the time needed to grow a void to a size that can lead to circuit opening. While the number of vacancies needed to form a critical nucleus is small, the supersaturation needed for nucleation can be quite high. Actually, because there are sinks of vacancies in the sample, a large number of vacancies may be consumed before the sinks can be deactivated, so it will take a while to build up the supersaturation, therefore the incubation time can be quite long.

We show below that for the growth alone, the factor n should be close to unity instead of 2. If we assume that the volume of the void needed to fail the circuit is V and if we ignore the back stress, we can write

$$V = \Omega J_v At = \Omega \left(C_v \frac{D_v}{kT} Z^* \rho j \right) At \tag{12.2}$$

Thus

$$t = \frac{VkT}{\Omega CDZ^* \rho j} = B\frac{1}{j} \exp \left(\frac{E}{kT} \right) \tag{12.3}$$

where B is the pre-factor and $B = VkT/\Omega CD_0 Z^* \rho$, Ω is the atomic volume of a vacancy, A is the cross-sectional area of the void if we assume that the shape of the void is rectangular or A is the surface area of a spherical void, C_v and D_v are concentration and diffusivity of vacancy in the sample, respectively, and we can take $C_v D_v = CD$, where C and $D = D_0 \exp(-E/kT)$ are the atomic concentration and atomic diffusivity, respectively, and hence E is the activation energy of atomic diffusion rather than the activation energy of motion of a vacancy. The other terms have the usual meaning. We see that $n = 1$.

Experimentally, it has been found that the time taken to grow a void to the size of failure is much shorter than the incubation time of nucleation of the void. This is the case in electromigration-induced failure in flip-chip solder joints owing to the nucleation and growth of a pancake-type of void in the cathode contact interface; the incubation time of nucleation is much longer than the growth time. Hence, we need to consider the nucleation for the explanation of $n = 2$.

To consider the nucleation of a void in electromigration, recall that it has been reported repeatedly that a void starts to form in the low current density region, rather than in the high current density region, as discussed in Section 12.3. In order for vacancies to move from the high current density region to the low current density region, there exists a force normal to the electron wind force. A current-density gradient force was proposed, which is normal to the electron wind force. Since force is a vector, the magnitude of the combined force, which drives vacancies to the low current density region to nucleate a void, is the square root of the sum of the square of magnitude of these two forces. Thus, we have the $n = 2$ dependence on current density. The above argument to show that $n = 2$ is from the point of view of driving force, while the argument presented by Shatzkes and Lloyd was from the point of view of kinetics.

However, a much simpler answer of $n = 2$ is that it is because of joule heating since joule heating is given as $j^2 \rho$ per unit volume. But intuitively, it seems that MTTF should be directly, instead of inversely, proportional to joule heating. This is because it seems that the higher the heating, the faster the failure or the shorter the MTTF. Indeed, joule heating is a very important factor in device failure. However, we need to consider the effect of joule heating on the nucleation and growth of a void in the cathode. We need to consider the time, the temperature, and the rate of nucleation and growth of the void. This is similar to the classical theory of phase transformations on the basis of the time-temperature-transformation (TTT) diagram. The TTT diagram indicates that the time or the overall rate of transformation is a function of under-cooling. The effect of under-cooling on nucleation is strong. Nucleation must have under-cooling. The smaller the under-cooling, the slower the nucleation. According to the TTT diagram, with a small under-cooling, the temperature is high, while diffusion is fast, nucleation is slow, so the overall rate of transformation is small. With a large under-cooling, while the nucleation is fast, the diffusion is slow because the temperature is lower, again the overall rate of transformation is slow. The maximum rate of transformation occurs when both nucleation and diffusion are relatively fast at a moderate under-cooling.

Joule heating affects under-cooling in the sample under electromigration. When joule heating is large, the temperature of the sample increases, so the under-cooling needed for void nucleation is reduced so that it will delay void nucleation and the MTTF is increased. When joule heating is small, the temperature of the sample is low, then diffusion is slow but void nucleation is easier. Since MTTF depends more on void nucleation than on growth, joule heating is inversely proportional to MTTF, so we have j^{-2} dependence. The effect of joule heating on diffusion is represented by the term $\exp(E/kT)$, which is directly proportional to MTTF. Thus, the effect of j^{-2} and the effect of $\exp(E/kT)$ on MTTF are opposite to each other.

12.5 Electromigration-induced failure in Cu interconnects

Although the Al(Cu) alloy has performed well as the interconnect conductor in Si micro-electronic technology for a long time, the trend of miniaturization has recently demanded a change for the following reasons: first, because of the resistance-capacitance (RC) delay in signal transmission in fine lines; second, because of the high cost of building multilay-ered Al interconnect structure of more than eight layers; third, because of the concern of electromigration. For the use of narrower and narrower lines, not only the line resistance increases, but also the capacitance between lines drags down signal propagation. If we choose to maintain the dimensions of Al interconnect without change, we must add more layers of Al, from six to eight or ten. To make more layers of metallization on Si is very undesirable because of cost. Moreover, if we have to add two more layers of interconnects, the additional processing steps can reduce yield. This is why the dual-damascene processing of Cu interconnects is attractive. It combines the steps making one level of lines and one level of vias together, so most of the via-making steps have been removed. Using the same number of processing steps, we can build more levels of Cu interconnect than Al interconnect.

Since Cu has a much higher melting point (1083 °C) than Al (660 °C), atomic diffusion should be much slower in Cu than Al at the same device working temperature. So, electromigration is expected to be much less in Cu interconnects. Surprisingly, this benefit is not as great as expected. As we stated in the beginning, electromigration in Cu occurs by surface diffusion which has lower activation energy than grain-boundary diffusion. Why has electromigration changed from grain-boundary diffusion in Al to surface diffusion on Cu? Ironically, this is because of the use of the damascene process to fabricate the Cu interconnect. In addition, it is because Cu intrinsically does not adhere to oxide surfaces. Why do we have to use damascene processes to produce Cu interconnects? This is because Cu cannot be etched or patterned by dry or reactive ion etching. Therefore, we have to form Cu lines by electrolytic plating of Cu into trenches in dielectric, followed by a wet process of chemical-mechanical polishing (CMP) to flatten the Cu with its surrounding dielectric. On the polished flat surface we repeat the damascene process and build the multilevel Cu interconnect shown in Fig. 12.7. In the

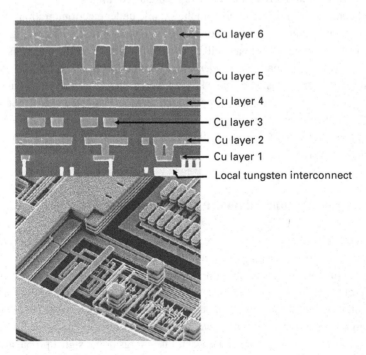

Fig. 12.7   Damascene process of multilevel Cu interconnects.

damascene process, via holes are etched together with trenches in the dielectric layer, followed by electroplating Cu into the trenches and via holes simultaneously. Compared to the process of making Al lines and W-plug vias, the dual damascene process has saved the via-making steps. In filling Cu into the trenches and via holes in the dielectric, we need to improve the adhesion of Cu to dielectric, so a liner such as Ta, TaN, or TiN is used to cover the bottom and sidewalls of the trenches and holes before the electroplating of Cu. The liner also serves as a diffusion barrier to prevent the diffusion of Cu into Si. To plate the Cu, a seeding layer of electroless Cu or vapor-phase deposited Cu is needed before the electrolytic plating.

Then, CMP is used to polish the top surface of Cu, which has no liner, followed by the deposition of a dielectric layer so that the dual damascene process of building another interconnect layer of vias and lines can be repeated. Thus, CMP produces a free surface of Cu on top of the Cu interconnect which does not adhere to the dielectric layer deposited over it. Atomic diffusion on the top surface of the Cu interconnect becomes the "built-in" highway for electromigration [12–17]. Furthermore, the liner between a Cu via and the Cu line below it is an interface of flux divergence. These two, the top surface and the via interface, are the two weak places for electromigration failure to occur. Which is the weaker one may depend on process control and can lead to early failure, resulting in a bimodal distribution of failures; the early failure and the wear-out failure. Nevertheless, the mechanism of surface diffusion-induced void formation in a via is intriguing. It may be that a certain amount of interfacial diffusion can occur between the Cu and its liner. Then, the nucleation of a via void could be stress-induced due to poor adhesion. Moreover, the third mode of failure is stress-induced extrusion at anodes, resulting in dielectric delamination or fracture, especially for long interconnects with ultralow k dielectric insulation.

12.5.1   Effect of microstructure on electromigration

Because electromigration in Al interconnects takes place along grain boundaries, the effect of microstructure is a key issue. The knowledge learned from the reliability study of Al interconnects was applied to Cu interconnects with the intention of improving electromigration resistance, but with less success. For example, no significant difference in electromigration failure was found between poly-granular and very long-grained bamboo Cu interconnects. The microstructure of electrolytic Cu has the unusual property that it undergoes abnormal grain growth or recrystallization near room temperature. It has a strong [111] texture in the as-plated state. Why some of the grain boundaries in the electrolytic Cu have high mobility near room temperature is unclear; it could be due to a minute amount of the organic and inorganic additives from the plating bath. Also, whether grain-boundary diffusivity along these grain boundaries is the same as that given in Table 11.2 is of interest. Both grain size and impurity may affect electrical conductivity, and good conductivity is the first requirement of interconnect metallization. So far, this is not an issue for electrolytic Cu. The minute amount of additives from the plating bath has little effect on the conductivity, and the grain size is typically about 0.1 μm before grain growth.

The room-temperature grain growth has led to a non-uniform distribution of large grains and clusters of small grains in the interconnect. The large grains are bamboo-like and their grain boundaries to smaller grains have triple points. Hence, the microstructure is undesirable if electromigration is dominated by grain-boundary diffusion, but it becomes irrelevant when electromigration occurs by surface diffusion. On a surface, the grain-boundary triple point is not an effective flux divergence point. The surface plane of the [111] oriented grains and the triple points of their twins are of interest since they affect surface diffusion.

The microstructure of the seeding layer of electroless or vapor-phase deposited Cu is also interesting, since its grain size is linearly proportional to its thickness. The grain growth during deposition is called flux-driven grain growth, yet after deposition the grains do not grow at room temperature as do those in the electrolytic Cu. Nevertheless, when electrolytic Cu is plated on the seeding layer, the microstructure of the latter is lost due to ripening. The interaction between these two types of Cu film is interesting, and it seems that the organic and inorganic additives in the electrolytic Cu might have diffused into the seeding layer to enhance its grain-boundary mobility.

The divergence at Al-line–W-via interfaces is no longer an issue in Cu interconnects due to the dual-damascene process. Nevertheless, the interface between a via and the line beneath it can still be an interfacial discontinuity and flux divergence due to the liner. Besides adhesion, the liner also serves as a diffusion barrier to prevent Cu from reaching Si. As a diffusion barrier, the thicker the better. Yet a thick liner will increase the resistance of interconnect, so actually, the thinner the better. While a cleaning of via holes is performed before depositing the liner, if the cleaning is not done properly, it affects the adhesion between a Cu via and the Cu line beneath it. Hence, besides being a flux divergence plane, the liner between a via and a line is of reliability concern due to poor adhesion under thermal stress.

12.5.2   Effect of solute on electromigration

Since adding 1 atomic % Cu to Al is beneficial to improve the electromigration resistance in Al interconnects, we ask if a similar beneficial effect of solute can be found in Cu interconnects and what is the solute? Again, if we try to find a solute to slow down the grain-boundary diffusion of Cu, it is irrelevant, since what we really need is a solute that can slow down the surface diffusion of Cu. There is no guideline to find such a solute until we realize that electromigration in Cu interconnects is dominated by surface diffusion. To sustain a continuous surface atomic flux of Cu, we must be able to release Cu atoms continuously from kink sites on surface steps. In other words, the mechanism and energy needed to dissociate Cu atoms from kink sites, as in desorption and low-temperature sublimation of atoms from a solid surface, are important in the consideration of surface electromigration.

After adding and testing many elements in Cu, it was found that Sn has shown a significant effect in resisting electromigration in Cu. Table 12.1 compares the conductivity of Al and Cu interconnects. Fig. 12.8 shows the resistivity of Cu (0.5 wt. % Sn) and Cu (1.0 wt. % Sn) alloys as a function of temperature using Van der Pauw test structures. Fig. 12.9

Table 12.1. A comparison of the conductivity of Al and Cu interconnects

Films	Resistivity at 20 °C ($\Omega\mu$-cm)
Sputtered Cu	2.1
Al (2 wt.% Cu)	3.2
Cu (0.5 wt.% Sn)	2.4
Cu (1 wt.% Sn)	2.9
W	5.3

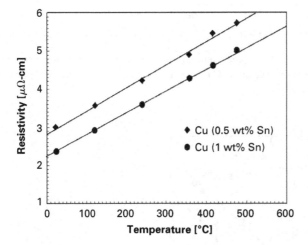

Fig. 12.8   Resistivity of Cu (0.5 wt. % Sn) and Cu (1.0 wt. % Sn) alloys as a function of temperature using Van der Pauw test structures.

shows the resistance change of Cu(Sn) and pure Cu thin films under electromigration at 250 °C and 10^6 A/cm^2. While the resistance of the Cu in testing changed quickly with time, the resistance of the Cu(Sn) remained unchanged. Fig. 12.10 shows a comparison of the measured edge displacement versus time for 150 μm-long and 5 μm-wide Cu and Cu(Sn) test strips with a current density of 2.1×10^6 A/cm^2 at 300 °C. The slope of the edge displacement versus time gives the average drift velocity for the Cu mass transport during electromigration. The average drift velocity of the Cu mass transport in the Cu(Sn) alloy strips is small at the beginning of testing, then increases slowly with time and eventually reaches a value comparable to that of pure Cu.

Why Sn is beneficial in resisting electromigration in Cu is unclear. Since electromigration in Cu interconnects is known to occur by surface diffusion, it is likely that, somehow, the surface diffusivity of Cu is slowed down by adding Sn and, more importantly, the supply of a surface flux of Cu atoms is reduced, in particular, if the dissociation of Cu atoms from the kinks on Cu surface steps is retarded due to a strong binding of Sn atoms to the kinks. Fig. 12.11 is a schematic diagram of surface steps and kinks on a Cu surface. The shaded atoms at the kinks represent Sn atoms. If we assume a strong binding

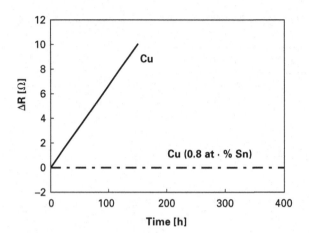

Fig. 12.9   Resistance change of Cu(Sn) and pure Cu thin films under electromigration at 250 °C and 10^6 A/cm². While the resistance of the Cu in testing changed quickly with time, the resistance of the Cu(Sn) remained unchanged.

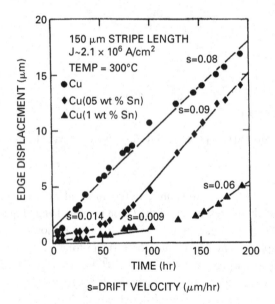

Fig. 12.10   Comparison of the measured edge displacement versus time for 150 μm-long and 5 μm-wide Cu and Cu(Sn) test strips with a current density of 2.1 × 10^6 A/cm² at 300 °C. The slope of the edge displacement versus time gives the average drift velocity for the Cu mass transport during electromigration.

there, the release of Cu atoms from the surface steps will be blocked. Moreover, when Sn segregates to Cu surfaces and forms an oxide bond with the top dielectric layer that improves the adhesion between the Cu and the top dielectric layer, it may further retard surface dissociation and diffusion of Cu.

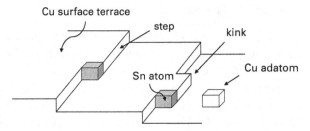

Fig. 12.11 Schematic diagram of surface steps and kinks on a Cu surface. The shaded atoms at kinks represent solute atoms. If we assume a strong binding there, the release of Cu atoms from the surface steps will be blocked.

The device working temperature is around 100 °C, which is only about 0.275 of the absolute melting temperature of Cu. So, at 100 °C we can ignore both lattice and grain-boundary diffusion in Cu. We may still be able to ignore both of them at a testing temperature of 250 °C. But at 350 °C, which is about 0.46 of the absolute melting temperature of Cu, we cannot ignore grain-boundary diffusion in Cu. Hence, the finding that there is no electromigration in Cu(Sn) up to 250 °C, yet some has occurred at 350 °C, tends to indicate that while Sn may be effective in retarding surface diffusion of Cu, it is less effective in retarding grain-boundary diffusion. More importantly, what is the proper temperature range for conducting an accelerated electromigration test for a Cu interconnect? On the basis of the fact that in devices the electromigration in Cu interconnects is dominated by surface diffusion, we may conduct the tests at 250 °C but not at 350 °C. The result at 350 °C may be misleading, since it may include grain-boundary diffusion.

12.5.3   Effect of stress on electromigration

The effect of applied compressive stress on electromigration in Cu is expected to be the same as that in Al; it retards electromigration at the anode. But the effect of back stress on electromigration in Cu is unclear. The first question is whether there is any back stress to be induced by the electromigration in Cu strips at a temperature below 250 °C. If a Cu strip is not covered by a liner such as Ta, or not confined by rigid walls, it is hard to envision how back stress can be generated under electromigration by surface diffusion. This is because a free surface is the most effective source and sink of vacancies, so stress will be relaxed easily and will not build up. Hence, definitive experimental measurements on back stress in Cu interconnects will be needed.

In devices, the Cu interconnect is embedded in an interlayer dielectric, so it is confined. Electromigration will induce extrusion or hillock formation at the anode end, even though the electromigration occurs by surface diffusion. The extrusion can deform the surrounding dielectric if it is soft and cause delamination if its adhesion is poor, or crack the dielectric if it is brittle. The effect of electromigration-induced stress in Cu/ultralow *k* interconnects is of concern. This will be a serious reliability issue when an ultralow dielectric constant insulator is integrated with Cu metallization.

12.5.4   Effect of nanotwins on electromigration

In 2004, there was a paper published in *Science* by Lu *et al.*, and Chen *et al.* on nano-twinned Cu [18, 19]. By using pulsed electro-deposition, Cu foils having a high density of nanotwins were prepared and the Cu foils were found to possess ultrahigh strength and at the same time to maintain a normal electrical conductivity. The nanotwinned Cu has a yield stress which is about 10 times higher than that of coarse-grained Cu. At the same time, the nanotwinned Cu also has a good ductility. The rare combination of excellent mechanical and normal electrical properties will be very valuable for the Cu to serve as an interconnect conductor. The ultrahigh strength which has yield stress 10 times higher than a normal Cu foil will be good for the mechanical property of the multilayered interconnect structure and also for chemical-mechanical polishing (less dishing) in manufacturing of the dual-damascene structure. More importantly, it may lead to a critical length 10 times longer in electromigration since the critical length is linearly proportional to the yield stress.

Furthermore, it has been reported that in-situ TEM observation of electromigration in the nanotwinned Cu showed a 10 times slower migration rate. The nanotwins intersect the grain boundaries and the free surface of Cu to form triple points. The triple points have modified the free surface and grain-boundary structure and properties of Cu. These triple points slow down the diffusion of Cu atoms driven by electromigration. In high-resolution TEM video recording, it was observed that when Cu atoms diffuse across a triple point, they stop and wait for a long while.

Fig. 12.12 is a set of high-resolution TEM images of the motion of a surface step on the zig-zag (111)–(422) free surface. As electron wind force drives Cu atoms to diffuse to the anode, the surface step migrates in the opposite direction to the cathode. It is the

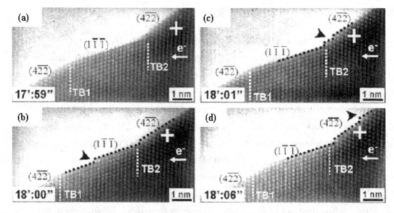

Fig. 12.12   (a)–(d) High-resolution TEM images of the (011) oriented Cu grain under electromigration as a function of time. The time of the captured image (in min and s) is given in the rectangular box at the lower left corner. The direction of electron flow is from right to left. The arrow head indicates the atomic steps on the free surface. Reaching a triple point, the step waited and took about 5 s to move over the triple point to continue its migration. The cross in each panel refers to a fixed reference point of easy inspection.

step motion that enables us to recognize the surface diffusion. The images were recorded by video tapes so that we can measure the velocity of the step. The surprising finding was that the surface step stops at every triple point and waits for a long while (5 s in the recording) before jumping over a triple point. This is because across a triple point, the atomic plane changes typically from (111) to (422) or vice versa. The stop is because of the incubation time needed in order to nucleate a new step on the new atomic plane, going from (111) to (422) or from (422) to (111). After it has jumped over a triple point, the stepwise motion is fast on the flat (111) plane or the flat (422) plane. The step velocity on the (422) plane is slightly slower than that on the (111) plane.

References

[1] F. M. d'Heurle and P. S. Ho, "*Thin Films: Interdiffusion and Reactions,*" eds J. M. Poate, K. N. Tu and J. W. Mayer (Wiley-Interscience, NY, 1978), 243.

[2] C. K. Hu and J. M. E. Harper, "Copper interconnects and reliability," *Mater. Chem. Phys.* **52** (1998), 5.

[3] R. Rosenberg, D. C. Edelstein, C. K. Hu and K. P. Rodbell, "Copper metallization for high performance silicon technology," *Annual Review Mater. Sci.* **30** (2000), 229.

[4] E. T. Ogawa, K. D. Li, V. A. Blaschke and P. S. Ho, "Electromigration reliability issues in dual-damascene Cu interconnections," *IEEE Trans. Reliability* **51** (2002), 403.

[5] K. N. Tu, "Recent advances on electromigration in very-large-scale-integration of interconnects," *J. Appl. Phys.* **94** (2003), 5451–73.

[6] J. S. Huang, H. K. Liou and K. N. Tu, "Polarity effect of electromigration in Ni2Si contacts on Si," *Phys. Rev. Lett.* **76** (1996), 2346–49.

[7] H. Okabayashi, H. Kitamura, M. Komatsu and H. Mori, "In-situ side-view observation of electromigration in layered Al lines by ultrahigh voltage transmission electron microscopy," *AIP Conf. Proc.* **373** (1996), 214 (see Figs. 2 and 4).

[8] S. Shingubara, T. Osaka, S. Abdeslam, H. Sakue and T. Takahagi, "Void formation mechanism at no current stressed area," *AIP Conf. Proc.* **418** (1998), 159 (see Table I).

[9] M. Hasunuma, H. Toyota, T. Kawanoue, S. Ito, H. Kaneko and M. Miyauchi, "A highly reliable Al line with controlled texture and grain boundaries," *Materials Reliability in Microelectronics V* **391** (1995).

[10] C. K. Hu, L. Gignac, S. G. Malhotra, R. Rosenberg and S. Boettcher, "Mechanisms for very long electromigration lifetime in dual-damascence Cu interconnections," *Appl. Phys. Lett.* **78** (2001), 904.

[11] M. Shatzkes and J. R. Lloyd, "A model for conductor failure considering diffusion concurrently with electromigration resulting in a current exponent of 2," *J. Appl. Phys.* **59** (1986), 3890.

[12] C. S. Hau-Riege and C. V. Thompson, "Electromigration in Cu interconnects with very different grain structures," *Appl. Phys. Lett.* **78** (2001), 3451.

[13] K. L. Lee, C. K. Hu and K. N. Tu, "In-situ scanning electron microscope comparison studies on electromigration of Cu and Cu(Sn) alloys for advanced chip interconnects," *J. Appl. Phys.* **78** (1995), 4428.

[14] M. Y. Yan, K. N. Tu, A. V. Vairagar, S. G. Mhaisalkar and Ahila Krishnamoorthy, "Confinement of electromigration induced void propagation in Cu interconnect by a buried Ta diffusion barrier layer," *Appl. Phys. Lett.* **87** (2005), 261906.

[15] A. V. Vairagar, S. G. Mhaisalkar, Ahila Krishnamoorthy, K. N. Tu, A. M. Gusak, M. A. Meyer and Ehrenfried Zschech, "In-situ observation of electromigration induced void migration in dual-damascene Cu interconnect structures," *Appl. Phys. Lett.* **85** (2004), 2502–4.

[16] C. W. Park and R. W. Vook, "Activation energy for electromigration in Cu films," *Appl. Phys. Lett.* **59** (1991), 175.

[17] J. R. Lloyd and J. J. Clement, "Electromigration in copper conductors," *Thin Solid Films* **262** (1995), 135.

[18] L. Lu, Y. Shen, X. Chen, L. Qian and K. Lu, "Ultrahigh strength and high electrical conductivity in copper," *Science* **304** (2004), 422.

[19] Kuen-Chia Chen, Wen-Wei Wu, Chien-Neng Liao, L. J. Chen and K. N. Tu, "Observation of atomic diffusion at twin-modified grain boundaries in copper," *Science* **231** (2008), 1066–9.

Problems

12.1 Ag is a better conductor than Cu and Al. Why don't we use Ag as interconnects?

12.2 To reduce RC delay, we will need a low dielectric constant insulator. Since air has a dielectric constant of 1 and if we use Cu interconnects with air-gap or even free-standing Cu interconnects, what will the reliability issues be? If we can use air-gap or free-standing interconnects, why not use Ag?

12.3 It is well known now that we can add 1 atomic % Cu into a Al interconnect to improve its resistance to electromigration. What element can we add to Cu interconnect to improve its resistance to electromigration? Explain the criteria of your selection.

12.4 In Fig. 9.3 in Chapter 9, we considered the flux divergence of a grain-boundary triple point under the driving force of concentration gradient. If we change the driving to electromigration and take the same configuration, temperature, and grain-boundary diffusivity, calculate the time to grow the same void by ignoring the nucleation event.

12.5 In Fig. 11.11, an Al interconnect overhang is shown. Calculate the time needed to form a void to deplete the entire overhang if it is pure Al and the length of the overhang is 5 μm. If the overhang is Al (1 atomic % Cu), what will happen?

12.6 Electric current crowding occurs in the multilevel interconnect structure when there is a turn in the path of the conduction. Besides turns, what other structural features can lead to current crowding?

12.7 A drift velocity of 2 μm/s was measured in a Blech test structure of Cu at 350 °C under a current density of 1×10^6 A/cm^2. Calculate the effective charge number Z^* of Cu by assuming surface diffusion and also grain-boundary diffusion of Cu.

13 Thermomigration

13.1 Introduction

When an inhomogeneous binary solid solution or alloy is annealed at constant temperature and constant pressure, it will become homogeneous to lower the free energy. Conversely, when a homogeneous binary alloy is annealed at constant pressure but under a temperature gradient, i.e. one end of it is hotter than the other, the opposite will happen: the alloy will become inhomogeneous, and the free energy increases. This de-alloying phenomenon is called the Soret effect, as mentioned in Chapter 10. It is due to thermomigration or mass migration driven by a temperature gradient [1–3]. Since the inhomogeneous alloy has higher free energy than the homogeneous alloy, thermomigration is an energetic process which transforms a phase from a low-energy to a high-energy state. It is unlike a conventional phase transformation which occurs by lowering Gibbs free energy.

In thermodynamics, under homogeneous external conditions defined by a constant temperature and constant pressure (for example, if T is fixed at $100\,°C$ and p is fixed at atmospheric pressure), a thermodynamic system will minimize its Gibbs free energy, and it will move toward the equilibrium condition at the given T and p. Both enthalpy and entropy are state functions, so the Gibbs free energy of the equilibrium state is defined when T and p are given. On the other hand, if the external conditions are inhomogeneous, for example, having different temperatures at the two ends of a sample in thermomigration, the equilibrium state of minimum Gibbs free energy is unattainable. Instead, if the deviation from homogeneity is small, the system will move toward a steady state instead of equilibrium. As discussed in Chapter 10, irreversible thermodynamics indicates that the entropy production inside the inhomogeneous system is caused by the flux of heat due to the temperature gradient.

Thermomigration should occur in a pure metal. One would expect that a kitchen utensil, such as a Cu kettle, should expand in size after years of use. In boiling water, the temperature inside the kettle is $100\,°C$ and outside is about $500–600\,°C$. If the thickness of the kettle is about 1 mm, the temperature gradient is very large, 4000 to $5000\,°C/cm$, and themomigration is expected. When the outside of the kettle is hotter than the inside, Cu atoms would have diffused from the outside to the inside and the latter should have expanded. Yet, this does not seem to happen! One reason is that the lattice diffusion in Cu occurs by vacancy mechanism. The outside of the kettle which is hotter will have a higher concentration of vacancies than the inside. The vacancy concentration gradient induces a counter-atomic flux which might have compensated nearly all the flux of Cu atoms driven by the temperature gradient. The net change may be too small to be noticed. Another reason is due to back stress. The temperature inside the kettle is too low for creep to take place. As thermomigration drives more and more Cu atoms into the cold side and builds up a high compressive stress there, the stress gradient will produce an atomic flux of Cu against the thermomigration. The equilibrium vacancy concentration is affected by the stress, which will be discussed in Chapter 14, Section 14.2.

Solder is a typically binary system, so the Soret effect can be found. Actually, the Soret effect has been reported to occur in PbIn alloy which forms a solid solution over a wide concentration range [4, 5]. On the other hand, eutectic solder has a two-phase microstructure, and the effect of thermomigration in a eutectic two-phase structure is different from that in a solid solution. At a constant temperature below the eutectic temperature, the chemical potential of the two phases in the eutectic region are equal and independent of composition of the two phases. Thus, the two phases can be redistributed without resistance because there is no chemical potential gradient due to composition redistribution. For this reason, the nature of thermomigration in solder joints is different from that in the Soret effect of a solid solution.

However, thermomigration has a temperature gradient, so it is not a constant-temperature process, yet the effect of chemical potential change in the temperature gradient on phase redistribution is small, provided that the temperature difference between the hot and the cold ends of the solder joint is not large, only a few degrees of centigrade.

It is worth noting that thermomigration in Al and Cu interconnects has seldom been studied. In comparison, it is much easier to have thermomigration in solder alloys, especially in flip-chip solder joints. This is because we can have a temperature gradient of 1000 °C/cm in the joint, which is large enough to cause thermomigration. Also at the device working temperature of 100 °C, lattice diffusion in solder joints is fast, especially if joule heating exists, so the kinetic effect of thermomigration can be observed.

Thermomigration in solder joints has been found to be a harder subject to study than electromigration for two reasons [6–9]. First, it is difficult to apply a temperature gradient across a small flip-chip solder joint. For a solder joint of dimension 100 μm in diameter, if we can apply a temperature difference of 10 °C across it, we have a temperature gradient of 1000 °C/cm, which is sufficient to induce thermomigration in the solder, to be discussed in Section 12.3. Therefore, a temperature difference of 10 °C or even a difference of a few degrees of centigrade in a solder joint is of concern. Second, the heat dissipation is hard to control because of the two interfaces in a joint. Therefore, it is difficult to simulate temperature distribution or temperature gradient in a solder joint because of the complicated boundary conditions of UBM on the chip side and bond-pad structure on the substrate side. We have to simplify the test structure of solder joints in order to study thermomigration. On the other hand, solder has a low melting point, so we can use the melting of the solder as an internal calibration. The condition of heat generation and dissipation in the melting experiment can be used to check the simulation.

Due to joule heating, electromigration has caused a non-uniform temperature distribution in a flip-chip solder joint, thus there may be a component of thermomigration in any electromigration experiment. In other words, electromigration in flip-chip solder joints is accompanied by thermomigration when a large current density is applied and when the current distribution is non-uniform due to current crowding. It is an advantage that we can combine electromigration and thermomigration in studying flip-chip solder joints.

In Section 13.2, we shall discuss the design of a test structure of flip-chip solder joints that will enable us to conduct thermomigration with and without electromigration. It will be shown that if a composite solder joint of high-Pb and eutectic SnPb is used, the redistribution of Sn and Pb in thermomigration can be recognized easily by using optical microscopy, even though the original compositional distribution is not homogeneous in the composite sample. Observation of thermomigration in eutectic SnPb flip-chip solder joints will be easy too.

In Section 13.3, the fundamentals of thermomigration will be presented and the driving force of thermomigration and transport heat will be discussed. In Section 13.4, thermomigration under DC or AC electromigration will be given. In Section 13.5, thermomigration in Pb-free flip-chip solder joints will be presented. In Section 13.6, the interaction between thermomigration and stress-migration will be discussed.

13.2 Thermomigration in flip-chip solder joints of SnPb

13.2.1 Thermomigration in unpowered composite solder joints

In Chapter 10, Figs. 10.8(a), (b), and (c) show respectively the schematic diagram of a flip-chip on a substrate, the cross-section of a composite of 97Pb3Sn and 37Pb63Sn flip-chip solder joints, and SEM images. The composite solder joints have been used to study thermomigration. As a control experiment, a constant temperature heating of the composite flip-chip samples was performed in an oven at constant temperature of 150 °C and at constant atmospheric pressure for a period of one, two, to four weeks. The microstructures of the cross-section were examined under optical microscope (OM) and SEM, as shown in Fig. 10.9. No mixing between the high-Pb and the eutectic was observed and the image was essentially the same as that shown in Fig. 10.8(c), the reason being that there is negligible chemical potential difference between the high-Pb and the eutectic SnPb at a constant temperature of 150 °C.

To conduct thermomigration in the composite solder joints, we use the temperature gradient induced from joule heating in electromigration. The set of flip-chip samples was depicted in Fig. 10.10(a); there were 24 bumps on the peripheral of the Si chip, and all the bumps have the original microstructure as shown in Fig. 10.8(c) before electromigration stressing. The darker region in the bottom area of each bump is the eutectic SnPb and the brighter region in the top part is 97Pb3Sn. After electromigration was conducted through only four pairs of bumps, the effect of thermomigration is clearly visible across all of the unpowered solder joints, as shown in Fig. 10.10(b), because in all of them Sn has migrated to the Si side, the hot end, and Pb has migrated to the substrate side, the cold end. The redistribution of Sn and Pb was caused by temperature gradient across the solder joints, since no current was applied to them.

13.2.2   In-situ observation of thermomigration

In-situ observation of thermomigration was conducted with flip-chip samples depicted in Fig. 13.1(a). The chip was cut and only a thin strip of Si was kept. The strip has one row of solder bumps connecting it to the substrate. The bumps were cut and polished to the middle so that the cross-section of each of the bumps was exposed for in-situ observation during thermomigration. Fig. 13.1(a) is a schematic diagram of the Si strip and a row of four of the cross-sectioned bumps. Due to the excellent thermal conduction of Si and the small strip used, when one pair of the bumps is powered under DC or AC current, the other pair of unpowered solder joints experiences almost the same thermal gradient as the powered pair. This set of samples can be used to conduct in-situ experiments by observing changes on the cross-sectioned surfaces directly during electromigration and thermomigration. The major difference between the first set as shown in Fig. 10.10 and this set is that in the latter, the bumps have a polished free surface during the test. Besides composition redistribution, surface bulging can occur if a large amount of materials is driven to the cold end by thermomigration, and the bulge can be observed easily.

(a)

(b)

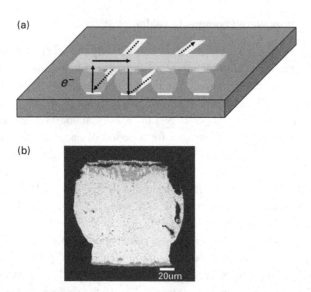

Fig. 13.1   (a) Schematic diagram of the Si stripe and a row of four of the cross-sectioned bumps. The chip was cut and only a thin strip of Si was kept. The strip has one row of solder bumps connecting it to the substrate. The bumps were cut to the middle so that the cross-section of the bump was exposed for in-situ observation during electromigration. (b) Cross-sectional SEM image of one of the bumps on the right after thermomigration.

The pair of joints on the left shown in Fig. 13.1(a) was powered at 2×10^4 A/cm^2 for 20 h at 150 °C; the pair of joints on the right had no electric current at all, yet all of them showed composition redistribution and damage. The pair on the right (unpowered) showed a uniform void formation at the interface on the top side, i.e. the Si side which is also the hot side. The SEM image of one of them is shown in Fig. 13.1(b). In the bulk of the joint, some phase redistribution can be recognized. The redistribution of elements of Sn, Pb, and Cu can be measured by electron microprobe from the cross-sections of this unpowered pair on the right. The Sn has migrated to the hot end and there is more Cu in the hot end too, and Pb has moved to the cold end.

If we assume that there was no temperature gradient in the pair of bumps on the right in Fig. 13.1, in other words, the temperatures were uniform in these bumps, then its thermal history is similar to isothermal annealing, and no phase redistribution or void formation should have been found since isothermal annealing has no effect on phase mixing or unmixing, as shown in Fig. 10.9 in Chapter 10. However, we may ask: can there be other kinds of driving force that can lead to the phase change as observed? Besides electrical and thermal forces, we could have mechanical force. Yet the mechanical force should have existed in isothermal annealing. The annealing does cause interfacial chemical reactions between solder and UBM on the chip side and between solder and bond-pad metal on the substrate side. The growth of IMC may generate stress owing to molar volume change. However, this effect should have existed in the sample which was isothermally annealed at 150 °C for four weeks, yet no noticeable change was detected, as shown in Fig. 10.9. Furthermore, solder has a very high homologous temperature at 150 °C; it is unlikely that stress will not be relaxed in four weeks.

Thus, we conclude that the composition redistribution and the damage (void formation) in the unpowered bumps are due to thermomigration. Then, which is the dominant diffusing species in thermomigration, or which species diffuses with the temperature gradient, is of interest. During electromigration at 150 °C, Pb has been found to be the dominant diffusing species. In thermomigration of the composite solder joint, the temperature gradient drives Pb from the hot side to the cold side and Sn atoms from the cold side to the hot side. Since void was found in the hot side, this indicates that Pb is the dominant diffusing species and the flux of Pb is greater than the flux of Sn. The high Pb forms Cu_3Sn after reflow, but the Cu_3Sn will transform to Cu_6Sn_5 after the diffusion of Sn to the hot side. The formation of void and Cu_6Sn_5 is rather uniform across the entire contact area to the Si side. Why Sn diffuses against temperature gradient in the composite solder joints is an interesting question. To answer it, we shall discuss the driving force and the flux motion in a two-phase microstructure under the constraint of constant volume. Furthermore, in Section 13.5, thermomigration in Pb-free flip-chip solder joints will be discussed, and Sn moves from the cold end to the hot end without Pb.

13.2.3 Random states of phase separation in the two-phase eutectic structure

Fig. 13.2(a), (b), and (c) show a set of cross-sectional SEM images of unpowered composite solder joints after thermomigration of 30 min, 2 h, and 12 h, respectively. Before

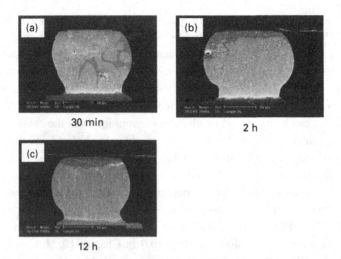

Fig. 13.2   (a), (b), and (c) show a set of the cross-sectional image of unpowered solder joints after thermomigration of 30 min, 2 h, and 12 h, respectively. Before thermomigration, the image is similar to that shown in Fig. 10.8(a). In Fig. 13.2(a) a random state of phase separation is observed. In Fig. 13.2(b) the eutectic is segregated towards the hot end. In Fig. 13.2(c) a near-complete phase separation is achieved.

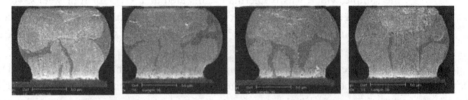

Fig. 13.3   Four images of random states in phase separation the two-phase microstructure in both DC and AC stressing are shown.

thermomigration, the image is similar to that shown in Fig. 10.8(c). In Fig. 13.2(a) a random state of phase separation is observed. In Fig. 13.2(b) the eutectic is segregated towards the hot end. In Fig. 13.2(c) a near-complete phase separation is achieved. Many images similar to Fig. 13.2(a) in both DC and AC stressing were obtained, and four of them are shown in Fig. 13.3 to illustrate the random state of phase separation in the two-phase microstructure before achieving the complete phase separation. It appears that a fluid-like motion occurs in the solid state phase separation.

When an electron microprobe was used to measure composition distribution across a polished cross-section of the flip-chip sample after thermomigration, a highly irregular or stochastic composition distribution of three scans was observed as shown in Fig. 13.4; no smooth concentration profile was observed. If the electromigration experiment was extended to several days, a clear phase separation of Sn and Pb in the unpowered joints was found.

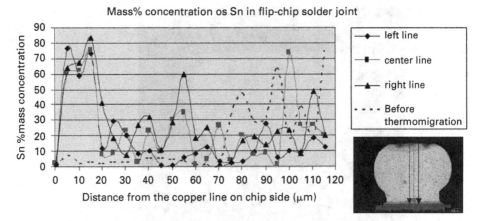

Mass% concentration os Sn in flip-chip solder joint

Fig. 13.4   Electron microprobe measurement of composition redistribution across a polished cross-section of the flip-chip sample after thermomigration. A highly irregular or stochastic composition distribution was observed; no smooth concentration profile was observed.

13.2.4   Thermomigration in unpowered eutectic SnPb solder joints

The eutectic 37Pb63Sn flip-chip solder joints, without the highPb, used for the thermomigration test were arranged similarly to that shown in Fig. 10.10(a), except that there were 11 bumps. The UBM thin films on the chip side were Al ($\sim$0.3 μm)/Ni(V) ($\sim$0.3 μm) /Cu ($\sim$0.7 μm) deposited by sputtering. The bond-pad metal layers on the substrate side were Ni (5 μm)/Au (0.05 μm) prepared by electroplating. The bump height between the UBM and the bond-pad is 90 μm. The contact opening on the chip side has a diameter of 90 μm. Fig. 13.5 shows the cross-sectional SEM image of a eutectic SnPb solder bump in the as-received state.

Only one pair of them, numbers 6/7, was current stressing with a DC current of 0.95 A at 100 °C for 27 h. The average current density at the contact opening was 1.5 $\times$ 10^4 A/cm^2. The unpowered bumps neighboring the powered pair were used to study thermomigration.

Fig. 13.6(a) depicts the arrangement of the 11 bumps. Fig. 13.6(b) displays a SEM image of the cross-section of all 11 bumps after the electromigration test. The lighter color in the SEM image represents the Pb-rich phase and the darker color represents the Sn-rich phase. Compared to the as-received sample shown in Fig. 13.5, the results show that the Pb-rich phase has moved to the substrate side (the cold side) in the unpowered bumps. Also, one of the unpowered neighboring bumps, shown in Fig. 13.6(b), has some dendritic crystallization structure of a liquid phase, indicating that it was partially melted in the test. It is worth noting that crystallization of a molten eutectic phase should show a eutectic microstructure. The dendritic structure indicates that phase separation has occurred before melting. The melting suggests that thermomigration in eutectic SnPb occurs at high temperatures close to the melting point.

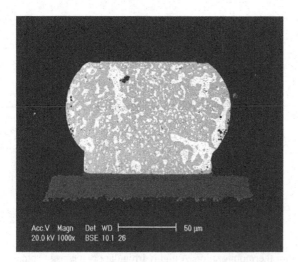

Fig. 13.5   Cross-sectional SEM image of an eutectic SnPb solder bump in the as-received state.

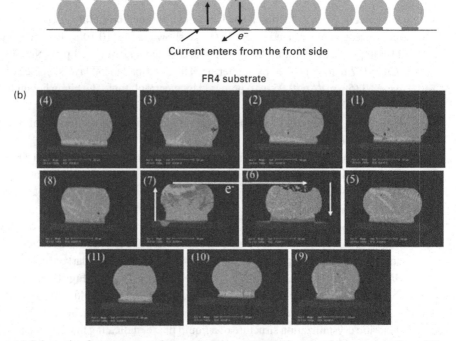

Fig. 13.6   (a) Schematic of arrangement of 11 solder joints. (b) SEM images of the cross-section of 11 solder bumps after only one pair of them, No. 6/7, underwent current stressing. The direction of electron flow is marked by arrows. The lighter color in the SEM image is the Pb-rich phase and the darker color the Sn-rich phase.

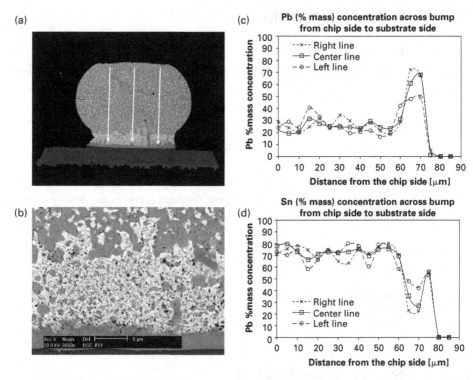

Fig. 13.7 (a) Enlarged SEM picture of bump no. 11. The redistribution of Sn and Pb can be seen. It shows the migration of a large amount of Pb to the substrate side (the cold side), yet there is no accumulation of Sn to the chip side (the hot side). (b) A higher magnification image of the Pb-rich phase in the cold end. (c) and (d) The concentration profiles across the bump by EPMA of Pb and Sn, respectively. Three profile lines across the bumps were scanned and every line is the average of three sets of data points. Each point was taken at every 5 μm step from the chip side to the substrate side.

Fig. 13.7(a) shows an enlarged SEM picture of an unpowered bump after thermomigration. The redistribution of Sn and Pb is shown by the accumulation of a large amount of Pb (lighter color) to the substrate side (the cold side), yet there is no accumulation of Sn to the chip side (the hot side) and the Sn distribution is quite uniform across the bump. The very surprising finding is not only that the microstructure in the bulk of the bump is quite uniform (except the accumulated Pb-rich phase), but also the lamellar structure is much finer, indicating the existence of many more interfaces in the microstructure after the phase separation, in turn a higher energy state. Recall that when a eutectic two-phase microstructure is annealed at constant temperature, coarsening, instead of refinement, of the two-phase lamellar microstructure should occur to reduce the surface energy. Fig. 13.7(b) shows an enlarged SEM image of the Pb-rich phase in the cold side. Figs. 13.7(c) and (d) show the concentration distribution of Pb and Sn, respectively, in the sample.

13.3   Analysis of thermomigration

In terms of the irreversible processes, the heat flow and mass flow in thermomigration can be expressed by the temperature gradients and chemical potential gradients as below.

$$J_Q = -L_{QQ}\frac{1}{T}\frac{dT}{dx} - L_{QM}T\frac{d}{dx}\left(\frac{\mu}{T}\right) \tag{13.1}$$

$$J_M = -L_{MQ}\frac{1}{T}\frac{dT}{dx} - L_{MM}T\frac{d}{dx}\left(\frac{\mu}{T}\right)$$

When a material is held in a temperature gradient until a concentration gradient is established to balance the temperature gradient, it comes to a steady state that the mass flow J_M will be zero. Taking $J_M = 0$, we have from the last equation,

$$L_{MQ}\frac{1}{T}\frac{dT}{dx} = -L_{MM}T\frac{d}{dx}\left(\frac{\mu}{T}\right)$$

Eliminating T/dx, we have

$$d\left(\frac{\mu}{T}\right) = -\frac{L_{MQ}}{L_{MM}}\frac{dT}{T^2}$$

Now, by differentiation we have

$$d\left(\frac{\mu}{T}\right) = \frac{1}{T}d\mu + \mu d\left(\frac{1}{T}\right) = \frac{1}{T}d\mu - \mu\frac{1}{T^2}dT$$

Using the thermodynamic relations,

$$d\mu = -SdT + Vdp \text{ and } \mu = H - TS$$

and substituting them into the previous equation, we obtain

$$d\left(\frac{\mu}{T}\right) = \frac{Vdp}{T} - H\frac{dT}{T^2} = -\frac{L_{MQ}}{L_{MM}}\frac{dT}{T^2}$$

thus

$$\frac{Vdp}{T} = \left(H - \frac{L_{MQ}}{L_{MM}}\right)\frac{dT}{T^2}$$

To gain an understanding of the meaning of L_{MQ}/L_{MM}, we consider the ratio of heat flow to mass flow under isothermal conditions, that is, when $dT/dx = 0$. We have

$$\frac{J_Q}{J_M} = \frac{L_{QM}}{L_{MM}} = \frac{L_{MQ}}{L_{MM}}$$

by using Onsager's relation that $L_{QM} = L_{MQ}$.

The term L_{MQ}/L_{MM} represents the energy flow associated with a mass flow. Defining $Q' = L_{MQ}/L_{MM}$, we have

$$\frac{Vdp}{T} = (H - Q')\frac{dT}{T^2} = Q^*\frac{dT}{T^2} \tag{13.2}$$

where we define the heat of transport, $Q^* = H - Q'$. It represents the difference between the energy associated with the materials that flows (Q') and the enthalpy of the materials (H) in the reservoir from which the flow starts. In iron-carbon alloys, Shewmon [1] showed that under temperature gradient, carbon moved to the hot side and a steady state was established. The value of Q^* for carbon in α-iron is about -24 kcal/mol near $700\,^\circ$C. The sign of Q^* will be discussed in the next section.

13.3.1 Driving force of thermomigration

In the thermoelectric effect, a temperature gradient can move electrons. Similarly, a temperature gradient can drive atoms. In essence, the electrons in the high-temperature region have higher energy in scattering or stronger interaction with diffusing atoms, hence atoms move down the temperature gradient. On the driving force of atomic diffusion, recall that the atomic flux driven by chemical potential can be given as

$$J = C < v > = CMF = C\frac{D}{kT}\left(-\frac{\partial \mu}{\partial x}\right) \tag{13.3}$$

where $< v >$ is drift velocity, $M = D/kT$ is mobility, and μ is chemical potential energy. Considering temperature gradient as the driving force, we have on the basis of conjugate force as shown in Eq. (10.14),

$$J = C\frac{D}{kT}\frac{Q^*}{T}\left(-\frac{\partial T}{\partial x}\right) \tag{13.4}$$

where Q^* is defined as heat of transport. Comparing the last two equations, we see that Q^* has the same dimension as μ, so it is the heat energy per atom. The definition of Q^* is the difference between the heat carried by the moving atom and the heat of the atom at the initial state (the hot end or the cold end).

To define the sign of Q^*, we consider J in Eq. (13.4) between two points; point 1 at (x_1, T_1) and point 2 at (x_2, T_2) in the Cartesian coordinates, and we assume that $T_1 > T_2$ and $x_1 < x_2$, and atomic flux moves from hot to cold, i.e. from point 1 to point 2. Then, $\Delta T/\Delta x$ is negative, so Q^* is positive. Recall that this is also the reason why all the flux equations as shown in Eq. (10.4) have a negative sign. Thus, for an element which moves from the hot end to the cold end, Q^* is positive. For an element moving from cold to hot, its Q^* is negative.

The driving force of thermomigration is given as

$$F = -\frac{Q^*}{T}\left(\frac{\partial T}{\partial x}\right) \tag{13.5}$$

To make a simple estimation, we take $\Delta T / \Delta x = 1000\,°C/cm$, and consider the temperature difference across an atomic jump and take the jump distance to be $a = 3 \times 10^{-8}$ cm. We have a temperature change of 3×10^{-5} K across an atomic spacing, so the thermal energy change will be

$$3k\Delta T = 3 \times 1.38 \times 10^{-23}\,(J/K) \times 3 \times 10^{-5}\,K \approx 1.3 \times 10^{-27}\,J$$

As a comparison, we shall consider the driving force, F, of electromigration at a current density of 1×10^4 A/cm^2 or 1×10^8 A/m^2 which we know has induced electromigration in solder alloys:

$$F = Z^* e E = Z^* e \rho j \tag{13.6}$$

We shall take $\rho = 10 \times 10^{-8}\,\Omega\,m$, Z^* of the order of 10, and $e = 1.602 \times 10^{-19}$ coulomb, and we have $F = 10 \times 1.6 \times 10^{-19}$ (C) $\times 10 \times 10^{-8}\,(\Omega\,m) \times 10^8$ A/m$^2 = 1.6 \times 10^{-17}$ C V/m $= 1.6 \times 10^{-17}$ N.

The work done by the force in a distance of atomic jump of 3×10^{-10} m will be $\Delta w = 4.8 \times 10^{-27}$ N m $= 4.8 \times 10^{-27}$ J. This value is close to the thermal energy change we have calculated in the above for thermomigration. Thus, if a current density of 10^4 A/cm^2 can induce electromigration in a solder joint, a temperature gradient of $1000\,°C/cm$ will induce thermomigration in a solder joint.

On heat of transport, note that Q^* can be positive or negative. In a Fe–C system, carbon was found to move to the hot end interstitially with a positive heat of transport. In alloys of SnPb, when thermomigration drives Pb to move from the hot zone to the cold zone, it moves down the temperature gradient. But the thermomigration drives Sn to move in the opposite direction; it moves against the temperature gradient. The Q^* for Pb is negative or the heat decreases, but for Sn, it seems that the Q^* is positive since it moves to the hot end and gains heat. This is because we have one temperature gradient in thermomigration for both species, unlike interdiffusion in a diffusion couple, in which the concentration gradient of the two interdiffusing species is in the opposite direction, so the chemical potential change in interdiffusion can be positive for both species.

To measure Q^*, if we know the atomic flux, we can use the flux equation, i.e. Eq. (13.3), to determine Q^* when diffusivity, the average temperature, and temperature gradient are known. The heat of transport of Pb in thermomigration discussed in Section 13.2.4 is estimated below by using the flux equation, Eq. (13.3).

By measuring the accumulation width of Pb (12.5 μm) on the substrate side from Fig. 13.7(a), the total volume of atomic transportation can be obtained from the product of the width and the cross-section of the solder joint. Taking the density of 27Sn73Pb as 10.25 g/cm^3, the molecular weight of 27Sn73Pb as 183.3 g/mole, the flux of $J_{TM} = 4.26 \times 10^{14}$ atom/cm^2 s is obtained. Assuming a temperature gradient of $1000\,°C/cm$, and a temperature of $180\,°C$ at the hot side, which is very close to the melting temperature of eutectic SnPb, and a diffusivity of $D_{Pb} = 4.41 \times 10^{-13}$ cm^2/s, the molar heat of transport Q^*_{Pb} is estimated to be $+79$ kJ/mole.

The determination accuracy of Q^* may be affected by the measurement of flux in Eq. (13.3) since the concentration distribution is non-uniform. The assumed temperature gradient may be incorrect. However, more serious is the basic assumption in the analysis that both Pb and Sn move with the temperature gradient. Actually, if Pb is the dominant diffusing species and moves from the hot side to the cold side, the Sn will be pushed back in the opposite direction if a constant volume process is assumed. The effect of reverse flux of Sn on the calculation of transport heat in a two-phase microstructure should be studied.

13.3.2   Thermomigration in eutectic two-phase alloys

Thermomigration in a eutectic alloy is unique and it is different from that in a solid solution. A eutectic alloy below the eutectic temperature is an alloy of two phases at equilibrium. In a two-phase mixture the change of composition at constant temperature does not mean any change of chemical potential; instead, it means just the change of local volume fractions of the two phases. Composition of each primary phase is determined by the thermodynamic equilibrium between them and is known from the equilibrium phase diagram. Thus, if some redistribution of the two phases in a eutectic solder is induced by thermomigration, it means a change of gradient of volume fractions, not a change of the gradient of chemical potentials, so the redistribution can be enormous due to the lack of a counteracting force.

Strictly speaking, an equilibrium phase diagram is obtained assuming constant temperature and constant pressure. Thus, the concept of a constant chemical potential between the two eutectic phases at a constant temperature cannot be applied to thermomigration as discussed in the last paragraph because the temperature is not constant. It is an approximation, provided that ΔT is small.

At the end of thermomigration in a eutectic two-phase structure, no steady state of a linear concentration gradient is achieved; instead, a near-complete segregation of the two eutectic phases occurs. Furthermore, since a gradient of volume fractions is not a driving force, the lack of counteracting force in the form of $\Delta C / \Delta x$ will not produce a smooth segregation, so a stochastic behavior tendency occurs in the thermomigration of the eutectic mixture, as shown experimentally in Fig. 13.3, Section 13.2.3. No smooth concentration gradient exists as in the Soret effect of a solid solution.

Consider a two-phase mixture of almost pure components; hereafter the indexes 1 and 2 correspond to phases as well as to species. The shape of the sample is assumed to be unchanged, so we have the constraint of constant volume in every part of the sample. It means that in the laboratory reference frame, the sum of volume fluxes of two species should be zero everywhere:

$$\Omega_1 J_1 At + \Omega_2 J_2 At = 0 \tag{13.7}$$

or

$$\Omega_1 J_1 = -\Omega_2 J_2 \tag{13.8}$$

where J_1, J_2 are fluxes of atoms per unit area per unit time, and Ω_1, Ω_2 are atomic volumes. A and t are the cross-section of the sample and reaction time, respectively. Under the assumption of constant volume, in the two-phase system, J_2 is in the reverse direction of J_1.

13.4   Thermomigration under DC or AC stressing in flip-chip solder joints

In DC electromigration, there is a polarity effect between the cathode and the anode. When a daisy chain of bumps is tested by DC electromigration, the void formation occurs on every alternative bump at the cathode contact to the Si. Therefore, it is very easy to recognize DC electromigration in flip-chip solder joints. However, we have to consider the contribution of thermomigration to DC electromigration. Thermomigration may accompany electromigration when the joule heating of the latter has induced a temperature gradient of the magnitude of 1000 °C/cm across the solder joint.

If we consider DC electromigration in the pair of bumps on the left in Fig. 13.1, and we assume that electrons flow up in the left side bump and down on the right-side bump; if we assume the temperature is hotter on the Si chip side, i.e. the top side in Fig. 13.1, thermomigration will drive the dominant diffusing species down, and it is in

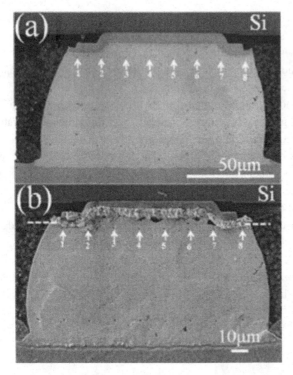

Fig. 13.8   SEM images of the cross-section sample (a) before and (b) after thermomigration respectively at a AC current density of 1×10^4 A/cm^2 and at 100 °C for 800 h.

the same direction as electromigration in the right-side bump with the downward flow electrons, so the effects of thermomigration and electromigration are added together. Both thermomigration and electromigration will drive vacancies to go to the Si contact, and void formation will occur near the contact. However, in the left-side bump of the pair, electromigration will drive atoms in the opposite direction and counteract thermomigration, i.e. the two effects tend to cancel each other. Since we can obtain different experimental results in a pair of bumps, we should be able to decouple the contribution of thermomigration and electromigration.

If we use pure Sn flip-chip samples, we have a simple case of diffusion of one element and we can use a marker to determine the net effect of fluxes. With two elements in eutectic SnPb bumps or in solid-solution PbIn bumps, the problem is more complicated. In these cases, besides marker motion, the concentration change of the Sn and Pb (or Pb and In) fluxes should be determined. Due to stochastic behavior, the analysis of eutectic SnPb is even more complicated than that of PbIn.

There is no difference in joule heating whether we apply AC or DC to stress a pair of flip-chip solder bumps. This is because the current distribution in the pair of powered bumps as shown in Fig. 13.1 is independent of the current direction or polarity; therefore, joule heating is the same whether we apply AC or DC in electromigration, except that could be a difference at a very high-frequency AC. However, unlike DC, it is generally assumed that AC does not induce mass flow. If the assumption is true that there is no electromigration-induced mass migration in a pair of bumps stressed by AC, we should expect only thermomigration in the pair of bumps powered by AC, provided that AC has generated a temperature gradient in the pair of bumps. This is the advantage of using AC to study thermomigration in a flip-chip solder joint; AC serves just as a heating source to generate a temperature gradient across the solder bump without electromigration.

We should be able to verify the assumption that AC does not induce mass migration by examining very carefully a pair of bumps that has been stressed by AC together with a neighboring bump that was unpowered. The arrangement has been depicted in Fig. 13.1, in which the pair of bumps on the left-hand side can be stressed by AC, but the neighboring pair on the right-hand side is a dummy pair and will carry no current. The joule heating generated by the AC-stressed left pair will cause the same thermomigration in both pairs. Marker displacement experiments should be conducted in both pairs to determine whether or not the mass migration is the same. Direct comparison with DC experiments should also be made.

13.5 Thermomigration in Pb-free flip-chip solder joints

Thermomigration in Pb-free SnAg3.5 solder joints was studied by a AC of 50 Hz at 1×10^4 A/cm^2 on a hotplate kept at 100 °C. On the cross-section of the solder joints, an array of markers of tiny holes of 10^4 nm^2 in area, made by focused ion beam, were used to determine the direction and magnitude of the atomic flux in thermomigration. Using infrared (IR), the temperature gradient across the solder bump was found to be around 2800 °C/cm. The temperature gradient, measured as $(T_{chip} - T_{subst})/h$, where h was the height of the solder joint of 100 μm, and T_{chip} (about 154 °C) and T_{subst} (about 125 °C) were the temperatures at the chip side and the substrate side, respectively, was determined by IR scan on a polished cross-section of the solder joint under power. Fig. 13.8(a) and (b) shows SEM images respectively of the cross-sectioned sample before and after thermomigration at a AC current density of 1×10^4 A/cm² and at 100 °C for 800 h. In Fig. 13.8(b), a large amount of hillocks is observed at the chip side which was the hot end. Also, some voids were formed just below the hillock. The composition of the hillock was determined to be Sn. The Ag in the solder was found to have migrated to the substrate side, which was the cold end. The markers were found to have moved toward the substrate side, indicating that the dominant diffusing species in the thermomigration had moved to the chip side or the hot end.

Note that the annealing time of 800 h is very long. The hillock formation and marker motion indicate that Sn has moved from the cold end to the hot end. The atomic flux as measured from the marker motion showed that the flux is about one order of magnitude smaller than that in the eutectic SnPb. The Sn transport heat is measured to be about +1.36 kJ/mole. The positive transport heat means that Sn atoms gain heat in thermomigration.

13.6   Thermomigration and creep in Pb-free flip-chip solder joints

The hillock formation at the hot end in Fig. 13.8(b) indicates that the hot end was under compression. The effects of stress and temperature on vacancy concentration in the hot end are mixed. According to the Nabarro–Herring creep model (see Section 14.3), the vacancy concentration in the compressive region is below the equilibrium concentration in the unstressed region. Then, from the temperature effect, the hot end should have a higher concentration of vacancy than the cold end. Since thermomigration has occurred and marker motion indicated that mass flux had moved to the hot end, the net vacancy flux was moving to the cold end, so the temperature effect is larger than the stress effect. However, why some voids were found at the hot end cannot be explained by the temperature effect.

In the following, we shall couple thermomigration and creep. Recall that in Chapter 10, Section 10.7.1, we have coupled electromigration and creep. Both electromigration and creep are constant-temperature processes, but thermomigration is not. To couple thermomigration and creep, the following analyses are considered.

First, we assume the solder composition to be pure Sn, so that there is no concentration gradient. This is a reasonable assumption, since in the SnAg solder, Ag diffuses interstitially in Sn, so most Ag will be driven to the cold end in the early stage of thermomigration. In the larger part of the 800 h period of thermomigration reported before, it was basically Sn diffusion in pure Sn.

Second, we should consider creep under a temperature gradient in order to couple it to thermomigration. For simplicity, while we have temperature as a variable, we shall use the concept of constant temperature creep, since the temperature difference across the solder joint is only a few degrees of centigrade, so $\Delta T/T_m$ is very small where T_m is the melting point of the solder.

Third, to estimate the driving force of stress migration in order to see if it is of the same order of magnitude as thermomigration, we consider a flip-chip solder joint of 100 μm diameter. At the anode, we assume a compressive stress at the yield-stress level of 30 MPa, and it is stress-free at the cathode. The atomic volume of a Sn atom is taken to be 27×10^{-24} cm^3. So we have

$$\sigma\Omega = 30 \times 10^7 (\text{dyn/cm}^2) \times 27 \times 10^{-24}(\text{cm}^3) = 810 \times 10^{-17}\text{erg}$$

The driving force will be

$$F = -\frac{\Delta\sigma\Omega}{\Delta x} = -\frac{0 - 8 \times 10^{-15}\text{erg}}{10^{-2}\text{cm}} = 8 \times 10^{-13}\text{erg/cm}$$

The work done by this force over an atomic jump distance of 0.3 nm is 2.4×10^{-27} joule, which is of the same order of magnitude as those calculated for thermomigration. Thus, the driving force will lead to stress-migration and can be coupled to thermomigration.

In Eq. (10.25) and Eq. (10.2) in Chapter 10, we replace the chemical potential, μ, by the stress potential, $\sigma\Omega$, and we couple creep and thermomigration by the following pair of equations:

$$J_M = C\frac{D}{kT}\left[-T\frac{d}{dx}\left(\frac{\sigma\Omega}{T}\right)\right] - C\frac{D}{kT}\frac{Q^*}{T}\frac{dT}{dx} \tag{13.20}$$

$$J_Q = L_{QM}\left[-T\frac{d}{dx}\left(\frac{\sigma\Omega}{T}\right)\right] - \kappa\frac{dT}{dx} \tag{13.21}$$

We can rewrite the last two equations as

$$J_M = C\frac{D}{kT}\left[-\frac{d\sigma\Omega}{dx} + \left(\frac{\sigma\Omega - Q^*}{T}\right)\frac{dT}{dx}\right] \tag{13.22}$$

$$J_Q = L_{QM}\left(-\frac{d\sigma\Omega}{dx}\right) + \left(L_{QM}\frac{\sigma\Omega}{T} - \kappa\right)\frac{dT}{dx} \tag{13.23}$$

In Eq. (13.22), only Q^* is unknown. This is because, experimentally, we can measure J_M from marker motion, we can assume σ to be the elastic limit, so we have $d\sigma/dx$, and we know dT/dx and T. On marker motion, we have

$$\Omega J_M At = A\Delta x$$

where Ω is the atomic volume of Sn, A is the cross-sectional area of the sample, t is stressing time, and Δx is the average displacement of markers. We found Q^* to be -7.4 kJ/mole.

In Eq. (13.22), if we let $J_M = 0$, it means that the back stress will balance the thermomigration, and there is no net atomic flux at a steady state, so we have

$$\frac{d\sigma\Omega}{dx} = \left(\frac{\sigma\Omega - Q^*}{T}\right)\frac{dT}{dx} \tag{13.24}$$

By rearrangement, we have

$$\frac{dT}{T} = \frac{d\sigma\Omega}{\sigma\Omega - Q^*}$$

By integration, we obtain

$$\frac{T_1}{T_2} = \frac{\sigma_1\Omega - Q^*}{\sigma_2\Omega - Q^*}$$

Then,

$$\frac{\Delta T}{\Delta\sigma} = \frac{T_1 - T_2}{\sigma_1 - \sigma_2} = \frac{T_2\Omega}{\sigma_2\Omega - Q^*} \tag{13.25}$$

Note that both σ and Q^* can be positive or negative. The above equation indicates the condition under which thermomigration can be balanced by creep or by stress migration.

In the introduction to this chapter, we discussed the lack of thermomigration in copper kettles. Inside the kettle, the temperature is $100\,^\circ\mathrm{C}$ with boiling water, so the temperature may be too low for stress relaxation by atomic diffusion. Back stress will build up to retard thermomigration even though the temperature gradient is very large.

On the other hand, in Pb-free flip-chip solder joints, when Sn is being driven to the hot end by thermomigration, stress relaxation can occur because of the high homologous temperature, so we observed hillock formation at the hot end, as shown in Fig. 13.8(b). The hillock growth requires atomic diffusion of Sn in the direction normal to the sample surface, which will require vacancies to diffusion in the opposite direction. These vacancies will form voids surrounding the hillocks as shown in Fig. 13.8(b).

Recall that in electromigration in stressed short strips of Al, a critical length was found below which there will be no electromigration. In thermomigration in stressed flip-chip solder joints, no critical length is found.

Experimentally, to study the interaction between thermomigration and applied mechanical stress in solder joints, we can also use the sample of Cu wire/solder-ball/Cu wire, the same as the study of the interaction between electromigration and applied mechanical stress.

References

[1] Paul Shewmon, "Diffusion in solids," Ch. 7 of "Thermo- and Electro-Transport in Solids" (TMS, Warrendale, PA, 1989).

[2] D. V. Ragone, "Thermodynamics of materials," Vol. II, Ch. 8 of *Nonequilibrium Thermodynamics* (Wiley, New York, 1995).

[3] R. W. Balluffi, S. M. Allen and W. C. Carter, "Irreversible thermodynamics: coupled forces and fluxes," Ch. 2 of *Kinetics of Materials* (Wiley-Interscience, Hoboken, NJ, 2005).

[4] W. Roush and J. Jaspal, "Thermomigration in Pb-In solder," IEEE Proc. **CH1781** (1982), 342–45.

[5] D. R. Campbell, K. N. Tu and R. E. Robinson, "Interdiffusion in a bulk couple of Pb-PbIn alloy," *Acta Met.* **24** (1976), 609.

[6] H. Ye, C. Basaran and D. C. Hopkins, "Thermomigration in Pb-Sn solder joints under joule heating during electric current stressing," *Appl. Phys. Lett.* **82** (2003), 1045–7.

[7] Y. C. Chuang and C. Y. Liu, "Thermomigration in eutectic SnPb alloy," *Appl. Phys. Lett.* **88** (2006), 174105.

[8] Hsiang-Yao Hsiao and Chih Chen, "Thermomigration in Pb-free SnAg solder joint under alternating current stressing," *Appl. Phys. Lett.* **94** (2009), 092107.

[9] Annie Huang, A. M. Gusak, K. N. Tu and Yi-Shao Lai, "Thermomigration in SnPb composite flip-chip solder joints," *Appl. Phys. Lett.* **88** (2006), 141911.

Problems

13.1 In electromigration in Al and Cu interconnects, why is thermomigration unimportant, or why is it seldom mentioned? Yet in electromigration in flip-chip solder joints, thermomigration is important; why?

13.2 In electromigration in Al interconnects, typically we observe electromigration when the applied current density is 10^5 A/cm^2 at 200 °C. How large is the driving force in terms of $Z^* eE$? If we need the force of the same magnitude in thermomigration, what will the temperature gradient be?

13.3 The heat of transport for most metals is negative, indicating that thermomigration is going from the hot end to the cold end; why?

13.4 If we anneal a diffusion couple of $Cu_{40}Ni_{60}$ and $Cu_{60}Ni_{40}$ at 500 °C, what will happen? For comparison, if we anneal a diffusion couple of $Sn_{40}Pb_{60}$ and $Sn_{60}Pb_{40}$ at 150 °C, what will happen? Then, if we anneal the SnPb couple at 200 °C, what will happen?

13.5 In electromigration there is a critical length because of back stress at the anode. Below the critical length, there will be no electromigration. In thermomigration, there is back stress at the cold end (or hot end) when atoms are being driven to the cold end (or hot end). Is there a critical length in thermomigration below which no thermomigration occurs?

13.6 To boil water in a copper kettle, we assume that the inside and outside temperatures are 100 °C and 600 °C, respectively, and the thickness of the wall of the copper kettle is 0.5 mm. How large is the thermomigration force? What are the self-diffusion coefficients of Cu near the inside and the outside walls (find the diffusivity of Cu in Chapter 4)?

Assuming that the water will be boiled in 10 min, what will the diffusion distances be in the copper in 10 min?

13.7 When we use thermocouples to measure temperature, there is a temperature gradient in the thermocouple. If the couple is made of alloys, do we worry about thermomigration in the thermocouple? Also, when we use a Peltier device to remove heat, there is a temperature gradient in the device. Is thermomigration a concern?

14 Stress migration in thin films

14.1 Introduction

Stress-induced atomic migration is creep as we have discussed in Chapter 10. We have emphasized that it is stress gradient not stress that can induce atomic diffusion. From the viewpoint of device reliability, we must ask the following questions. First, from where is the stress coming? Second, how does a stress gradient develop in an interconnect? Third, how can the elastic stress gradient induce atomic migration? Fourth, what is the mechanism of creep that leads to void or whisker formation to cause failure in interconnects? Finally what is the rate of creep [1–4]?

On the first question, typically the answer is thermal stress which occurs due to different thermal expansion coefficients in the interconnect structure. The most obvious one is that between Al (or Cu) metallic wire and the interlayer dielectric insulator. Another one comes from the chip-packaging interaction in flip-chip technology because of the large difference in thermal expansion between the chip and the packaging substrate. Then, electromigration can introduce back-stress in interconnects as discussed in Chapter 11, Section 11.6. Mechanical stress due to externally applied force is rare in electronic devices. However, we should mention impact-induced stress due to the dropping of a handheld device to the ground. The impact is a high rate shear in a very short time, about 1 millisecond, or a shear rate of 1×10^3 cm/s. Since impact failure is not a long-time event as in creep, we will not cover it here.

On the second question of stress gradient, it is worth mentioning that three-dimensional FEA software is available commercially. Therefore, it is not difficult to obtain a simulation of stress distribution in interconnect structures so that stress gradient can be examined. Nevertheless, the most common stress gradient is that between a free surface and a stressed region or a point of stress concentration. Since a void has a free surface, the stress gradient between a void and its surrounding tensile region is the most important one from the point of view of stress-migration, since it leads to vacancy diffusion to the void and the void will grow. This is the so-called stress-induced voiding. In this chapter, a description and explanation of the stress gradient driving force and kinetics of stress-induced voiding will be given. As ultralow dielectric constant materials are being introduced into multilayered interconnect structures, stress-induced fracture is becoming more and more important because of the poor mechanical and thermal properties, and weaker interfacial adhesion of the ultralow k materials. The failure can be cracking and delamination, in addition to void formation. We will not cover crack propagation.

On the third question, the failure of an interconnect structure caused by stress migration is a phenomenon of wear-out, which means that it is time-dependent and takes a long time to fail. The typical remedy is by extending the MTTF beyond the device lifetime so that the device may not fail within its lifetime. Alternatively, a diffusion barrier is added to block the vacancy diffusion to the void so that the reliability against stress-migration can be improved. In future, a closer interaction among device designers and processing and reliability engineers is needed in order to ensure that stress-migration is of lesser reliability concern.

Note that a steady-state creep may not lead to failure, provided that vacancies can be maintained at equilibrium everywhere in the sample. A typical example is the sagging of lead (Pb) pipes by their own weight in some very old houses over hundreds of years; it does not fail. Room temperature is a relatively high homologous temperature for lead (Pb), which melts at 327 °C, therefore, atomic diffusion at room temperature is sufficiently fast for creep to occur. Also, it is worth mentioning that a modern application of creep is in the use of pure and well-annealed copper O-rings as pressure seals in ultrahigh vacuum systems. Fig. 14.1 depicts a schematic diagram of the cross-section of a Cu O-ring between two steel fringes. The screws are tightened to enable the teeth of the fringes to bite into the soft Cu O-ring to form grooves under a high-stress gradient. At room temperature, creep can occur in the O-ring at atomic scale to close atomic size gaps in the groove seal to maintain the ultrahigh vacuum pressure.

As to the fourth question, generally speaking, creep is a high-temperature phenomenon except where grain-boundary diffusion becomes dominant, and then only a moderate temperature is required. To cause failure, we have emphasized in Chapter 12 that it is atomic flux divergence and the absence of lattice shift that lead to void or hillock formation. We should consider the site of atomic flux divergence for stress-induced voiding. Furthermore, similarly to current crowding in electromigration, we have to consider the effect of stress concentration on void formation.

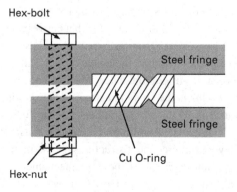

Fig. 14.1   Schematic diagram of the cross-section of a Cu O-ring between two steel fringes.

As to the final question of the rate of creep, it is limited by lattice diffusion or the activation energy of lattice diffusion.

14.2 Chemical potential in a stressed solid

Consider a pure metal bar under a uni-axial constant tensile stress or constant tensile load, within the elastic limit. In the initial elastic deformation, there is no flux of any kind, yet the bar will elongate slowly with time if the stress or the load is maintained. The time-dependent deformation is called diffusional creep, when there is a free surface to allow a stress gradient or a chemical potential gradient to form within the bar. On the other hand, under a hydrostatic compression or tension, there is no creep.

We consider the chemical potential in a stressed solid, below. In thermodynamics, the change of Helmholtz free energy F is given as

$$dF = -SdT - pdV$$

If the change occurs at constant temperature as in room-temperature creep, we eliminate the first term on the right-hand side and rewrite

$$p = -\frac{\partial F}{\partial V} \tag{14.1}$$

The last equation can be interpreted to mean that pressure (stress) is an energy density (i.e. energy per unit volume).

We should explain why we take Helmholtz free energy instead of Gibbs free energy in the above. This is because to consider chemical potential change in a stressed solid, the stress is assumed to be a hydrostatic compression or tension. We evaluate the chemical potential energy change or the change of free energy in each atom under a uniform hydrostatic compression or tension. Hence, we need the pressure to be constant, so we have to use Helmholtz free energy. However, under a hydrostatic compression or tension, while the chemical potential energy of every atom has been changed and the rate of random walk will change, there will be no directional flow or flux of atoms. In order to have a diffusional flux or a directional flux, a stress potential gradient is required. It can be established between a tensile region and a compressive region, or between a stressed region and a free surface.

For a given volume, the energy change equals the energy density times the given volume. Therfore, for an atomic volume Ω, we have

$$p\Omega = -\frac{\partial F}{\partial V}\Omega = -\frac{\partial F}{\partial \left(\frac{V}{\Omega}\right)} = -\frac{\partial F}{\partial N} \tag{14.2}$$

where N is the number of atoms in volume V. The last term is by definition the chemical potential, where the negative sign is used to indicate that a decrease in volume by pressure results in an increase in energy. Pressure is a compressive stress which is negative. The chemical potential change in a stressed solid can be given as

$$\mu = \pm\sigma\Omega \tag{14.3}$$

where the positive and negative signs refer to tensile and compressive hydrostatic stress respectively, following the sign convention given in Chapter 6. In other words, we can express the Helmholtz free energy of a stressed solid as

$$dF = -SdT - (p \pm \sigma)dN\Omega \tag{14.4}$$

where p is the ambient pressure and σ is the external applied stress.

To gain a quantitative feeling of $\sigma\Omega$, we shall consider a piece of Al stressed at the elastic limit (i.e. strain is 0.2%). Young's modulus for Al is $Y = 6 \times 10^{11}$ dyne/cm^2, so that the stress is

$$\sigma = Y\varepsilon = 1.2 \times 10^9 \text{dyne/cm}^2 = 1.2 \times 10^9 \text{erg/cm}^3$$

Since Al has a fcc lattice with a lattice parameter of 0.405 nm, there are four atoms in a unit cell of (0.405 nm)^3, or 0.602×10^{23} atom/cm^3. Then

$$\sigma\Omega = \frac{1.2 \times 10^9 \text{ erg}}{0.602 \times 10^{23} \text{ atom}} = 2 \times 10^{-14} \text{ erg/atom} = 0.0125 \text{ eV/atom}$$

It is interesting to compare this value to that of elastic strain energy per atom calculated by Eq. (6.9) in Section 6.3. The latter is much smaller with values around 10^{-5} eV/atom. The elastic strain energy per atom is the energy needed to deform an atom in the solid (by increasing or decreasing interatomic distance) due to the applied stress. The chemical potential energy of $\sigma\Omega$ is the energy change in removing one atom from or adding one atom to the stressed solid.

Yet the driving force of diffusion of an atom under stress is not the stress potential; rather, it is the stress potential gradient (to be discussed in Section 14.3 below), and the force is quite small. An example of the experimental measurement of stress gradient and a calculation of the magnitude of the force will be presented in Section 14.5.4, when we discuss the stress-migration-induced Sn whisker growth by synchrotron radiation micro-beam X-ray diffraction.

For thermally activated processes such as diffusion, the chemical potential, $\sigma\Omega$, enters as an exponential factor. For Al stressed to the elastic limit at 400 °C, we have

$$kT = 0.058 \text{ eV and}$$

$$\exp\left(\frac{\sigma\Omega}{kT}\right) = \exp\left(\frac{0.0125}{0.058}\right) = 1.23$$

Usually, creep occurs at a much lower stress ($\sigma\Omega \ll kT$), so we can linearize the exponential term by

$$\exp\left(-\frac{\sigma\Omega}{kT}\right) \cong 1 - \frac{\sigma\Omega}{kT} \tag{14.5}$$

However, we have to be careful in using the linearization. For example, we consider a different case where we deposit an Al thin film on a thick fused quartz substrate kept at 400 °C. Then we lower the temperature to 100 °C and observe the relaxation of the Al film under a tensile stress. The tensile stress in the Al film is due to the much smaller thermal expansion of the quartz substrate. The linear thermal expansion coefficients of Al and quartz in the temperature range of 100 °C are $\alpha = 25\times$ and $0.5 \times 10^{-6}/°C$, respectively. The thermal strain is

$$\varepsilon = \Delta\alpha\Delta T = 25 \times 10^{-6} \times 300 = 0.75\%$$

which is greater than the typical elastic limit. Then, the thermal stress is $\sigma = Y\varepsilon = 4.5 \times 10^9$ dyne/cm^2 and so $\sigma\Omega = 0.045$ eV. On the other hand, $kT = 0.032$ eV at 100 °C, so that we have $\sigma\Omega > kT$ in this case of a high-stress and low-temperature creep.

14.3 Diffusional creep (Nabarro–Herring equation)

In Fig. 14.2, we consider a hexagonal grain in a polycrystalline material which is under a shear stress. We can imagine that the grain is acted upon by a combination of tensile and compressive stresses as shown. The effect of the elastic stress is to deform the grain from its original shape delineated by the solid lines, to that delineated by the broken lines. If the stress persists, the grain can change shape by transporting the part of material in the shaded area from the compressive region to the tensile region in order to release the stress. The transport is by atomic diffusion as indicated by the curved arrows. To analyze

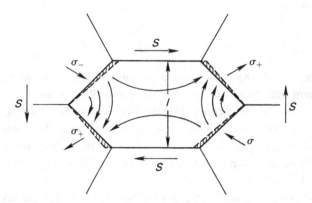

Fig. 14.2 Hexagonal grain in a polycrystalline material which is under a shear stress.

this problem, we shall follow the Nabarro-Herring creep model in which it is assumed that grain boundaries are effective sources and sinks of vacancies which mediate the mass transport, and that diffusion is via the vacancy mechanism in the lattice.

In the tensile region very close to the grain boundary, the chemical potential deviates from the equilibrium value μ_0, due to the stress, by an amount $\mu_1 - \mu_0 = \sigma\Omega$.

Similarly, in the compressive region, $\mu_2 - \mu_0 = -\sigma\Omega$, and thus the chemical potential difference in going from the compressive region to the tensile region is

$$\Delta\mu = \mu_2 - \mu_1 = -2\sigma\Omega \tag{14.6}$$

This potential difference will drive atoms to diffuse from the compressive regions to the tensile regions, as depicted by the arrows in the grain in Fig. 14.2. The force acting on the diffusing atoms is a stress-potential gradient,

$$F = -\frac{\Delta\mu}{\Delta x} = \frac{2\sigma\Omega}{l} \tag{14.7}$$

where l is the grain size. The flux of the diffusing atoms, according to Eq. (4.7) and (4.8) in Chapter 4 is

$$J = C\frac{D}{kT}F = C\frac{D}{kT}\frac{2\sigma\Omega}{l} = \frac{2D\sigma}{kTl} \tag{14.8}$$

where $C = 1/\Omega$ in a pure metal. The number of atoms transported by the flux in a period of time t and through an area A is $N' = JAt$, or the volume accumulated is

$$\Omega N' = \Omega JAt \tag{14.9}$$

The strain is then

$$\varepsilon = \frac{\Delta l}{l} = \frac{\Omega N'/A}{l} = \frac{\Omega Jt}{l} \tag{14.10}$$

so that the strain rate is

$$\frac{d\varepsilon}{dt} = \frac{\Omega J}{l} = \frac{2\sigma\Omega D}{kTl^2} \tag{14.11}$$

This is the well-known Nabarro–Herring creep equation. It has an inverse dependence on the square of grain size; the rate is much faster for a small grain size.

The last equation was derived by considering the flux of atoms. Since atomic diffusion occurs by having vacancies diffusing in the opposite direction, we should be able to obtain the same equation by considering the flux of vacancies. We shall illustrate this later for a comparison. First, we deal with the concentration of vacancies in the tensile and the compressive regions.

We have argued that the chemical potentials in the tensile and the compressive regions have changed by the amounts $\sigma\Omega$ and $-\sigma\Omega$, respectively, from the equilibrium value.

Since chemical potential is free energy per atom, this means that if we wish to remove an atom from these stressed regions (i.e. to create a vacancy), the work needed to do so is changed by the same amounts. When we consider a vacancy in the stressed solid, the formation energy (which is actually the potential energy) is changed by $\pm\sigma\Omega$, assuming that Ω is the volume of a vacancy. The positive and the negative signs are now reversed and refer to the compressive stress and the tensile stress, respectively. In other words, the formation energy of a vacancy in the tensile regions is reduced by the amount $\sigma\Omega$, and in the compressive regions it is increased by $\sigma\Omega$. This means that in the compressive region, it takes more energy to form a vacancy, and in a tensile region, it takes less, so that we have more vacancies in the tensile region and fewer in the compressive region at a given temperature. There is a gradient of vacancies between these two regions, and the vacancies will diffuse from the tensile to the compressive region. According to Eq. (4.10), we can express the concentration of vacancies as

$$C_v^\pm = C \exp[(-\Delta G_f \pm \sigma\Omega)/kT] \tag{14.12}$$

where C_v^+ and C_v^- correspond to the vacancy concentrations in tensile and compressive regions, respectively. Assuming that $\sigma\Omega \ll kT$, we have

$$C_v^\pm = C_v\left(1 \pm \frac{\sigma\Omega}{kT}\right) \tag{14.13}$$

where $C_v = C\exp(-\Delta G_f/kT)$ is the concentration of vacancies in the equilibrium state. Then the concentration difference is

$$\Delta C_v = C_v^+ - C_v^- = 2\sigma\Omega\frac{C_v}{kT} \tag{14.14}$$

The flux of vacancies going from the tensile to the compressive region is

$$J_v = -D_v\frac{\Delta C_v}{\Delta x} = \frac{-2\sigma\Omega D_v C_v}{kTl}$$

where D_v is the diffusivity of a vacancy. Then the atomic flux is

$$J = \frac{2\sigma\Omega DC}{kTl} = \frac{2\sigma D}{kTl} \tag{14.15}$$

where we have taken $DC = -D_v C_v$ since the atomic flux J is opposite to the vacancy flux J_v.

Equation (14.15) is the same as Eq. (14.8), so whether we consider the atomic flux or the vacancy flux, the creep equation is the same. We present both of them here because when we consider void formation we use vacancy flux, but for hillock or whisker growth we use atomic flux.

The creep relation in Eq. (14.11) shows that if we plot ln $(Td\varepsilon/dt)$ versus $1/kT$, we determine the activation energy of creep which is the same as the activation energy of lattice diffusion. Many high-temperature creep data for pure metals have been analyzed, and the measured activation energies indeed agree well with those of lattice diffusion; see Fig. 14.3. However, lower temperature creep data show a smaller activation energy. This may be due to grain-boundary diffusion or to creep-induced dislocation motion. As shown in Fig. 14.2, the shaded volume in the compressive regions can be transported along the grain boundary to the tensile regions. In this case, the creep rate becomes

$$\frac{d\varepsilon_{gb}}{dt} = A\frac{\sigma\Omega D_{gb}\delta}{kTl^3} \tag{14.16}$$

where A is a constant and D_{gb} and δ are grain-boundary diffusivity and grain-boundary width, respectively. Comparing Eq. (14.11) with Eq. (14.16), we see that the difference is in replacing D by $D_{gb}\delta/l$. The factor $1/l$ can be regarded as the density of cross-section of

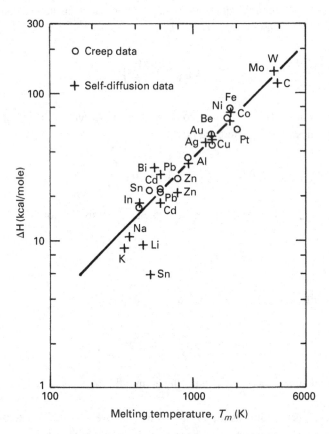

Fig. 14.3   A plot of measured activation energy against melting temperature. Many high-temperature creep data for pure metals have been analyzed, and the measured activation energies agree well with those of lattice diffusion.

grain boundaries per unit area, hence δ/l is the cross-sectional area of grain boundaries per unit area. Creep by grain-boundary diffusion has a stronger dependence on grain size than creep by lattice diffusion; it is known as Coble creep. When both lattice diffusion and grain-boundary diffusion occur simultaneously, we have

$$\frac{d\varepsilon}{dt} = \frac{2\sigma\Omega D}{kTl^2} \left(1 + \frac{A}{2}\frac{D_{gb}\delta}{Dl}\right) \tag{14.17}$$

For a thin film on a substrate, diffusional creep leads to relaxation rather than deformation when the relaxation is uniform or homogeneous. Yet if the relaxation is inhomogeneous (i.e. localized, when the thin film has a protective oxide surface, for example), it induces void formation or hillock growth, which can be a serious reliability issue.

We have shown in Section 14.2 that elastic strain energy is much smaller than chemical energy. Hence, the former is unimportant in most of the chemical reactions such as silicide formation, yet we show in this section that stress can influence vacancy concentration and affect diffusion. The difference lies in the period of time involved in the reactions; in silicide formation the reaction finishes typically in minutes or hours, whereas in creep it usually lasts for months. We ignore the long-term effect in short-term events.

14.4 Void growth in Al interconnects driven by tensile stress

Thin films of Al are known to have very good adhesion to SiO_2 and other oxide surfaces. Good adhesion means that the interface is not an effective sink and source of vacancies. In addition, lattice shift will be difficult because of the good interfacial adhesion, so stress relaxation is hard, except at high temperatures. In device operation, thermal stress in Al thin films is typically tensile. Thus, stress-induced void formation is a reliability issue. For a void to form and grow, it must nucleate first. Since the free surface of the void is stress-free, a stress gradient exists between the void surface and the tensile region, so vacancy will be driven from the tensile region to the void and the void grows.

Fig. 14.4(a) shows a schematic diagram of a cross-section of a spherical void in a piece of Al under unidirectional tensile stress of σ. The growth of the void occurs because it serves as a sink for vacancies.

The radius of the void is limited by the elastic limit of 0.2%. If we consider in a given volume of V, a void of the size ΔV is formed, then the bulk strain of $\Delta V/V$ cannot be larger than 0.2%, since there is no driving force to grow the void beyond that. Similarly, in an Al interconnect of length l, if we assume the width of the void to be Δl, the ratio of the width to the interconnect length, $\Delta l/l$, cannot be larger than 0.2%. So, for a line 100 μm long, the maximum width (or radius) of the void is only about 0.2 μm. Thus, we do not expect to find a large spherical or cylindrical void in a line caused by stress migration; instead a slit-type void is commonly found in stress migration, as depicted in Fig. 14.4(b). It is possible that a slit-type void could be a crack formed by fracture under tensile stress. If so, it is not a time-dependent problem.

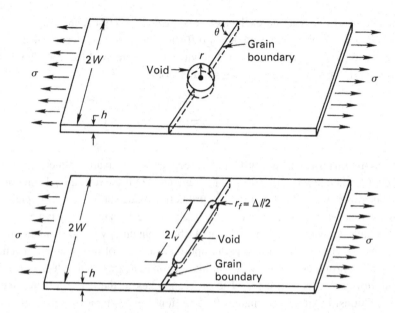

Fig. 14.4   (a) Schematic diagram of a cross-section of a spherical void on a piece of Al under hydrostatic tension of σ. (b) Schematic diagram of the growth of a slit-type void from the edge of an Al line.

First, we shall consider the growth of a spherical void of radius r as depicted in Fig. 14.4(a). Actually, the kinetics of void growth is similar to the growth of a spherical precipitate as treated in Chapter 5. In the initial growth stage, the radius is small. Owing to the Gibbs–Thomson potential, the atoms at the circumference of the void have a potential of

$$pdV = \frac{2\gamma}{r}\Omega$$

where γ is the surface energy per unit area of the void. The vacancy concentration in the neighborhood of the tip will be

$$C_{v1} = C \exp\left(-\frac{(\Delta G_f - (2\gamma/r)\Omega)}{kT}\right)$$

where ΔG_f is the formation energy of a vacancy in the unstressed Al. In the faraway region under the tensile stress, the vacancy concentration will be given as

$$C_{v2} = C \exp\left(-\frac{(\Delta G_f - \sigma\Omega)}{kT}\right)$$

To grow the void, we assume that

$$\sigma\Omega > \frac{2\gamma}{r}\Omega$$

The difference in vacancy concentration is

$$C_{v2} - C_{v1} = \Delta C_v = C_v \left[\exp \frac{\sigma \Omega}{kT} - \exp \frac{2\gamma \Omega}{rkT} \right] = C_v \left(\frac{\sigma \Omega}{kT} - \frac{2\gamma \Omega}{rkT} \right) \quad (14.18)$$

where $C_v = C \exp(-\Delta G_f / kT)$ and we assume that both $\sigma \Omega$ and $\gamma \Omega / r$ are much smaller than kT. For simplicity we can assume a linear concentration gradient of vacancy and obtain the vacancy flux arriving at the void surface. Then we can use the law of conservation of mass and take $\Omega J_v 4\pi r^2 dt = 4\pi r^2 dr$ to obtain the growth rate of the sphere.

Alternatively, we can assume a steady-state process to solve the continuity equation in spherical coordinates as presented in Chapter 5 and obtain the flux of vacancies arriving at the void surface by Fick's first law. The void growth is then obtained by the law of conservation of void volume with the total volume of vacancies diffusing to the void.

Next, we shall consider the growth of a slit-type void from the edge of an Al line as depicted in Fig. 14.4(b). The growth kinetics is quite similar to the growth of a whisker to be presented in the next section, except that the growth of a whisker occurs at the bottom of the whisker instead of at the tip as in the growth of the slit-type void. Also, the driving force of whisker growth is a compressive stress gradient rather than a tensile-stress gradient for void growth. In Chapter 15, the growth of a pancake-type void under electromigration is presented in Section 15.4.3.

14.5   Whisker growth in Sn/Cu thin films driven by compressive stress

14.5.1   Morphology of spontaneous Sn whisker growth

Spontaneous whisker growth on beta-tin (β-Sn) is a surface relief phenomenon of creep. It is driven by a compressive-stress gradient and occurs at room temperature. Spontaneous Sn whiskers are known to grow on a matte Sn finish on Cu. Today, because of the wide applications of Pb-free solders on Cu conductors used in the packaging technology of consumer electronic products, Sn whisker growth has reappeared as a serious reliability issue because the Sn-based Pb-free solders are very rich in Sn. The matrix of most Sn-based Pb-free solder is almost pure Sn, so the well-known phenomena of tin such as tin-cry, tin-pest, and tin-whisker are again of concern.

The Cu lead frames in surface mount technology of electronic packaging are finished with a layer of solder for surface passivation and for enhancing wetting during the joining of the lead frames to printed circuit boards. When the solder finish is eutectic SnCu or matte Sn, whiskers are often observed. Some whiskers can grow to several hundred microns in length, which are long enough to become electrical shorts between neighboring legs of a lead frame. The trend in consumer electronic products is to integrate more and more systems in packaging, so that the device elements and components are closer and closer together, and the probability of shorting by whiskers becomes greater. A broken whisker can fall between two electrodes and become a short.

However, the serious concern about whisker is not for handheld consumer electronic products because these products are cheap; when they fail, they will be replaced without concern. For high reliability devices, such as satellites, they cannot be replaced easily, so even the growth of a single whisker is of concern.

Cross-sectional scanning and TEM have been used to examine Sn whiskers, with samples prepared by focused ion beam thinning and polishing. Also, X-ray micro-diffraction in synchrotron radiation has been used to study the structure, phase formation, and stress distribution around the root and vicinity of a whisker grown on eutectic SnCu and matte Sn.

In Fig. 14.5(a), an enlarged SEM image of a long whisker on the eutectic SnCu finish is shown. The whisker in Fig. 14.5(a) is straight and its surface is fluted. The crystal structure of Sn is body-centered tetragonal with the lattice constant $a = 0.58311$ nm and $c = 0.31817$ nm. The direction of whisker growth, or the axis along the length of the whisker, has been found mostly to be the c-axis, but growth along other axes such as [100] and [311] has also been found.

On the pure or matte Sn finish surface, short whiskers or hillocks were observed as shown in Fig. 14.5(b). The surface of the whisker in Fig. 14.5(b) is faceted. Besides the difference in morphology, the rate of whisker growth on the pure Sn finish is much slower than that on the SnCu finish. The direction of growth is more random too.

(a)

(b)

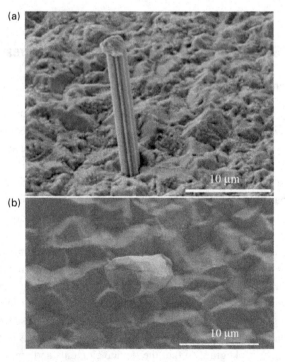

Fig. 14.5   (a) SEM image of a long whisker on the eutectic SnCu finish. (b) SEM image of a short whisker on matte Sn finish.

Comparing the whiskers formed on eutectic SnCu and pure Sn, it seems that the Cu in eutectic SnCu enhances Sn whisker growth. Although the composition of eutectic SnCu consists of 98.7 atomic % of Sn and 1.3 atomic % of Cu, the small amount of Cu seems to have a profound effect on whisker growth on the eutectic SnCu finish.

In Fig. 14.6(a), a cross-sectional SEM image of a lead frame leg with a SnCu finish is shown. The rectangular Cu lead frame core is surrounded by an approximate 15 μm-thick SnCu finish. A higher magnification image of the interface between the SnCu and the Cu, prepared by focused ion beam, is shown in Fig. 14.6(b). An irregular layer of Cu_6Sn_5 compound can be seen between the Cu and SnCu. No Cu_3Sn was detected at the interface. The grain size in the SnCu finish is about several microns. More importantly, there are Cu_6Sn_5 precipitates in the grain boundaries of SnCu. The grain-boundary precipitation of Cu_6Sn_5 is the source of stress generation in the CuSn finish. It provides the driving force of spontaneous Sn whisker growth. We shall address this critical issue of stress generation later.

In Fig. 14.6(c), a cross-sectional SEM image of matte Sn finish on a Cu lead frame is shown, prepared by focused ion beam. While the layer of Cu_6Sn_5 compound can

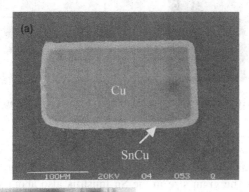

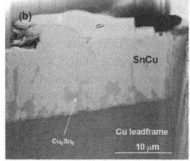

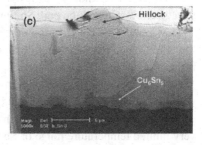

Fig. 14.6 (a) Cross-sectional SEM image of a lead frame leg with SnCu finish. The rectangular core of Cu lead frame is surrounded by an approximate 15 μm-thick SnCu finish. (b) A high magnification image of the interface between the SnCu and the Cu, prepared by focused ion beam. IMC of Cu_6SN_5 grows at the Cu–SnCu interface and in the grain boundaries of SnCu finish. (c) A high magnification image of the interface between matte Sn and Cu. Much less Cu_6Sn_5 is shown.

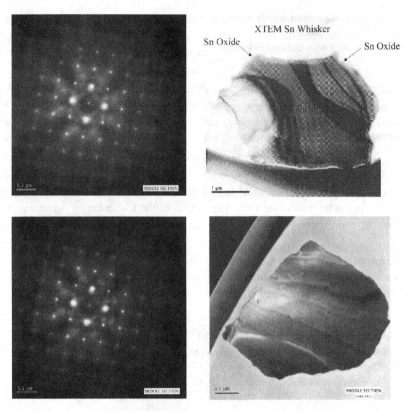

Fig. 14.7   TEM images of the cross-section of whiskers, normal to their length, together with electron diffraction pattern.

be seen between the Cu and the Sn, there are fewer Cu_6Sn_5 precipitates in the grain boundaries of Sn. The grain size in the Sn finish is also about several microns. The lacking of grain-boundary Cu_6Sn_5 precipitates is the most important difference between the eutectic SnCu and the pure Sn finish with respect to whisker growth.

TEM images of the cross-section of whiskers, normal to their length, are shown in Fig. 14.7(a) and (b) together with an electron diffraction pattern. The growth direction is the c-axis. There are a few spots in the images which might be dislocations.

How to suppress Sn whisker growth, and how to perform systematic tests of Sn whisker growth in order to understand the driving force, the kinetics, and the mechanism of growth are challenging tasks in the electronic packaging industry today. Due to the very limited temperature range of Sn whisker growth, from room temperature to about 60 °C, accelerated tests are difficult. This is because if the temperature is lower, the kinetics is insufficient due to slow atomic diffusion, and if the temperature is higher, the driving force is insufficient because of stress relief by lattice diffusion owing to the high homologous temperature of Sn.

The Sn whisker growth is spontaneous, indicating that the compressive stress needed for the growth is self-generated; no externally applied stress is required. Otherwise, we expect that the growth slows down and stops when the applied stress is exhausted, if it is not applied continuously. Therefore, it is of interest to ask from where the self-generated compressive stress is coming, how the driving force can maintain itself to sustain the spontaneous whisker growth, and also, how large is the compressive stress gradient needed to grow a whisker?

Spontaneous whisker growth is a unique creep process in which both stress generation and stress relaxation occur simultaneously at room temperature. The three indispensable conditions of Sn whisker growth are (a) the fast room temperature diffusion in Sn, (b) the room temperature reaction between Sn and Cu to form Cu_6Sn_5 which generates the compressive stress in Sn, and (c) the cracking of the protective surface oxide on Sn. The last condition is needed in order to produce a compressive stress gradient for creep. When the oxide is broken at a weak spot, the exposed free surface is stress-free, so a compressive stress gradient is developed, and creep or the growth of a whisker can occur to relax the stress.

While whisker growth occurs at a constant temperature, it does not occur under a constant pressure; therefore, we cannot use minimum Gibbs free energy change to describe the growth. Rather, it is an irreversible process of the interaction between the flux of Cu to form Cu_6Sn_5 and the flux of Sn to grow the whisker.

The growth of Sn whiskers is from the bottom, not from the top, since the morphology of the whisker tip does not change with whisker growth [5]. Many Sn whiskers are long enough to short two neighboring legs of the lead frame, shown in Fig. 14.5(a). It is possible that when there is a high electrical field across the narrow gap between the tip of a whisker and the point of contact on the other leg, just before the tip of the whisker touching the other leg, a spark may ignite fire. The fire may result in failure of the device or a satellite.

Since it only needs one whisker to fail a device which requires high reliability, this has been the most challenging issue in preventing whisker growth. While the basic mechanism of whisker growth is clear in terms of its driving force and kinetics, it is difficult to guarantee that we can have no whisker growth at all since the growth is a localized phenomenon. We cannot guarantee that there is no local microstructure variation over the entire solder finish.

14.5.2 Stress generation (driving force) in Sn whisker growth

The origin of the compressive stress can be mechanical, thermal, and chemical, but the mechanical and thermal stresses tend to be finite in magnitude, so they cannot sustain a spontaneous or continuous growth of whiskers for a long time. The chemical force is essential for spontaneous Sn whisker growth, but is not obvious. The origin of the chemical force is due to the room temperature reaction between Sn and Cu to form the IMC of Cu_6Sn_5 [6–8]. The chemical reaction provides a sustained driving force for the spontaneous growth of whiskers as long as the reaction keeps going between unreacted Sn and Cu.

Compressive stress is generated by the interstitial diffusion of Cu into Sn and the formation of Cu_6Sn_5 in the grain boundary of Sn. When the Cu atoms from the lead frame

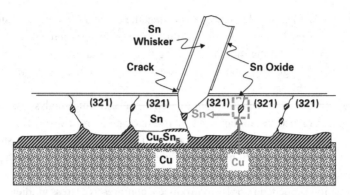

Fig. 14.8   (a) FIB images of a group of whiskers on SnCu finish. (b) Same images when the oxide on a rectangular area of the surface of the finish was sputtered away by using glancing incidence ion beam to expose the microstructure beneath the oxide. (c) A high magnification of the sputtered area is shown, in which the microstructure of Sn grains and grain-boundary precipitates of Cu_6Sn_5 are clear.

diffuse into the finish to grow the grain boundary of Cu_6Sn_5, as shown in Fig. 14.6(b), the volume increase due to the IMC growth will exert a compressive stress to the grains on both sides of the grain boundary. In Fig. 14.8, we consider a fixed volume V in the Sn finish that contains an IMC precipitate, as shown by the dotted square. The growth of the IMC due to the diffusion of a Cu atom into this volume to react with Sn will produce a stress,

$$\sigma = -B\frac{\Omega}{V}$$

where σ is the stress produced, B is the bulk modulus, and Ω is the partial molecular volume of a Cu atom in Cu_6Sn_5 (we ignore the molar volume change of Sn atoms in the reaction, for simplicity). The negative sign indicates that the stress is compressive.

In other words, we are adding an atomic volume into the fixed volume. To absorb the added atomic volume by the fixed volume of V in the finish due to the in-diffusion of Cu as considered in Fig. 14.8, we must add lattice sites in the fixed volume. Furthermore, we must allow Kirkendall shift or allow the added lattice plane to migrate, otherwise compressive stress will be generated. When more and more Cu atoms, say n Cu atoms, diffuse into the volume, V, to form Cu_6Sn_5, the stress in the above equation increases by changing Ω to $n\Omega$.

Since Sn has a native and protective oxide on the surface, the interface between the oxide and Sn is a poor source and sink for vacancies. Furthermore, the protective oxide ties down the edge of lattice planes in Sn and prevents them from moving. This is the basic mechanism of stress generation in spontaneous Sn whisker growth.

For the oxide to be effective in tying down lattice plane migration, the SnCu or Sn finish cannot be too thick. In a very thick finish, say over $100\,\mu m$, there are more sinks in the bulk of the finish to absorb the added Cu volume. Note that a whisker is a surface relief phenomenon. When bulk relief mechanism occurs, a whisker will not grow. There is a dependence of whisker formation on the thickness of finish. Since the average diameter of whiskers is a few microns, a whisker will grow more frequently on a finish having a thickness from a few microns to a few times its diameter.

Sometimes it is puzzling to find that Sn whiskers seem to grow on a tensile region of a Sn finish. For example, when a Cu lead frame surface was plated with SnCu, the initial stress state of the SnCu layer right after plating was tensile, yet whisker growth was observed. If we consider the cross-section of a Cu lead frame leg coated with a layer of Sn as shown in Fig. 14.6(a), the lead frame experienced a heat-treatment of reflow from room temperature to 250 °C and back to room temperature. Since Sn has a higher thermal expansion coefficient than Cu, the Sn should be under tension at room temperature after the reflow cycle. Yet with time, a Sn whisker grows, so it seems that a Sn whisker grows under tension. Furthermore, if a leg is bent, one side of it will be in tension and the other side in compression. It is surprising to find that whiskers grow on both sides, whether the side is under compression or tension. These phenomena are hard to understand until we recognize that the thermal stress or the mechanical stress, whether it is tensile or compressive, is finite. It can be relaxed or overcome quickly by atomic diffusion at room temperature. After that, the continuing chemical reaction will develop the compressive stress needed to grow whiskers, so the chemical force is dominant and persistent. When we consider the driving force of spontaneous whisker growth on Sn or SnCu solder finish on Cu, the compressive stress induced by chemical reaction at room temperature is essential. Room temperature reaction between Sn and Cu was studied by using thin-film samples.

The idea of compressive stress induced by the growth of a grain-boundary precipitate of Cu_6Sn_5 has a few variations. One of them is the wedge model in which the Cu_6Sn_5 phase between the Cu and Sn has a wedge shape growing into the grain boundaries of Sn. The growth of the wedge will exert a compressive stress to the two neighboring Sn grains, similar to splitting a piece of wood by a wedge. So far, very few wedge-shaped IMCs have been observed in XTEM; for example, see Fig. 14.6(b).

14.5.3 Effect of surface Sn oxide on stress-gradient generation

To discuss the effect of surface oxide on Sn whisker growth, we shall refer to the same effect of surface oxide on Al hillock growth. In an ultrahigh vacuum, no surface hillocks were found on a Al surface under compression. Hillocks grow on Al surfaces only when the Al surface is oxidized, and Al surface oxide is known to be protective. Without surface oxide in an ultrahigh vacuum, the free surface of Al is a good source and sink of vacancies, so a compressive stress can be relieved uniformly on the entire surface or on the surface of every grain of the Al, on the basis of the Nabarro–Herring model of lattice creep or the Coble model of grain boundary creep. When the relief is homogeneous, no hillock will form, since hillock formation is an inhomogeneous or localized relief phenomenon.

Note that a whisker or hillock is a localized growth on a surface. To have a localized growth, the surface cannot be free of oxide, and the oxide must be a protective oxide so that it effectively blocks all the vacancy sources and sinks on the surface. Furthermore, a protective oxide also means that it pins down the lattice planes in the matrix of Sn (or Al), so that no lattice plane migration can occur to relax the stress in the volume, V, considered in Fig. 14.8. Only those metals which grow protective oxides, such as Al and Sn, are known to have serious hillock or whisker growth. When they are in thin-film or thin-layer form, the surface oxide can pin down the lattice planes near the surface easily. On the other hand, it is obvious that if the surface oxide is very thick, it will physically block the growth of any hillock and whisker. No hillocks or whiskers can penetrate a very thick oxide or a thick coating. No break means no free surface and no stress gradient. Thus, a necessary condition of whisker growth is that the protective surface oxide must not be too thick so that it can be broken at certain weak spots on the surface to form free surfaces, and whiskers grow to relieve the stress from these spots.

In Fig. 14.9(a), a focused ion beam image of a group of whiskers on the SnCu finish is shown. In Fig. 14.9(b), the oxide on a rectangular area of the surface of the finish was sputtered away by using a glancing incidence ion beam to expose the microstructure beneath the oxide. In Fig. 14.9(c), a higher magnification image of the sputtered area is shown, in which the microstructure of Sn grains and grain-boundary precipitates of Cu_6Sn_5 are clear. Due to the ion channeling effect, some of the Sn grains appear darker than the others. The Cu_6Sn_5 particles distribute mainly along grain boundaries in the Sn matrix, and they are brighter than the Sn grains owing to less ion channeling and more ion backscattering. The diameter of the whiskers is a few microns, which is comparable to the grain size in the SnCu finish.

In ambience, we assume that the surface of the finish and the surface of every whisker are covered with oxide. The growth of a hillock or whisker is an eruption from the oxidized surface; it has to break the oxide. When the Sn matrix is under compression, its oxide is under tension, so the oxide breaks under tension. The stress that is needed to break the oxide may be the minimum stress needed to grow whiskers. It seems that the easiest place to break the oxide is at the base of the whisker. Then, to maintain the growth, the break must remain open so that it behaves like a stress-free free surface and vacancies can be supplied continuously from the break and can diffuse into the Sn layer to sustain the long-range diffusion of the Sn atoms needed to grow the whisker. In addition, the free surface is stress-free and it creates the stress gradient needed for stress migration to occur.

In case a part of the break at the base of the whisker is healed by oxide, the growth of the whisker becomes uneven and will lead to a turn in whisker growth direction towards the healed side; as a consequence, a bent whisker is formed.

In Fig. 14.8, we depict a surface of the whisker that is oxidized, except the base. The surface oxide of the whisker serves the very important purpose of confinement so that the whisker growth is essentially a one-dimensional growth. The surface oxide of the whisker prevents it from growing in a lateral direction; thus it grows with a constant cross-section and has the shape of a pencil. Also, the oxidized surface may explain why the diameter of a Sn whisker is just a few microns. This is because the gain in strain-energy reduction in whisker growth is balanced by the surface formation of the whisker. By balancing the strain energy against the surface energy in a unit length of the whisker,

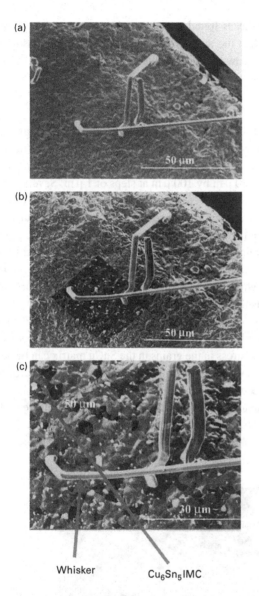

(a)

50 µm

(b)

50 µm

(c)

80 µm

30 µm

Whisker Cu_6Sn_5 IMC

Fig. 14.9   Low magnification picture of an area of CuSn finish wherein a whisker is seen and scanned.

$\pi R^2 \varepsilon = 2\pi R \gamma$, we find that

$$R = \frac{2\gamma}{\varepsilon} \qquad (14.19)$$

where R is radius of the whisker, γ is the surface energy per unit area, and ε is the strain energy per unit volume. Since strain energy per atom is about four to five orders of magnitude smaller than the chemical bond energy or surface energy per atom of the oxide, the diameter of a whisker is found to be several microns, which are about four orders of magnitude larger than the atomic diameter of Sn. For this reason, it is very difficult to have spontaneous growth of nanodiameter Sn whiskers.

14.5.4   Measurement of stress distribution by synchrotron radiation micro-diffraction

The micro-diffraction apparatus in ALS, at Lawrence Berkeley National Laboratory, was used to study Sn whiskers grown on a SnCu finish on a Cu lead frame at room temperature [9, 10]. The white radiation beam was 0.8 to 1 μm in diameter and the beam step-scanned over an area of 100 μm by 100 μm at steps of 1 μm. Several areas of the SnCu finish were scanned and those areas were chosen so that in each of them there was a whisker, especially the areas that contained the root of a whisker. During the scan, the whisker, and each grain in the scanned area, can be treated as a single crystal to the beam. This is because the grain size is larger than the beam diameter. At each step of the scan, a Laue pattern of a single crystal is obtained. The crystal orientation and the lattice parameters of the Sn whisker and the grains in the SnCu matrix surrounding the root of the whisker were measured by the Laue patterns. The software in ALS is capable of determining the orientation of each of the grains, and of displaying the distribution of the major axis of these grains. Using the lattice parameters of the whisker as a stress-free internal reference, the strain or stress in the grains in the SnCu matrix can be determined and displayed. Fig. 14.10 shows a low-magnification picture of an area of the SnCu finish wherein a whisker is seen and scanned.

Figure 14.11 shows an in-plane orientation map of the angle between the (100) axis of Sn grains and the X-axis of the laboratory frame. An image of the whisker is seen. The X-ray micro-diffraction study shows that in a local area 100 μm $\times$ 100 μm, the stress is

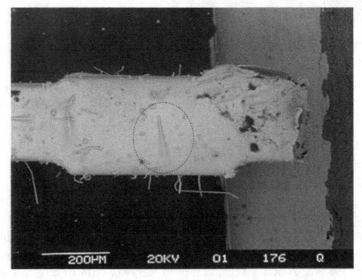

Fig. 14.10   In-plane orientation map of the angle between (100) axis of Sn grains and x-axis of the laboratory frame.

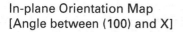

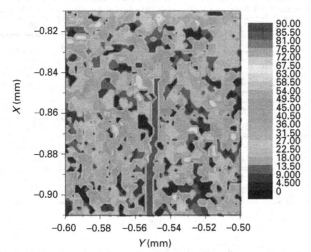

Fig. 14.11 Numerical value, and the distribution of stress, where the root of the whisker is at the coordinates of $X = -0.8415$ and $Y = -0.5475$ as shown in Fig. 14.11.

highly inhomogeneous with variations from grain to grain. The finish is therefore only under an average biaxial stress. This is because each whisker has relaxed the stress in the region surrounding it, but the stress gradient around the root of a whisker does not have a radial symmetry. The numerical value and the distribution of stress are shown in Fig. 14.12, where the root of the whisker is at the coordinates of $X = -0.8415$ and $Y = -0.5475$ as shown in Fig. 14.11. Overall, the compressive stress is quite low, of the order of several MPa; however, we can still see the stress gradient going from the whisker root area to the surroundings. It means that the stress level just below the whisker is slightly less compressive than the surrounding area. This is because the stress near the whisker has been relaxed by whisker growth. In Fig. 14.12, the light-colored arrows indicate the directions of local stress gradient. Some circles next to each other in Fig. 14.13 show a similar stress level, which most likely means that they belong to the same grain.

Figure 14.13 shows a plot of $-\sigma'_{zz}$, which is the deviatoric component of the stress along the normal surface. The total strain tensor is equal to the sum of the deviatoric strain tensor and the dilatational strain tensor. The latter is measured from the energy of the Laue spot using a monochromatic beam and the former is measured from deviation in the crystal Laue pattern using a white radiation beam. Here

$$\varepsilon_{ij} = \varepsilon_{\text{deviatoric}} + \varepsilon_{\text{dilatational}}$$

$$= \begin{vmatrix} \varepsilon'_{11} & \varepsilon_{12} & \varepsilon_{13} \\ \varepsilon_{21} & \varepsilon'_{22} & \varepsilon_{23} \\ \varepsilon_{31} & \varepsilon_{32} & \varepsilon'_{33} \end{vmatrix} + \begin{vmatrix} \delta & 0 & 0 \\ 0 & \delta & 0 \\ 0 & 0 & \delta \end{vmatrix} \quad (14.20)$$

where the dilatational strain $\delta = (1/3)(\varepsilon_{11} + \varepsilon_{22} + \varepsilon_{33})$ and $\varepsilon_{ii} = \varepsilon'_{ii} + \delta$.

|1.5μm| | | | | | | | | |(Unit: MPa)|
	-0.5400	-0.5415	-0.5430	-0.5445	-0.5460	-0.5475	-0.5490	-0.5505	-0.5520	-0.5535	-0.5550
-0.8340	-2.82	-3.21	-2.26	0.93	0.93	-0.23	-8.17	2.22	1.49	1.6	-0.03
-0.8355	-2.26	-2.64	-2.64	-1.04	1.37	1.37	-1.31	0.87	0.87	0.87	-0.7
-0.8370	-2.53	-3.21	-3.21	-2.64	-1.04	3.61	0.75	0.87	0.7	0.7	-0.19
-0.8385	-7.37	-9.62	-6.57	-2.64	3.61	4.52	3.61	0.29	-1.31	0	-4.79
-0.8400	-7.37	-8.22	-6.57	-1.18	0.75	4.23	0.75	-2.25	-2.27	-2.91	-6.91
-0.8415	-4.17	-4.84	-4.17	-1.81	-0.67	0.00	-1.96	-1.96	-3.74	-5.08	-5.08
-0.8430	-4.17	-4.17	-3.63	-1.81	-1.81	-2.29	-2.29	-1.96	-1.96	-3.27	-3.27
-0.8445	-4.14	-4.17	-3.86	-3.63	2.79	-4.64	-4.78	-0.84	-1.4	1.49	-3.27
-0.8460	-3.14	-3.63	-3.86	-3.63	-3.13	-4.78	-4.78	0.04	0.04	-1.41	-2.33
-0.8475	-4.14	-4.49	-4.49	-4.64	-3.86	-6.04	-1.72	3.55	3.55	-0.41	-2.33
-0.8490	-3.33	-5.67	-6.29	-6.29	-2.66	-2.08	-1.72	-1.79	0	-1.79	-3.73

Whisker

Fig. 14.12   Plot of $-\sigma'_{zz}$, which is the deviatoric component of the stress along the surface normal.

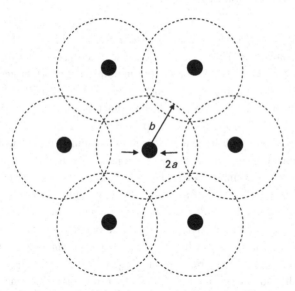

Fig. 14.13   Distribution of whiskers is assumed to have a regular arrangement so that each has a diameter $2a$ and occupies a diffusional field of diameter of $2b$.

We explain in the following the measurements of these two strain tensors. The deviatoric strain tensor is calculated from the deviation of spot positions in the Laue pattern with respect to their "unstrained" positions. The latter is obtained from an "unstrained" reference. By assuming that the whisker is strain-free, we used the Sn whisker itself as the unstrained reference and calibrated the sample-detector distance and the detector tilt with respect to the beam. The geometry is fixed. From the Laue spot positions of the strained sample, we can then measure any deviation of their positions from the calculated positions if the sample has zero strain. The transformation matrix which relates the unstrained to the strained Laue spot positions is then calculated and the rotational part is taken out. The deviatoric strain can then be computed from this transformation matrix. The more spots we have in the Laue pattern, the more accurate will be the deviatoric strain tensor determined. Note that the deviatoric strain is related to the change in the shape of the unit cell, but the unit cell volume is assumed to be constant and consists of five independent components. The sum of the three diagonal components should be equal to zero.

To obtain the total strain tensor, we must add the dilatational strain tensor to the deviatoric strain tensor. The dilatational component is related to the change in volume of the unit cell and consists of a single component of expansion or shrinkage, δ, in the last equation. In principle, when the deviatoric strain tensor is known, only one additional measurement is needed, i.e. the energy of a single reflection is required to obtain this single dilatational component. We can use the monochromatic beam to do so. From the orientation of the crystal and the deviatoric strain, we can calculate for each reflection what would be the energy of E_0 for zero dilatational strain. We scan the energy by rotating the monochromator around this energy E_0 and watch the intensity of the peak of interest on the CCD camera. The energy which maximizes the intensity of the reflection is the actual energy of the reflection. The difference in the observed energy and the E_0 gives the dilatational strain.

Since $\sigma'_{xx} + \sigma'_{yy} + \sigma'_{zz} = 0$ by definition, $-\sigma'_{zz}$ is a measure of the in-plane stress (note that for a blanket film, with free or passivated surface, on average the total normal stress $\sigma_{zz} = 0$), from that σ_b (biaxial stress) $= (\sigma_{xx} + \sigma_{yy})/2 = (\sigma'_{xx} + \sigma'_{yy})/2 - \sigma'_{zz} = -3\sigma'_{zz}/2$. This relation is always true on average. A positive value of $-\sigma'_{zz}$ indicates an overall tensile stress, whereas a negative value indicates an overall compressive stress. However, the measured stress values, corresponding to a strain of less than 0.01%, are only slightly larger than the strain/stress sensitivity of the white beam Laue technique (the sensitivity of the technique is 0.005% strain).

No very long-range stress gradient has been observed around the root of a whisker, indicating that the growth of a whisker has released most of the local compressive stress in the distance of several surrounding grains. In Fig. 14.12, the whisker part is removed in order to observe more clearly the stress around the whisker root. The absolute stress value in the whisker is higher than that in the surrounding grains. If we assume the whisker to be stress-free, the SnCu surface finish is under compressive stress.

In Chapter 13, we compared the driving force of electromigration and thermomigration in solder joints. In terms of the work done in moving an atom of an atomic distance in electromigration or in the thermal energy across an atom in a temperature gradient in thermomigration, they are in the order of 10^{-27}. We can calculate the driving force and work done in stress-migration-induced Sn whisker growth to see if they agree. In Fig. 14.11, we consider the stress gradient between the origin ($x = 0$, $y = 0$) and the lower left corner point at ($x = -0.5400$, $y = 0.8490$). We found that $\Delta\sigma \approx 4$ MPa and

$\Delta x \approx 10 \mu$m, and we take the atomic volume of a Sn atom to be 27×10^{-24} cm^3.

$$F = -\frac{\Delta \sigma \Omega}{\Delta x} = \frac{4 \times 10^6 (\text{N/m}^2) \times 27 \times 10^{-24} (\text{cm}^3)}{10^{-3}(\text{cm})}$$

$$= \frac{4 \times 10^7 (\text{dyn/cm}^2) \times 27 \times 10^{-24} (\text{cm}^3)}{10^{-3} \text{cm}}$$

$$= \frac{108 \times 10^{-17} \text{erg}}{10^{-3} \text{cm}} \approx 10^{-12} \text{erg/cm}$$

The work done by this force over an atomic jump distance of 0.3 nm is 3×10^{-27} J, which is of the same order of magnitude as those calculated for electromigration and thermomigration. Thus, the driving force will lead to stress-migration. It is worth mentioning that if we calculate a mechanical force of σA, where σ is yield stress and A is the cross-sectional area of a Sn atom, the calculated force will be several orders of magnitude larger. This is because the stress difference across an atom in a stress gradient is quite small.

14.5.5 Stress relaxation by creep: broken oxide model in Sn whisker growth

Whisker growth is a unique creep phenomenon in which stress generation and stress relaxation occur simultaneously. Therefore, we must consider two kinetic processes of stress generation and stress relaxation and their coupling by irreversible processes. About the two processes in whisker growth, the first is the diffusion of Cu from the lead frame into the Sn finish to form grain-boundary precipitates of Cu_6Sn_5. This kinetic process generates the compressive stress in the finish. The second is the diffusion of Sn from the stressed region to the stress-free region at the root of a whisker to relieve the stress. The distance of diffusion in the second process is much longer than the first and also the diffusivity in the second process is slower, so the second process tends to control the rate in whisker growth.

Room temperature is a relatively high homologous temperature for Sn, which melts at $232\,^\circ$C; hence, the self-diffusion of Sn along Sn grain boundaries is fast at room temperature. Therefore, the compressive stress in the Sn induced by the chemical reaction at room temperature can be relaxed at room temperature by atomic rearrangement via grain-boundary self-diffusion.

The growth of a whisker occurs at the root; it is being pushed out. We ask: what is the growth mechanism? When Sn atoms diffuse to the root of a whisker, how can they be incorporated into the root of a whisker? The growth can be regarded as grain growth because the whisker is a single crystal and grows longer with time. In the classical model of normal grain growth, the basic process is grain-boundary migration against its curvature by atoms jumping from one grain across a grain boundary to the grain on the other side of the grain boundary. Yet in whisker growth, it is unclear if there is grain-boundary migration at the root. Using a series of cross-sectional SEM images, the microstructure of a whisker root and its surrounding grains has been observed. This suggests that most likely there is no migration of the grain boundaries between the whisker and the surrounding grains during the growth of the whisker. Whisker growth is a grain growth with very little grain-boundary migration at the root of the whisker. It seems that Sn atoms arrive at the root region along grain boundaries and they can be incorporated into the root of a whisker without jumping across a grain boundary as in normal grain growth. This is because the Sn atoms are already diffusing in grain boundaries. Hence, no grain-boundary migration is needed. The atomistic model of incorporation of atoms into the whisker for its growth requires more study; it may take place at the kink sites on the bottom side of the whisker, similar to step-wise growth on a free crystal surface in the epitaxial growth of thin films. We must mention that there are vacancies coming in from the surface crack at the root area driven by a stress gradient to assist the growth.

To analyze the growth kinetics of a whisker, we assume a two-dimensional model in cylindrical coordination. The distribution of whiskers is assumed to have a regular arrangement so that each occupies a diffusional field of diameter of $2b$, as shown in Fig. 14.14. We assume that the whisker has a constant diameter of $2a$ and a separation of $2b$ and that it has a steady-state growth in the diffusional field which can be described by a two-dimensional continuity equation in cylindrical coordinates. Recall that stress can be regarded as an energy density, and a density function obeys the continuity equation:

$$\nabla^2 \sigma = \frac{\partial^2 \sigma}{\partial r^2} + \frac{1}{r}\frac{\partial \sigma}{\partial r} = 0 \qquad (14.21)$$

The boundary conditions are

$$\sigma = \sigma_0 \text{ at } r = b, \text{ and } \sigma = 0 \text{ at } r = a$$

The solution is $\sigma = B\sigma_0 \ln(r/a)$, where $B = [\ln(b/a)]^{-1}$ and σ_0 is the stress in the Sn film. Knowing the stress distribution, we can evaluate the stress gradient,

$$X_r = -\frac{\partial \sigma \Omega}{\partial r} \qquad (14.22)$$

Then the flux to grow the whisker is calculated at $r = a$,

$$J = C\frac{D}{kT}X_r = \frac{B\sigma_0 D}{kTa} \qquad (14.23)$$

Note that in a pure metal, $C = 1/\Omega$, the volume of materials transported to the root of the whisker in a period of dt is

$$JAdt\Omega = \pi a^2 dh \qquad (14.24)$$

where $A = 2\pi as$ is the peripheral area of the growth step at the root, s is the step height, and dh is the increment of height of the whisker in dt. Therefore, the growth rate of the whisker is

$$\frac{dh}{dt} = \frac{2}{\ln(b/a)}\frac{\sigma_0 \Omega s D}{kTa^2} \qquad (14.25)$$

To evaluate the whisker growth rate, we need to know the self-diffusivity of Sn. The self-lattice diffusivities in the direction parallel and normal to the c-axis are slightly different and are given as

$$D_{//} = 7.7 \times \exp(-25.6 \text{ kcal/kT}) \text{cm}^2/\text{s}$$

$$D_\perp = 10.7 \times \exp(-25.2 \text{ kcal/kT}) \text{ cm}^2/\text{s}$$

The lattice diffusivities at room temperature are about 10^{-17} cm^2/s. It means that in one year, or $t = 10^8$ s, the diffusion distance calculated by using $x^2 \cong Dt$ is approximately 1 μm. Therefore, the lattice diffusivities are too slow to be responsible for whisker growth at room temperature. Self grain-boundary diffusion of Sn has not been determined. If we assume that the large-angle grain-boundary diffusivity requires one-half of the activation energy of lattice diffusion given above, we obtain a self grain-boundary diffusivity about 10^{-8} cm^2/s.

We take $a = 3\,\mu$m, $b = 0.1$ mm, $\sigma_0\Omega = 0.01$ eV (at $\sigma_0 = 0.7 \times 10^9$ dyne/cm^2), $kT = 0.025$ eV at room temperature, $s = 0.3$ nm, and $D = 10^{-8}$ cm^2/s (the self grain-boundary diffusivity of Sn at room temperature), we obtain a growth rate of 0.1×10^{-8} cm/s. At this rate, we expect a whisker of 0.3 mm after one year, which agrees well with the observed result. Since we assume grain-boundary diffusion, note that there are only several grain boundaries connecting the base of a whisker to the rest of the Sn matrix. Hence, in taking the total atomic flux which supplies the growth of a whisker to be $JAdt\Omega$, where $A = 2\pi as$, we have assumed that the flux goes to the entire peripheral of the whisker "$2\pi a$" but only for a step height of "s" for its growth. In the above calculation of whisker growth rate, if we take the value of b to be a few grain diameters as shown in Fig. 14.14 and the stress σ_0 to be about 10 MPa or 10^8 dyne/cm^2, the result is the same.

References

[1] C. Herring, "Diffusional viscosity of a polycrystalline solid," *J. Appl. Phys.* **21** (1950), 437.

[2] B. Chalmers, *Physical Metallurgy* (Wiley, New York, 1959).

[3] A. S. Nowick and B. S. Berry, *Anelastic Relaxation in Crystalline Solids* (Academic Press, New York, 1972).

[4] M. F. Ashby and D. R. H. Jones, *Engineering Materials I* (Pergamon Press, Oxford, 1980).

[5] Fan-Yi Ouyang, Kai Chen, K. N. Tu and Yi-Shao Lai, "Effect of current crowding on whisker growth at the anode in flip chip solder joints," *Appl. Phys. Lett.* **91** (2007), 231919.

[6] K. N. Tu, "Interdiffusion and reaction in bimetallic CuSn thin films," *Acta Met.* **21** (1973), 347.

[7] K. N. Tu and R. D. Thompson, "Kinetics of interfacial reaction in bimetallic CuSn thin films," *Acta Met.* **30** (1982), 947.

[8] K. N. Tu, "Irreversible processes of spontaneous whisker growth in bimetallic CuSn thin film reactions," *Phys. Rev.* **B49** (1994), 2030–4.

[9] G. T. T. Sheng, C. F. Hu, W. J. Choi, K. N. Tu, Y. Y. Bong and Luu Nguyen, "Tin whiskers studied by focused ion beam imaging and transmission electron microscopy," *J. Appl. Phys.* **92** (2002), 64–9.

[10] W. J. Choi, T. Y. Lee, K. N. Tu, N. Tamura, R. S. Celestre, A. A. MacDowell, Y. Y. Bong and
L. Nguyen, "Tin whisker studied by synchrotron radiation micro-diffraction," *Acta Mat.* **51**
(2003), 6253–61.

Problems

14.1 Both electromigration and thermomigration are cross-effects in irreversible
processes. How come stress-migration is not a cross-effect?

14.2 Upon applying an elastic stress to a piece of metal, what is the primary flux or
flow in the metal?

14.3 When we couple stress-migration and electromigration, a critical length is
obtained below which there is no electromigration. In comparison, when we couple
stress-migration and thermomigration, can we obtain a critical length or not? If not, why
not?

14.4 We have modeled Nabarro–Herring creep on the basis of lattice diffusion and
also Coble creep on grain-boundary diffusion. Can we have creep on the basis of surface
diffusion?

14.5 In zero creep as discussed in Chapter 3, we consider an Au wire of diameter of
1 mm and length of 10 cm and having a grain size of 2 mm in length. What is the stress
potential gradient in the sample? How fast is the creep rate at 800 °C?

14.6 We conduct a zero creep experiment using an Au nanowire of diameter of 100 nm
and length of 100 μm at 800°C. What will happen?

15   Reliability science and analysis

15.1   Introduction

We should define what reliability science is. When a device is manufactured to provide a unique function in applications, it is generally expected that the microstructure in the device will be unchanged in its lifetime of use. Unfortunately, this is not true. In electronic device applications, we have to apply an electric field or current. Under a high-current density, electromigration induces changes in microstructure and leads to circuit failure due to opening by void formation or shorting by whisker extrusion. The high-current density also causes joule heating and the temperature rise will lead to thermal stress between different materials having a different thermal expansion coefficient in the device. The stress and temperature gradients will induce atomic diffusion, phase change, and microstructure instability. What is unique in these microstructure changes is that they occur in the domain of non-equilibrium thermodynamics or they are irreversible processes. The basic science to provide an understanding of phase changes in irreversible processes that leads to device failure is reliability science. From the point of view of applications, physical and statistical analyses on the basis of reliability science should be able to predict the lifetime of a device [1–4].

The traditional metallurgical phase changes occur between two equilibrium states, and they are defined under constant temperature and constant pressure, for example the phase change in a piece of solder of eutectic SnPb going from 200 °C to 100 °C at ambient pressure. The phase change in the solder is from a molten state to a solid state. We can minimize Gibbs free energy change to describe the process and use the equilibrium phase diagram of Sn–Pb to define the composition of the two solid phases in the eutectic microstructure. Nevertheless, in this special case it is a near-equilibrium phase change, not the final equilibrium, because we cannot define the lamellar spacing in the eutectic structure.

However, when we have a temperature gradient or a pressure (stress) gradient, we do not have the boundary condition of constant temperature and constant pressure, so the change cannot be described by minimum Gibbs free energy. Instead, we have irreversible processes, and the kinetics of phase change is in the domain of non-equilibrium thermodynamics. A classical example is the Soret effect of thermomigration, in which a homogeneous alloy becomes inhomogeneous under a temperature gradient. Since the inhomogeneous state has a higher free energy than the homogeneous state, it is a process of increasing free energy. Another case is a eutectic solder joint under electromigration, which can lead to a complete separation of the two eutectic phases, in which there is no lamellar microstructure.

Nevertheless, these irreversible processes may not lead to device failure. Take electromigration as an example: if the atomic flux induced by electromigration is uniform in the conductor and if the cathode and the anode are a very large source and sink of atoms, respectively, there is no failure when we define failure as void or hillock formation in the conductor. Hence, for failure to occur, we shall require divergence of the mass flux in the irreversible processes. For example, void formation has been found in triple points of grain boundaries in Al and Cu interconnects, where atomic flux divergence occurs.

Still, the condition of divergence is necessary but insufficient for void or hillock formation if it is a constant lattice site or constant volume process. Note that while the total number of atoms is conserved, the total number of lattice sites has to change when there is void formation or hillock and whisker growth. To do so, we must require no lattice shift in the region of flux divergence, so we have a non-constant lattice site or non-constant volume process.

In this chapter, we shall discuss first the constant volume and non-constant volume processes. When the irreversible process is accompanied by lattice shift, the total number of lattice sites is conserved and it is a constant volume process. There is no void or hillock formation, and no failure. When lattice shift is absent, it is a non-constant volume process and failure occurs. To analyze failure, we shall discuss both physical analysis and statistical analysis, and we use electromigration in flip-chip solder joints as an example. The physical analysis will enable us to understand the mode and mechanism of failure. The statistical analysis will give us MTTF which is needed in predicting the lifetime of a device. Finally, the link between the two analyses will be briefly discussed.

15.2 Constant volume and non-constant volume processes

Recall the classic case of the Kirkendall effect in interdiffusion between A and B, as discussed in Section 5.3.2. The concentration of A in B or B in A changes with time and position, thus there is divergence of atomic flux in the interdiffusion. However, in Darken's analysis of interdiffusion, there is no stress and no void formation, since vacancy concentration has been assumed to be at equilibrium everywhere in the sample, although vacancy divergence exists. Because of the equilibrium, there is no supersaturation of vacancies, and so no nucleation of void. When the structure has no void formation, we can assume that there is no open failure.

Implicitly, the assumption of equilibrium vacancy requires that vacancies can be absorbed or created as they are needed in the structure, for example, by dislocation climb in the bulk of the sample. The interdiffusion under Darken's analysis leads to lattice plane migration or lattice shift, in turn leading to marker motion when markers are embedded in the lattice. In lattice shift, the number of total lattice sites is constant. If it is assumed that the partial molar volume of A and B in the AB alloys is the same, a constant total lattice site means that the total volume remains constant. Furthermore, there is no stress or no strain if there is no volume change. We can regard the interdiffusion as occurring under the "constant volume" condition and write $j_A + j_B = -j_V$, using the moving lattice frame as given in Eq. (5.37), where j_A, j_B, and j_V are the fluxes of A atoms, B atoms, and vacancies, respectively.

On the other hand, many interdiffusion cases have Kirkendall or Frenkel void formation. This is because lattice shift is incomplete or absent; in turn, vacancy is not at equilibrium everywhere in the sample. The total volume is not constant in the process. If we assume that there are more A atoms diffusing into B than B into A, more lattice sites are needed in B in order to accommodate the excess A atoms. Also, more lattice sites are needed in A in order to accommodate the excess vacancies. The excess A atoms may accumulate on the free surface in order to reduce strain, so we can have hillock or whisker growth. When the excess vacancies in A are super saturated, Kirkendall or Frenkel void can nucleate. It is worth mentioning that the amount of excess A atoms in B may not be equal to the amount of excess vacancies in A; each of them can be an independent event. When whisker or void grows, we have failure. The essence in reliability failure is a non-constant volume process or the absence of lattice shift. We have explained the absence of lattice shift for the case of Al interconnect in Section 5.3.2.3.

15.3   Effect of lattice shift on divergence of mass flux in irreversible processes

15.3.1   Initial distribution of current density, temperature, and chemical potential in a device structure before operation

Before electric power is turned on to operate a device, there will be no current flow and no joule heating, so there is no current density distribution in the conductors. In a p-n junction, there is dopant distribution, but there is no flow of charges until a bias voltage is applied. This is because the dopants are assumed to be frozen in the semiconductor lattice at near room temperature, and the build-in potential of the p-n junction is at an equilibrium state until a bias is applied. Without power, the temperature distribution in the device is uniform as the ambient. Furthermore, for simplicity, we may assume that initially the chemical potential, including stress potential, in the device is also uniform. This may be an approximation and may not be true; we shall discuss it later. With this very simple initial condition, we postulate that when we apply a current to operate the device, it induces a current distribution in the device structure; in turn, because of joule heating, it induces a temperature distribution in the device. Electron flow and heat flow can induce atomic diffusion, so we shall induce a chemical potential gradient in the device which will lead to microstructure and phase change with time, and hence the reliability issues occur. Joule heating may induce thermal stress too. Obviously, before we apply electric power, there is no reliability concern.

Knowing the design of the device structure and the interfaces between different materials, we know the composition distribution and the composition gradient in the interfacial reactions. In principle, we can obtain the chemical potential distribution as well as the stress potential in the device structure.

In Chapter 14, we mentioned stress potential ($\sigma\Omega$) as a part of chemical potential. Hence, the stress potential gradient ($d\sigma\Omega/dx$) is a driving force of atomic motion. Since different materials will have different thermal expansion coefficients, we can have thermal stress between different parts in the device during the processing of the device. When the thermal stress is large across an interface, it can induce crack formation and propagation. This is a problem of yield rather than reliability. On the other hand, it is the thermal stress gradient which produces atomic flow, similar to creep under a mechanical stress gradient, and it becomes a reliability issue. However, thermal stress and its gradient can be reduced by annealing.

In device manufacturing, both high yield and good reliability are emphasized. If there is no yield, there is no need to consider reliability. An example is the integration of a low dielectric constant material (low k) and Cu to form the multilevel damascene interconnects. When porous and polymeric materials as ultralow k dielectric are integrated with Cu, the thermal stress between them tends to crack the dielectric. Even if the integration is successful, when the chip is joined by flip-chip solder joints to its packaging substrate, the thermal stress between the chip and the substrate may lead to crack formation in the low k because the low k is mechanically weaker than the solder. This is called chip-packaging interaction and affects the yield of the device.

With high yield, then to achieve product assurance of good reliability, industry introduced a "burn-in" process to remove the early and easy failure parts or elements. An annealing at a given temperature must be performed in order to meet the reliability criteria. For example, solder joints in electronic packaging require an annealing at 150 °C for 1000 h. Basically, the annealing is to achieve a near-homogeneous chemical potential in the device. It is expected to remove any pre-existing stress and chemical potential gradient. After that, we consider reliability when we turn on the power to operate the device. The annealing at 150 °C for 1000 h may seem long, yet its effect on interfacial IMC formation is the same as that in reflow of 1 min when the solder is molten.

Since devices are built by layered thin-film structures, there are interfaces and interfacial reactions. For example, in solder joint formation, the solder will react with the Cu UBM on the chip side as well as with the Cu bond-pad on the substrate side. Since the solder reaction in the solid state is four orders of magnitude slower than that in the liquid state, even the annealing at 150 °C for 1000 h will not be able to consume all the Cu and the solder completely, so there will be unreacted Cu and solder in the joint. Actually it is better to have some non-reaction Cu and solder so that multiple reflows and reworks can be performed. Hence, there is a chemical potential gradient in the flip-chip structure because of the unreacted Cu and solder. Nevertheless, the IMC formed between the Cu and solder becomes the diffusion barrier to slow down the Cu/solder reaction, and the compound is so thick that the rate of chemical reaction to form more Cu–Sn compounds will be very slow at the device operation temperature. Yet the reaction will be a reliability concern when it is enhanced by electrical or thermal force.

15.3.2   Change of the distributions during device operation

Among mass flow, heat flow, and charge carrier flow, it is the last that initiates the distribution change of current density, temperature, and chemical potential in the device. When the structure design of a device is given, we know the dimension and material (resistance) from the design, so we can use FEA to simulate the "current distribution" in the device structure when an applied current is present. There are commercial computer programs, such as ANSYS, available for three-dimensional simulation of current distribution, that are reliable. In the simulation, it is important to know the non-uniform distribution of current density, where current crowding occurs.

The next step will be to obtain temperature distribution in the device. Knowing the current distribution, we can calculate joule heating, as given in Chapter 10. While we will need to know heat dissipation in the device, in order to obtain its temperature distribution, we can nevertheless assume a set of reasonable heat loss parameters to represent the heat dissipation in the device, since heat conduction has been a subject very well studied. Then we can use an infrared sensor or other temperature measurement techniques to determine the temperature distribution in the device, so we can check whether or not our assumed parameters are accurate. We can tailor the parameters to match the measured data. From the temperature distribution, we will know the hot spots and the temperature gradient in the device. The melting of the solder can be used as a check-point.

Besides current density distribution and temperature distribution, we need to know chemical potential distribution induced and affected by electron flow and heat flow. There is a transient state in the effect of electron flow and heat flow on atomic flow before a steady state can be established. The effect of electron flow and heat flow on atomic flow is given by electromigration and thermomigration, respectively. In these irreversible processes, it is typically assumed to be at a steady state, which usually takes time to reach. The transient state of electromigration in Al interconnects has been analyzed by using the continuity equation or the divergence. In the steady state, the key is to find out if there is an atomic flux divergence in the structure where failure tends to occur.

About flux divergence in atomic diffusion, the concentration C can change with time and position, i.e. $C = C(t, x)$, so the concentration gradient changes with time and position, and we have Fick's second law. But in electromigration, the current density or the driving force can be constant in time and position so the induced atomic flux can be constant; hence, there is no divergence except where microstructure changes. For example, at a triple point of grain boundaries or at an abrupt interface, we can have atomic flux divergence even if the current density is constant. The atomic flux change across the interface may be abrupt, rather than gradual. When there is no lattice shift, void or whisker can grow at the interface. The growth of a void, for example, is supplied by a constant flux of vacancies, and we can use the growth equation as discussed in Chapter 5 to model the growth.

When non-uniform current density occurs due to current crowding in a device structure, the driving force of electromigration is not constant in position, so we need to consider the atomic flux divergence driven by a non-uniform force of electromigration. We can solve the divergence equation when the boundary and initial conditions are given.

From the solution, we can obtain mass flux at a given position for the growth of a void or a whisker.

15.3.3 Effect of lattice shift on divergence of mass flux

We have discussed the essence of electromigration, thermomigration, and stress-migration in terms of driving force and flux in the previous chapters. Their interaction can be represented by the following 3×3 matrix equation in irreversible processes:

$$J_i = L_{ij}X_j \tag{15.1}$$

where J, L, and X represent flux, phenomenological coefficient, and driving force, respectively. They combine the interactions among Fick's law, Fourier's law, and Ohm's law. However, from the point of view of failure, we are interested in the flux of mass and the related cross-effects.

In Chapter 11, it was discussed that if the atomic flux under electromigration is uniform, there will be no failure. Therefore, knowing the flux is not enough; failure requires divergence of mass flux so that void formation or whisker growth can occur. Analogous to Fick's first and second laws, we can regard Eq. (15.1) as the first law, and we need to consider the divergence of J_i for failure, where J_i is the atomic flux given by Eq. (15.1). Assuming one dimension, we have

$$J_x = C\frac{D}{kT}\left[-\left(\frac{\partial\sigma\Omega}{\partial x}\right) + Z^*ej\rho + \frac{Q^*}{T}\left(-\frac{\partial T}{\partial x}\right)\right] \tag{15.2}$$

where on the right-hand side, the three terms correspond to atomic flux due to stress-migration, electromigration, and thermomigration, respectively. Assuming that diffusion occurs by vacancy mechanism, there is a reverse flux of vacancies, $J_v = J_x$. The concentration of vacancy, C_v, as a function of time and position can be obtained by solving the continuity equation,

$$\frac{\partial C_v}{\partial t} = -\frac{\partial J_v}{\partial x} + \delta \tag{15.3}$$

where δ is a source/sink term which allows for the creation and annihilation of vacancies in the lattice.

J. J. Clement solved Eq. (15.3) under both conditions of $\delta = 0$ and $\delta \neq 0$. Simulation of the solutions was obtained. The solutions will not be repeated here [6].

15.4 Physical analysis of electromigration failure in flip-chip solder joints

Physical analysis is required in order to understand the mode and mechanism of reliability failure. Typically, this is performed by conducting electromigration of a test structure at accelerated conditions, for example, at higher temperature and higher current density than those at the working condition of the device. The failed device will be cross-sectioned and the failure site will be examined by SEM, TEM, XRD, micro-probe, etc. Below, we shall illustrate it by the example of electromigration-induced failure in flip-chip solder joints.

15.4.1   Distribution of current density in a pair of joints

An example about the study of reliability of flip-chip solder joints is given here [7–10]. We begin with the simulation of a current distribution in a pair of solder joints. We need a pair of solder joints in the analysis in order to complete the electric charge flow in and out of the pair of solder joints. Fig. 15.1(a) depicts a three-dimensional simulation of a pair of flip-chip solder joints between a Si chip on the top side and a polymer substrate on the bottom side. On the top side, the two bumps were connected by an Al interconnect. The arrows indicate the electron flow direction. On the bottom side, they are connected separately on the substrate each by a Cu bond-pad to the outside. Fig. 15.1(b) depicts the cross-section of the pair of solder joints. Between the Al interconnect and the solder bump, there is an UBM to define the contact opening. As indicated by the arrows in Fig. 15.1(a), electron current can be applied from one of the bond-pads to flow through one solder bump, the Al interconnect, the other solder bump, and to get out from the other bond-pad. The dimension and resistance of the Al interconnect, the UBM, the solder bump, and the Cu bond-pad are known. We can simulate the current distribution as well as the joule heating in the flip-chip structure, as shown in Fig. 15.2(a) and (b), respectively.

From the current distribution simulation, two features are found to be important; first, the Al interconnect has the highest current density, and second, the current density

(a)

(b)

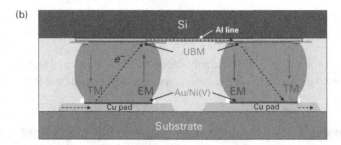

Fig. 15.1   (a) Three-dimenstional schematic diagram of a pair of flip-chip solder joints. The arrows indicate the electron flow direction. (b) Schematic diagram depicting the cross-section of a pair of solder joints between a Si chip on the top side and a polymer substrate on the bottom side. Directions of electromigration and thermomigration are indicated.

(a)

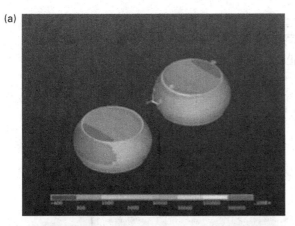

(b)

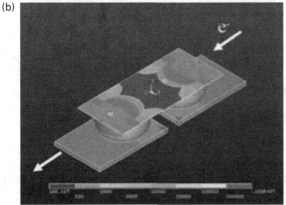

Fig. 15.2   (a) Simulation of the current distribution in the flip-chip structure of a pair of solder bumps with a Al interconnect between them and the Cu trace below them. (b) Simulation of joule heating to show that the Al interconnect is the heat source.

distribution in the solder bump is non-uniform. The first implies that the Al is the hottest part during device operation because joule heating is proportional to the square of current density, as in $j^2 \rho$. There will be a temperature gradient across the bump since the joule heating in the Cu bond-pad will be lower. In Fig. 15.1(b), it indicates the difference in the combined effects of electromigration and thermomigration in the two bumps. In the one on the right-hand side, they are in the same direction. In the one on the left-hand side, they are in the opposite direction. The second means that there is current crowding in the solder joint, where electrons enter the solder bump from the Al interconnect or exit the bump into the Al. Electromigration will be serious in the current crowding region.

15.4.2   Distribution of temperature in a pair of joints

Fig. 15.3 shows the temperature increase due to joule heating at the chip side and substrate side as a function of current density, and the furnace temperature was kept at 150 °C. It

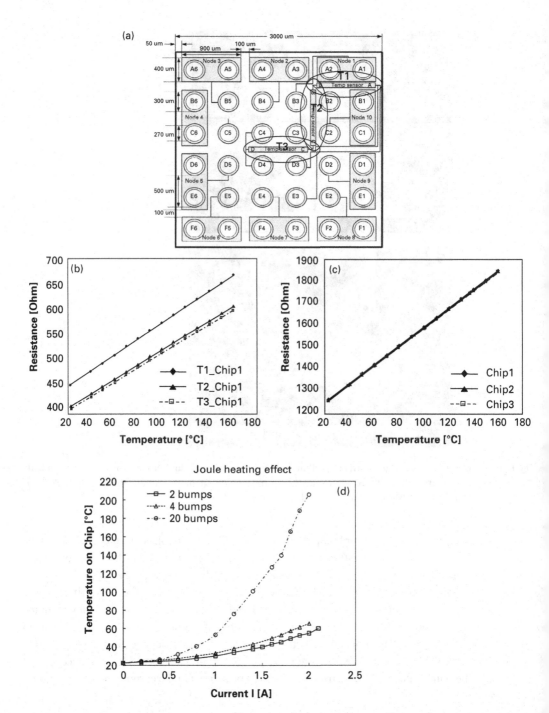

Fig. 15.3   (a) Schematic diagram of three Pt sensors on the chip side; (b) calibration curve of each of the Pt sensors at ambient temperature; (c) calibration curve of three Pt sensors together at ambient temperature; (d) The temperature increase due to joule heating at the chip side and substrate side as a function of current density and the furnace temperature was 150 °C.

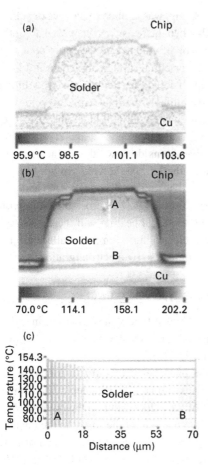

Fig. 15.4 Temperature distribution on the cross-section of one of the solder joints when the applied current is 0.5 amp, measured by infrared sensor. (a) Infra-red image before current stressing. (b) After stressing by 0.5 amp. (c) Temperature gradient across the solder joint.

was measured by temperature sensor of serpentine Pt wires deposited on the chip and on the substrate. For example, when the temperature at the chip side was found to be about 13° higher than the substrate side, this temperature difference indicates a temperature gradient of 1300 °C/cm across the solder bump having a height of 100 μm, which is sufficient to induce thermomigration.

Fig. 15.4 shows the temperature distribution on the cross-section of a solder joint when the applied current is 0.5 A, measured by IR sensor. The temperature is not uniform; the chip side is hotter than the substrate side, and the temperature difference across the solder bump of 100 μm in height is about 5 °C, so the temperature gradient is approximately 500 °C/cm. The temperature gradient suggests that thermomigration can occur in the solder bump. Therefore, electromigration in this bump will be accompanied by thermomigration. The temperature distribution can also be measured from the top side of the chip using infrared camera. The IR camera can penetrate the chip and detect the temperature increase. It can show that the Al interconnect has the highest temperature.

When a current density of 1×10^4 A/cm^2 was applied to the pair of solder joints kept at 150 °C, failure was found after 50 h when the resistance increased abruptly. The mode of failure is the growth of a pancake type of void across the entire contact interface on the chip side of the right-hand side solder bump of the pair shown in Fig. 15.1. In this bump, electrons flow down from the upper-left corner, and current crowding occurs when the electrons enter the solder bump from the Al interconnect. The void was initiated at the upper-left corner and grew to the left. The void will block the flow of electrons, so electrons have to move forward along the Al interconnect and enter the bump in front of the void. Electromigration will drive atoms downward and vacancies upward. The vacancies will feed the void to grow to the left, so the void growth has the shape of a pancake. The growth of such a pancake-type void has been modeled and simulated. The solder joint failed abruptly when the void had grown across the entire contact. When the conducting path in Al is increased, the joule heating will increase too. While there is thermomigration accompanying the electromigration and while there may be some stress-migration, the rate of failure is dominated by electromigration.

15.4.3   Effect of current crowding on pancake-type void growth

Fig. 15.5(a) shows SEM image of a pancake-type void formation in a daisy-chain of flip-chip solder joints. Enlarged images of the pancake-type void at the cathode contact of bumps 5, 3, and 1 are shown in Fig. 15.5(b).

Fig. 15.6(a) depicts the growth of a pancake-type void at a contact interface due to divergence of vacancies fluxes. The black solid arrow lines depict the atomic flux driven by the current crowding from the top of the solder bump to the bottom. Meanwhile, the reverse flux of vacancies goes back from the bulk of the solder to the interface indicated by the dotted arrow line. Fig. 15.6(b) depicts that due to the growth of a pancake-type void, the electron flux has to migrate to the front of the void, and Fig. 15.6(c) shows the measured change in voltage or resistance of the solder joint during the growth of the pancake-type void. There was little change until the void eclipsed the entire contact area at the cathode.

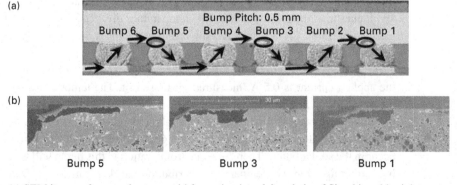

Fig. 15.5   (a) SEM image of a pancake-type void formation in a daisy chain of flip-chip solder joints. (b) Enlarged images of the pancake-type void as the cathode contact of bumps 5, 3, and 1 are shown.

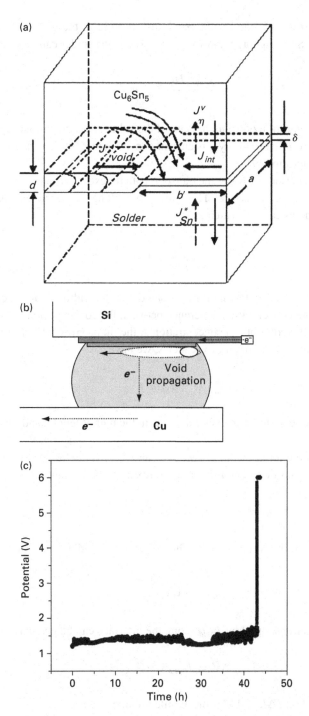

Fig. 15.6   (a) Schematic diagram depicting the growth of a pancake-type void at a contact interface due to divergence of vacancies fluxes. (b) Schematic diagram of the growth of a pancake-type void and the migration of electron flux to the front of the void (c) The corresponding change in voltage or resistance change of the solder joint during the growth of the pancake-type void. There was little change until the void eclipsed the entire contact area at the cathode.

To analyze the growth kinetics of the pancake-type void, if the vacancy flux in the interfacial IMC Cu_6Sn_5 is ignored, the vacancy fluxes in the solder can be written as

$$J_{Sn}^v = \frac{C_{Sn}^{bulk}D_{Sn}}{kT}Z_{Sn}^*e\rho_{Sn}j \tag{15.4}$$

where C is the concentration of Sn in the bulk of the solder, D is the diffusivity, e is the charge of an electron, ρ is resistivity, j is current density, and Z^* is the effective charge number of electromigration [9].

The solder–IMC interface provides the transport path for excess vacancies and enables them to diffuse along the interface. The lateral flux along the interface due to the divergence of vacancies can be written as

$$J_{int}^v = -D_{int}\frac{\Delta C}{\Delta x} \approx D_{int}\frac{\Delta C}{b'} \tag{15.5}$$

where D_{int} is the diffusivity in the interface, b' stands for the width of current crowding region, and ΔC is the concentration difference between the concentration in the higher current density and the equilibrium concentration at the tip or growth front of the void. In terms of conservation of number of vacancies, we have

$$J_{int}^v a\delta t = J_{Sn}^v ab't \tag{15.6}$$

where δ is the effective width of interface and a is in the unit of length, and t is a period of time.

It is assumed that the initial width of void is d, and J_{void} is the flux of vacancy at the tip of the void. The condition of conservation of flux is applied again:

$$J_{int}^v a\delta = J_{void}ad \tag{15.7}$$

Substituting Eq. (15.6) into Eq. (15.7), the flux to grow the void can be written as

$$J_{void} = (J_{Sn}^v)\frac{b'}{\delta} \tag{15.8}$$

The volume of mass transported by J_{void} along the interface can be given as

$$\Delta V = J_{void}A\Delta t\Omega \tag{15.9}$$

where $A = a\delta$, $\Delta V = ad\Delta l$, and Ω is the atomic volume.

Inserting Eq. (15.8) into Eq. (15.9), the growth velocity of void becomes

$$v = \frac{\Delta l}{\Delta t} = (J_{Sn}^v)\frac{b'}{d}\Omega \tag{15.10}$$

If it is assumed that $C_v^{bulk}\Omega = 1$, we obtain

$$v = \frac{ej}{kT}(D_{Sn}\rho_{Sn}Z_{Sn}^*)\frac{b'}{d} \tag{15.11}$$

To verify the mechanism of void propagation, the two key parameters are the width of current crowding region, b', and the width of the void, d. The Gibbs Thomas effect may play an important role in forming the tip of the void,

$$C_r = C_0\exp\left(\frac{\gamma}{r}\frac{\Omega}{kT}\right) \tag{15.12}$$

where γ is the surface energy per area.

By applying the linear approximation, we have the void width as

$$d = 2r = \frac{C_0}{\Delta C}\frac{4\gamma\Omega}{kT} \tag{15.13}$$

Since the model is two-dimensional, the void width is assumed to be constant. On the other hand, from Eq. (15.5) and Eq. (15.6), the width of current crowding is

$$b' = \left(\frac{\Delta C}{C_0}\frac{kTD_{gb}\delta}{ejD_{Sn}Z_{Sn}^*\rho_{Sn}}\right)^{1/2} \tag{15.14}$$

In the two-dimensional simulation shown in Fig. 15.6(a), the contact window length of a eutectic SnAgCu solder joint is taken to be 224 μm and the current crowding region is taken to be approximately 15% of the whole length, so the current crowding region, b', is estimated to be about 33.6 μm. From Fig. 15.5 the void width, d, is measured as 2.44 μm. The test temperature is 146 °C, the electric current density is about 3.67×10^3 A/cm^2, the void length is 33 μm, the void propagation has spent 6 h, thus the void growth velocity is about 5 μm/h.

In the other case of a eutectic SnPb solder bump stressed at 2.25×10^4 A/cm^2 and 125 °C as discussed in Section 9.4.1, the window length is 140 μm, the width of current crowd region is about 9 μm. The voids are formed at 38 h and failed at 43 h; therefore, the void growth velocity is about 28 μm/h.

The diffusivity is taken to be $D_{Sn} = 1.3 \times 10^{-10}$ cm^2/s for Sn, and the diffusivity of interface is taken to be 4.2×10^{-5} cm^2/s. The effective charge is $Z_{Sn}^* = 17$ for Sn. The resistivity of Sn is $\rho_{Sn} = 13.25$ $\mu\Omega$ cm. The surface energy $\gamma = 10^{15}$ eV/cm^2 and Ω is taken as 2.0×10^{-23} cm^3. The effective interfacial width is about 0.5 nm. The only unknown parameter is the ratio of ΔC and C_0. In order to obtain reasonable results, we choose the range of $\Delta C/C_0$ from 1% to 3%.

Using these parameters and experimental conditions, the theoretical values of current crowding length b' have been calculated from Eq. (15.14), void width d from Eq. (15.13), and void growth velocity v from Eq. (15.11). The comparison between theoretical values and experimental results are in reasonable agreement as listed in Table 15.1.

Table 15.1. Comparison of measured and calculated values of growth rate of pancake type void in electromigration

	Theory	Experiment
b'	25.49–44.15 μm	37.5 μm
D	0.81–2.42 μm	2.44 μm
V	1.24–6.44 μm/h	4.4 μm/h

One puzzling finding about the pancake-type void growth is that it grows below the dielectric which defines the contact opening. In cross-sectional SEM images of the pancake-type void, the void is always found to extend all the way to the boundary of the bump below the dielectric. Yet there is very little current under the dielectric.

Another question about the pancake-type void growth is its nucleation. It is unclear where the site of nucleation is. X-ray tomography using synchrotron radiation may be able to map the three-dimensional image of the pancake void, so more information about the nucleation and growth of the void can be obtained.

15.5   Statistical analysis of electromigration failure in flip-chip solder joints

In industrial manufacturing, since a large quantity of consumer electron products is manufactured, product assurance requires reliability tests. Besides physical analysis of failure mode and failure mechanism as discussed in the previous sections, statistical analysis of failure is needed in order to predict the lifetime of a product. The statistical analysis can provide two important data about the reliability of a product. First, what is the MTTF of the product under a given condition of use in the field, for example, the use of a flip-chip device under the hood of an automobile? Second, in the future application of an existing device by increasing its function, for example, the applied current density may have to be increased, then it is important to know: what is the maximum current density which can be applied to an existing device so that the device can still perform without failure in a required lifetime?

To conduct statistical analysis, there are two important elements. First, we must have the equipment that can measure the failure of a large number of devices as a function of time, temperature, and a specific driving force that stresses the device such as the applied current density. Fig. 15.7 shows such an equipment for statistical analysis of failure of flip-chip solder joints under electromigration; it consists of two furnaces, four power sources (for safety reasons, the applied current is limited to 2 A), a multi-channel control unit, and a personal computer for recording. Second, we must have the test vehicle or sample. For example, for flip-chip reliability, we must have multiple flip-chips on a board for accelerated tests or we must be able to connect electrically several single flip-chips together for the test so that a meaningful size of data can be obtained in a reasonable

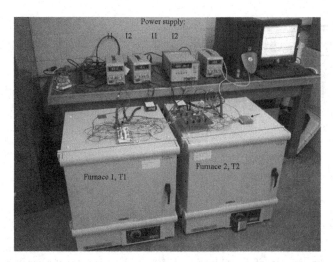

Fig. 15.7   Optical image of electromigration measurement equipment consisting of two furnaces, four power sources, a multi-channel control unit, and a personal computer for recording.

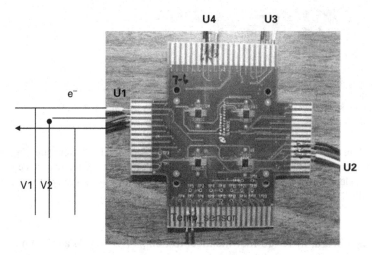

Fig. 15.8   Optical image of a test board having four chips for an electromigration test.

amount of time. Fig. 15.8 shows an optical image of such a test board having four chips on a board for electromigration test. The layout of the solder joints between the chip and the board in one of the four chips is shown in Fig. 15.9. The dimension of the chip is 0.3 mm × 0.3 mm, which has 36 solder joints. The diameter of the solder bump is 250 μm. The contact opening at UBM is 200 μm in diameter. The chips are mounted on the printed circuit board using Pb-free solder bumps of composition of Sn-1.2%Ag-0.5%Cu. The UBM is thin-film Al/Ni(V)/Cu in which the thickness of Al, Ni(V), and Cu is 1 μm, 0.3 μm, and 1 μm, respectively. A temperature sensor of Pt serpentine wire on the top surface of the Si chip is shown.

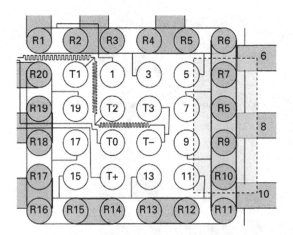

Fig. 15.9   Layout of the test chips on the board. The dimension of the chip is 0.3 mm × 0.3 mm, which has 36 solder balls. The diameter of the solder ball is 250 μm. The diameter of contact opening at UBM is 200 μm.

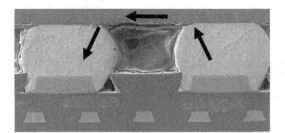

Fig. 15.10   In each chip, only two solder bumps (R1 and R2) are connected for current stressing in electromigration. A four-point probe structure is designed to monitor the resistance increase during the electromigration test. The arrows show the conduction path of electrons through the solder balls.

On each board, there are four identical test chips (C1 to C4) as shown in Fig. 15.8, which are connected electrically in series. In each chip, only one pair of solder bumps, as shown in Fig. 15.10, are connected for current stressing in electromigration. Four-point probe structure is designed to monitor the resistance increase during the electromigration test. By stressing only one pair of solder bumps on a chip, large joule heating that induces an extra temperature increase can be reduced. More importantly, compared to a daisy-chain which connects multiple solder bumps, the resistance increase for the one-pair test structure is directly correlated to one particular mode of electromigration failure, which can be examined physically and easily. This is important for modeling on the basis of Black's equation of failure and for correlating it to the mode of failure in electromigration. In electromigration using a daisy-chain of multiple bumps, it can capture all possible failure locations and modes. However, the interpretation of time to failure recorded by resistance increase is more complicated, and it is hard to identify the void at a certain interface that has caused the resistance increase.

Fig. 15.10 shows the path of electrons through the solder balls. When any one of the chips among the four fails (open circuit), a hook wire will be soldered to reconnect the working circuit. For example, if chip C2 fails, the time to failure will be recorded and the hook wire will be soldered to short the chip C2. In the subsequent electromigration test, C1, C3, and C4 will be tested, but C2 is not subject to current stressing and yet remains to the end until the other three fail too (see Fig. 15.8).

As an example, we show here the MTTF measurement of electromigration of one set of such flip-chip samples. They were tested at two current densities of 5×10^3 and 1×10^4 A/cm^2 and at two different temperatures of 125 and 150 °C. The test system as shown in Fig. 15.7 has been used to run tests simultaneously on a total of 64 chips (16 chips for each condition at one current density and one temperature, and four conditions for the set), on a total of 128 solder bumps being tested. The resistance change, as monitored by voltage change on each sample was automatically measured approximately every 5 min by the computer and the data recorded. The data were plotted by voltage versus time, from which we can determine the time to failure for each individual sample.

To see the array of solder bumps beneath the chip and to examine which bump has failed in the electromigration test, we can use X-ray tomography. Fig. 15.11 shows a three-dimensional image of X-ray tomography of the array of solder bumps corresponding to the schematic diagram as shown in Fig. 15.9. This sample was tested by a current density of 7.5×10^3 A/cm^2 at 125 °C. In the test, only one pair of bumps at the upper left corner was tested, and failure was found to occur after 4042 h. The X-ray tomographic image shows that melting has occurred in the failed bump, which had the electron current entering from the chip side. Since we can use software to obtain two-dimensional images from the three-dimensional image, we can obtain any two-dimensional cross-sectional image of the failed bumps in a row of bumps.

15.5.1 Time-to-failure and Weibull distribution

The data of time-to-failure of 16 samples tested at 1×10^4 A/cm^2 and 150 °C are listed in Table 15.2 and plotted using the Weibull distribution function in Fig. 15.12. MTTF at 50% of failure was determined. The parameters in Black's MTTF equation: the prefactor, the activation energy E_a, and the current density power factor n will be determined. The Weibull distribution function is given below

$$F(t) = 1 - \exp\left[-\left(\frac{t}{\eta}\right)^\beta\right]$$

(15.15)

where $F(t)$ is percentage or fraction of the failed sample as a function of time, η is the characteristic lifetime, and β is the shape factor or the slope of the Weibull plot. A larger slope indicates a narrower distribution of time-to-failure.

Table 15.2. Time-to-failure (TTF) of 16 pairs of flip-chip solder bumps tested at 10^4 A/cm^2 and 150 °C

Sample	TTF	Sample	TTF
8_T1_I1_U1	47	8_T1_I3_U1	59.5
8_T1_I1_U2	53	8_T1_I3_U2	68.5
8_T1_I1_U3	55.5	8_T1_I3_U3	43.5
8_T1_I1_U4	103	8_T1_I3_U4	58
8_T1_I2_U1	47.5	8_T1_I4_U1	68.5
8_T1_I2_U2	81.5	8_T1_I4_U2	25
8_T1_I2_U3	45.5	8_T1_I4_U3	41
8_T1_I2_U4	48	8_T1_I4_U4	53.5

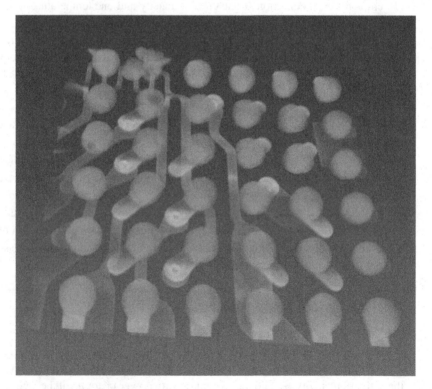

Fig. 15.11   Synchrotron radiation X-ray tomographic image of flip–chip solder joints. The second one on the top left failed by electromigration.

A simple explanation of the Weibull distribution function of failure is to assume $v(t_1)$ to be a frequency (probability per unit time) of failure, which means that $v(t_1)dt_1$ is the probability of failure within the very short time interval between t_1 and $t_1 + dt_1$. Then the probability of no failure in the very short period of dt_1 is

$$1 - v(t_1)dt_1 = \exp[-v(t_1)dt_1] \tag{15.16}$$

This is because $1 - x = \exp(-x)$ when x is very small. In turn, the probability of no failure in the interval from $t = 0$ to $t = t$ is the product of probabilities at each subinterval,

$$P_0(t) = \prod_{i=1}^{N} \exp[-\nu(t_1)dt_1] = \exp\left[-\sum_{i=1}^{N} \nu(t_1)dt_1\right] = \exp\left[-\int_0^t \nu(t_1)dt_1\right] \quad (15.17)$$

The Weibull or failure distribution function is just $1 - P_0(t)$. Physically, in the case of distribution of failure of a large number of flip-chip solder joints, we assume that a single event of failure consists of the nucleation of a void near one corner of the cathode contact of a solder joint having a frequency of $\nu_n(t_n)$ and the growth of the void into a pancake-type void across the entire cathode contact has a frequency of $\nu_g(t_g)$. Then, the failure frequency $\nu(t_1)$ as given in Eq. (15.16) should be a convolution of these two frequencies:

$$\nu(t_1) = \int_0^{t_1} \nu_n(t_n)\nu_g(t_1 - t_n)dt_n \quad (15.18)$$

To simplify the case, we assume that the growth takes much less time than the nucleation, and we take $\nu_g(t_g) = \delta(t_g)$, where δ is the Dirac function. Then,

$$\nu(t_1) = \int_0^{t_1} \nu_n(t_n)\delta(t_1 - t_n)dt_n = \nu_n(t_1) \quad (15.19)$$

For the nucleation event, we assume the characteristic frequency of nucleation is $1/\eta$, where η is the characteristic time of nucleation or the incubation time. Thus, $\nu(t_1) = \nu_n(t_1) = 1/\eta$, and we have

$$F(t) = 1 - P_0(t) = 1 - \exp\left[-\int_0^t \nu(t_1)dt_1\right] = 1 - \exp\left(-\frac{t}{\eta}\right)^\beta \quad (15.20)$$

where $\beta = 1$ in this case. Note that β can have other values depending upon the failure mode. By plotting $\ln\{-\ln[1 - F(t)]\}$ versus $\ln t$, the slope of the curve is β.

15.5.2 To calculate the parameters in Black's MTTF equation

In Black's MTTF equation, there are three parameters: pre-factor A, current density power factor n, and activation energy E_a:

$$\text{MTTF} = A\left(j^{-n}\right)\exp\left(\frac{E_a}{kT}\right) \quad (15.21)$$

To determine these parameters experimentally, we need to conduct at least the electromigration tests at two temperatures and two current densities. Fig. 15.12 shows the Weibull distribution plots of time-to-failure tested at two current densities of 5×10^3 and 1×10^4 A/cm^2 and at two different temperatures of 125 and 150 °C.

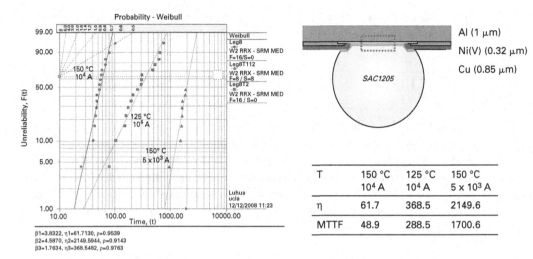

Fig. 15.12   Weibull distribution lifetime plots at the two current densities and two temperatures.

To find the activation energy, we use the data at the same current density and the relation below:

$$\frac{\text{MTTF}_2}{\text{MTTF}_1} = \frac{\exp\left(\dfrac{E_a}{kT_2}\right)}{\exp\left(\dfrac{E_a}{kT_1}\right)}$$

$$E_a = \left(\frac{1}{kT_2} - \frac{1}{kT_1}\right) \log\left(\frac{\text{MTTF}_2}{\text{MTTF}_1}\right) \qquad (15.22)$$

Then, to determine n, the power factor of current density, similarly we can use the data that have the same temperature at two different current densities and the relation below:

$$\frac{\text{MTTF}_2}{\text{MTTF}_1} = \left(\frac{j_1}{j_2}\right)^n \qquad (15.23)$$

However, due to joule heating, we have to be very careful in applying the above two relations. When we calculate the activation energy using Eq. (15.22), we have to take into account the effect of joule heating on temperature by changing kT to $k(T + \Delta T)$ where ΔT is due to joule heating at the applied current density and has been measured, so kT_1 becomes $k(T_1 + \Delta T)$ and kT_2 becomes $k(T_2 + \Delta T)$ in Eq. (15.22). This correction is straightforward.

When we calculate n using Eq. (15.21), it is more complicated. Since joule heating is different at different current densities, the temperature can hardly be assumed to be the same for two different current densities measured at the same furnace temperature. The actual temperature is not the same, so the problem is non-trivial. We need to know the ΔT increase at different current densities in order to make the temperature correction.

Figure 15.3 shows the temperature increase of the chip and the substrate of the test sample as a function of applied current density by using the temperature sensor of Pt thin-film serpentine lines. When we have measured ΔT_1 and ΔT_2 for j_1 and j_2, respectively, we need to adjust the furnace temperature of T_1 and T_2 so that we obtain the constant temperature condition of

$$(T_1 + \Delta T_1) = (T_2 + \Delta T_2) \tag{15.24}$$

In other words, when $j_2 > j_1$, we have $\Delta T_2 > \Delta T_1$, the furnace temperature T_2 for j_2 needs to be lower than T_1 for j_1. Furthermore, the ΔT increase on the chip side is different from the substrate side, so which one we should use is a choice. Since the mode of failure is a pancake-type void formation in the cathode contact on the chip side, we choose the ΔT on the chip side. Therefore, without the careful correction, it will make the calculated value of n different from 2, and this has been the cause of controversy in the literature about the question whether or not $n = 2$.

The activation energy is obtained to be 1.15 eV/atom, on the basis of the data shown in the last section. We found that $n = 2.08$. Then we can determine the constant $A = 1.5 \times 10^{-1}$ s-A^2/cm^4. Without correction of joule heating, we could have obtained $n = 5$ in this case.

On the basis of the obtained MTTF equation, the prediction or extrapolation of lifetime at the condition of field use can be calculated. Knowing the parameters in Black's equation, we extrapolated that at the chip temperature of 100 °C and solder joint current density of 2000 A/cm^2, the MTTF would be about 13.2 years for the samples tested. However, this does not mean that we can expect most of the solder joints to work for so long without failure. Instead, from the failure distribution, we see that the first 10% will fail at a much earlier time, especially if the slope of the Weibull distribution is small.

15.5.3 Modification of Black's equation for flip-chip solder joints

In 1969, Black provided his well-known equation to analyze failure in Al interconnects caused by electromigration. Nevertheless, whether Black's equation can be applied to MTTF in flip-chip solder joints deserves a careful examination.

Electromigration failure in flip-chip solder joints is different from that in Al interconnects because of lattice diffusion, current crowding at the cathode contact, and large joule heating from the Al interconnect on the chip side. The combined effect of electromigration and thermomigration is much larger in flip-chip solder joints than in Al interconnects. Typically, the failure mode in flip-chip solder joints is the nucleation and growth of a pancake-type void along the contact interface at the cathode. It was found that the bulk part of the time to failure is controlled not by the growth of a void across the contact interface, but by the incubation time of void nucleation. The latter takes about 90% of the time of failure. The propagation of the void across the entire contact takes only about 10% of the time.

Table 15.3. Mean-time-to-failure of eutectic SnPb flip-chip solder joints

	1.5 A (1.9×10^4 A/cm^2)		1.8 A (2.25×10^4 A/cm^2)		2.2 A (2.75×10^4 A/cm^2)	
	Calculated (h)	Measured (h)	Calculated (h)	Measured (h)	Calculated (h)	Measured (h)
100 °C	—	—	380	97	265	63
125 °C	108	573[a]	79.6	43	55.5	3
140 °C	46	121	34	32	24	1

[a] Not failed

Since the effect of current crowding on pancake-type void formation is crucial, it cannot be ignored in the MTTF analysis. The major effects of current crowding are to greatly increase the current density at the entrance of the solder joint and also to increase the local temperature due to joule heating. A solder joint has IMC formation at both the cathode and the anode interfaces, current crowding affects IMC formation and dissolution, and in turn IMC formation and dissolution may affect failure time and mode. Black did point out the importance of temperature gradient on interconnect failure, although it is unclear, at least it is not explicit, whether he has taken them into account in his equation.

Brandenburg and Yeh used Black's equation with $n = 1.8$ and $Q = 0.8$ eV/atom for eutectic SnPb flip-chip solder joints. The equation has been found to have greatly over-estimated the MTTF of flip-chip solder joints at high current densities. Table 15.3 compares the calculated and measured MTTF of eutectic SnPb flip-chip solder joints at three current densities and three temperatures. At the low-current density of 1.9×10^4 A/cm^2, the measured MTTF is slightly longer than the calculated value, but at 2.25×10^4 A/cm^2 and 2.75×10^4 A/cm^2, the measured MTTF is much shorter than the calculated. This is also true for the eutectic SnAgCu flip-chip solder joints. These findings show that the MTTF of flip-chip solder joints is very sensitive to a small increase of current density but drops rapidly when the current density is about 3×10^4 A/cm^2. However, it is worth mentioning that the current density of 3×10^4 A/cm^2 is very high for electromigration in flip-chip solder joints. If it is applied in a furnace of 150 °C, it can lead to melting. Furthermore, at such high-current density, the rate of nucleation and growth of void may be different from that at lower current density, or the fraction in time to failure spent in void nucleation and void growth will be different, meaning that the incubation time of nucleation could be very short. Then melting could occur.

To determine the activation energy in Black's equation, accelerated tests at high temperatures are performed. Attention must be paid to the temperature range in which lattice diffusion might overlap grain-boundary diffusion and also grain-boundary diffusion might overlap surface diffusion. For eutectic SnPb solder, it is more complicated because of the change of dominant diffusion species between Pb and Sn above and below 100 °C.

Black's equation can be modified to include the effect of current crowding joule heating, and the effect of stress,

$$\text{MTTF} = A \frac{1}{(cj)^2} \exp\left[\frac{E_a \pm \sigma\Omega}{k(T + \Delta T)}\right] \tag{15.25}$$

where the parameter c is due to current crowding and has a magnitude of 10, ΔT is due to Joule heating and can be as high as 100 °C, and $\sigma\Omega$ is stress potential that changes the vacancy concentration in the stressed region. Both the parameters c and ΔT will reduce the MTTF from the Black equation, i.e. make the solder joint fail much faster. Since ΔT depends strongly on j, the modified equation is much more sensitive to the change of current density than the original Black equation. Recall that the value of ΔT will depend on the design of flip-chip solder joint and interconnect, because of heat generation and heat dissipation. However, note that the modified Black's equation in essence is the same as the original Black's equation; there is no change in the basic form of the equation.

15.5.4 Weibull distribution function and JMA equation of phase transformations

Why do we use the Weibull distribution function rather than other distribution functions to analyze failure? It seems that it is based on the close similarity in the mathematical form between the Weibull distribution function and Johnson–Mehl–Avrami's (JMA's) canonical equation of phase transformation. Electromigration-induced failure of circuit open due to void formation can be regarded as phase transformation in the cathode end of an Al line or a flip-chip solder joint where a void nucleates and grows.

In classical phase transformations, for example the crystallization of an amorphous phase to a crystalline phase without change in composition, the fraction of the transformed phase is expressed by the JMA equation as

$$X_T = 1 - \exp(-X_{ext}) = 1 - \exp\left[-\left(\frac{t}{\lambda}\right)^n\right] \tag{15.26}$$

where X_T is the fraction of the volume transformed and X_{ext} is the fraction of the extended volume, which is defined as

$$X_{ext} = \int_{\tau=0}^{\tau=t} \frac{4\pi}{3} R_N R_G^3 (t - \tau)^3 d\tau \tag{15.27}$$

where R_N is the nucleation rate and R_G is the growth rate of spherical transformed particles, and t is the time of transformation. In essence, the physical meaning of X_{ext} is the sum of the volume of all spherical particles which have nucleated and grown in the period from $\tau = 0$ to $\tau = t$ without consideration of growth impingement and phantom nucleation. If the nucleation is random and continuous at a constant rate and if the growth is isotropic and linear with time, we have from the integration of the last equation,

$$X_{ext} = \frac{\pi}{3} R_N R_G^3 t^4 = K t^4 \tag{15.28}$$

where $K = \frac{\pi}{3} R_N R_G^3$. We see that in Eq. (15.28), we have $n = 4$ and $\lambda = (1/K)^{1/4}$. If we consider the growth to be one- or two-dimensional and linear with time, we have $n = 2$ or $n = 3$. Then, if the growth rate is diffusion-limited, we can have $n = 2.5$, etc.

Since the mathematical form of the Weibull distribution function, Eq. (15.15), and the JMA equation of phase transformation, Eq. (15.26), is similar, it is of interest to compare them so that we may obtain an explanation of the Weibull distribution of TTF from the point of view of phase transformation.

We consider electromigration-induced failure in a large number of Al strips in a multi-level interconnect structure. Between levels, the Al strips are interconnected by W vias. The wear-out failure mode is known in which failure is due to void formation and it occurs at the interface between the Al line and W via at the cathode ends. We can regard the wear-out as a phase transformation of forming a void in Al at the cathode ends, and the void formation requires nucleation and growth. The nucleation at all the cathode ends is a random process. Whether or not the nucleation rate is constant in time is unclear. The growth of a void can be taken to be linear under a constant current density. The fraction of failure as a function of time, $F(t)$, in Weibull distribution can be regarded as the fraction of the cathode ends that has been transformed to a void larger than the size of the W via to cause circuit open. While the void at a cathode end will not grow larger and larger after the circuit is open at that point, a phantom growth can be assumed theoretically. In the meantime, more and more voids will form in other cathode ends and the fraction of failure will increase. We can compare it to the fraction of transformed volume, X_T, in the JMA equation.

On the basis of the JMA equation, the TTT diagram and the S-curve of phase trans-formations can be constructed. Using the TTT diagram, the effect of undercooling on nucleation and growth and on the overall rate of transformation can be interpreted. Also, the activation energy of nucleation and activation energy of growth can be separated, provided that the kinetics of growth can be measured independently so that the activation energy of growth can be determined alone.

In Black's equation of MTTF, the physical meaning of the activation energy has not been defined clearly. It could be the activation energy of nucleation of the void or the growth of the void or both. If nucleation is the rate-limiting process, the activation energy should belong to the nucleation of the void, which is most likely a heterogeneous event. It is a subject which needs more study.

15.5.5   Physical analysis of statistical distribution of failure

By using cross-sectional FIB/SEM examination, single pancake-type void propagation in the cathode on the chip side can be studied and was found to be the key physical failure mode of flip-chip solder joints under electromigration. By using synchrotron radiation X-ray tomography, we can examine non-destructively a large number of flip-chip solder joints between a Si chip and a substrate ex-situ or in-situ, and we should be able to observe the distribution of failure and partial failure in all the solder bumps.

15.6   Simulation

In multi-scale modeling of simulation, according to length scale or time scale in the calculation, we can go from atomic dimension of quantum mechanics or first principles computation to classical molecular dynamics, Monte Carlo simulation of an aggregate of a fairly large number of atoms, dislocation dynamics of mechanical properties, statistical mechanics, and to continuum mechanics of a very large sample size using FEA.

The details of these simulations are beyond the scope of this book. Nevertheless, simulation is indispensable in reliability analysis and more and more progress will be made in this direction [11, 12].

References

[1] M. Ohring, *Reliability and Failure of Electronic Materials and Devices* (Academic Press, San Diego, 1998).

[2] W. J. Bertram, "Yield and reliability," Ch. 14 of *VLSI Technology*, ed. S. M. Sze (McGraw-Hill, New York, 1983).

[3] J. R. Black, "Mass transport of Al by momentum exchange with conducting electrons," *Proc. IEEE Int. Rel. Phys. Symp.* (1967), 144–59.

[4] M. Shatzkes and J. R. Lloyd, "A model for conductor failure considering diffusion concurrently with electromigration resulting in a current exponent of 2," *J. Appl. Phys.* **9** (1986), 3890–3.

[5] R. Rosenberg and M. Ohring, "Void formation and growth during electromigration in thin films," *J. Appl. Phys.* **42** (1971), 5671–9.

[6] J. J. Clement, "Electromigration modeling for integrated circuit interconnect reliability analysis," *IEEE Trans. on Device and Materials Reliability* **1** (2001), 33–42.

[7] S. Brandenburg and S. Yeh, *Proceedings of Surface Mount International Conference and Exhibition*, SMI98, San Jose, CA, Aug. 1998, pp. 337–44.

[8] Everett C. C. Yeh, W. J. Choi, K. N. Tu, P. Elenius and H. Balkan, "Current-crowding-induced electromigration failure in flip chip solder joints," *Appl. Phys. Lett.* **80** (4) (2002), 580–2.

[9] Lingyun Zhang, Shengquan Ou, Joanne Huang, K. N. Tu, Stephen Gee and Luu Nguyen, "Effect of current crowding on void propagation at the interface between intermetallic compound and solder in flip chip solder joints," *Appl. Phys. Lett.* **88** (2006), 012106.

[10] S. W. Liang, Y. W. Chang, T. L. Shao, Chih Chen and K. N. Tu, "Effect of three-dimensional current and temperature distribution on void formation and propagation in flip chip solder joints during electromigration," *Appl. Phys. Lett.* **89** (2006), 022117.

[11] T. V. Zaporozhets, A. M. Gusak, K. N. Tu and S. G. Mhaisalkar, "Three-dimensional simulation of void migration at the interface between thin metallic film and dielectric under electromigration," *J. Appl. Phys.* **98**, 103508 (2005).

[12] Y.-S. Lai, S. Sathe, C.-L. Kao and C.-W. Lee, "Integrating electrothermal coupling analysis in the calibration of experimental electromigration reliability of flip-chip packages," in *Proceedings of ECTC 2005 (55th Electronic Components and Technology Conference)*, Lake Buena Vista, FL, USA, 2005, pp. 1421–6.

Problems

15.1   What is the difference between yield and reliability in microelectronic technology?

15.2   What is burn-in and why do we do it?

15.3   Electromigration has been a major reliability concern in microelectronic devices. How come we have almost never found that our personal computer, laptop, or cell phone has failed due to electromigration?

15.4   What is an accelerated test in reliability? For electromigration in nanoscale Cu interconnects in the future, what will be the reasonable conditions in terms of temperature, time, and current density for accelerated tests?

15.5   In flip-chip technology, typically we use Al lines on the chip side to interconnect two solder bumps. Why not use Cu lines to connect them? Actually, on the substrate side, we do use Cu bond-pads to connect the solder bumps.

15.6   In Black's equation, why is the sign in front of the activation energy positive?

15.7   In Black's equation, we have measured the activation energy to be $E = 1$ eV/atom, the power factor of current density is $n = 2$ and the pre-exponential factor $A = 10^{-2}$. Calculate the MTTF when the device is used at $100\,^{\circ}$C with a current density of 2×10^3 A/cm^2. Then, for the next generation devices, when the current density will be increased to 1×10^4 A/cm^2, how do we calculate the MTTF?

15.8   In flip-chip technology, the line-to-bump configuration is unique, which means that electromigration can occur in the Al line as well as in the solder bump. Assume that the current density in the Al line is 10^6 A/cm^2 and the current density in the solder bump is 10^4 A/cm^2 due to current crowding. Which will fail first at $100\,^{\circ}$C due to void formation?

Appendix A   A brief review of thermodynamic functions

We begin by reviewing some general thermodynamic relations of closed systems having a fixed number of particles. The four energy functions, internal energy E, enthalpy H, Helmholtz free energy F, and Gibbs function G are related to one another and the four thermodynamic variables, pressure p, volume V, entropy S, and temperature T by

$$
\begin{aligned}
H &= E + pV \\
F &= E - TS \\
G &= E - TS + pV = F + pV = H - TS
\end{aligned}
\tag{A.1}
$$

The first law of thermodynamic relates changes in heat, dQ, and work done to the system, dW, to the change in internal energy, dE:

$$
dE = dQ + dW
\tag{A.2}
$$

The first law is given as a definition of the internal energy change of a closed system. If the change in heat is reversible, the second law of thermodynamic defines the increase in entropy to be

$$
dS = \frac{dQ}{T}
\tag{A.3}
$$

The second law is given as a definition of the change of entropy. If only mechanical work is done to the system, we have $dW = -pdV$ where the work is positive when the volume of the system decreases. Then,

$$
dE = TdS - pdV
\tag{A.4}
$$

Starting from Eq. (A.4), the differentials of H, F, and G can be obtained from Eq. (A.1) as

$$
\begin{aligned}
dH &= TdS + Vdp \\
dF &= -SdT - pdV \\
dG &= -SdT + Vdp
\end{aligned}
\tag{A.5}
$$

The four variables appear in Eq. (A.4) and Eq. (A.5) in pairs: p and V, T and S. Experimentally, it is easier to carry out a process at constant pressure or constant temperature, but harder at constant volume or constant entropy. A process occurring at constant entropy is called an adiabatic process, meaning that heat is isolated. Note that while it is hard to measure dS, we can measure $dQ = TdS$ at constant volume or at constant pressure; there are heat capacity measurements.

From Eq. (A.4), we define the heat capacity at constant volume,

$$c_v = \left.\frac{\partial E}{\partial T}\right|_v \tag{A.6}$$

And from the first equation in Eq. (A.5), we define the heat capacity at constant pressure,

$$c_p = \left.\frac{\partial H}{\partial T}\right|_p \tag{A.7}$$

The value of c_p is easier to measure than c_v for solids. Knowing c_p, we can evaluate the change of enthalpy and entropy:

$$\Delta H = \int_{T_1}^{T_2} c_p dT \tag{A.8}$$

$$\Delta S = \int_{T_1}^{T_2} \frac{c_p}{T} dT \tag{A.9}$$

Since we have defined $G = H - TS$, it means that we can obtain the change of Gibbs function at constant temperature and constant pressure when we have measured c_p. The Gibbs function has T and p as the two independent variables. They are easy to control experimentally, and do not depend on the size of the system. Hence, in considering equilibrium phase changes at constant temperature and pressure, we use Gibbs function. The Gibbs function is often called the Gibbs free energy. Knowing Gibbs free energy, the equilibrium state at a given temperature and pressure is defined. This is why heat capacity is important in thermodynamics, and why Einstein and Debye had a theory on heat capacity.

When we consider a chemical process of liquid or solid phases which occurs at one atmospheric pressure, the pV term is negligibly small, so the Gibbs free energy and Helmholtz free energy are practically the same. To calculate the pV term, we recall that 1 atm corresponds to a pressure of 1.013×10^6 dyne/cm^2 and to an energy density of 1.013×10^6 erg/cm^3. If we assume that there are approximately 3×10^{22} Ω atoms in 1 cm^3 where Ω is the atomic volume, we have

$$p\Omega = \frac{1.013 \times 10^6}{3 \times 10^{22} \times 1.6 \times 10^{-12}} \text{ eV/atom}$$

$$= 0.21 \times 10^{-4} \text{ eV/atom}$$

This value is much smaller than a typical binding energy of about 0.1 to 1 eV/atom in liquid or solid.

For surfaces there may be another contribution to the work, namely

$$dW_S = \gamma \, dA \qquad\qquad (A.10)$$

where γ is surface energy per unit area of A. This is the increase in the work against surface forces, by increasing the area of the surface.

Appendix B   Defect concentration in solids

We introduced the subject of diffusion in the beginning of Chapter 4 by stating that diffusion in crystalline solids is mediated by defects in the solid, especially the vacancy and interstitial point defects. Indeed, the point defect mechanism of atomic diffusion in metals is well developed. A key question is: what is the defect concentration in the solid, since the quantity will no doubt control the flux of diffusion? Point defect concentration is a thermodynamic equilibrium quantity, unlike that of complex defects such as dislocations, twins, and grain boundaries. These complex defects may serve as sources and sinks of point defects but their presence and their concentration in a sample depend on the sample history of thermal and mechanical treatments. However, a well-annealed sample will contain the equilibrium or fixed concentration of vacancies. Since a thermal equilibrium quantity can be calculated statistically, we give a simple example in the following.

We begin by considering the entropy of a lattice of N lattice sites in which N_v sites are vacant (vacancies) and $N - N_v$ sites are taken by atoms. The configurational entropy of the lattice is given by

$$S = k \ln \Gamma \tag{B.1}$$

where k is Boltzmann's constant and Γ is the number of possible states of the lattice or the number of ways to arrange the vacancies and the atoms in the lattice:

$$\Gamma = \frac{N\,(N-1)......(N-N_v+1)}{N_v!}$$

$$= \frac{N!}{(N-N_v)!N_v!} \tag{B.2}$$

Using Stirling's approximation, $\ln x! = x \ln x$ when x is large,

$$S = k\,[N \ln N - (N - N_v)\ln(N - N_v) - N_v \ln N_v] \tag{B.3}$$

The Gibbs function of the lattice having N_v vacancies at temperature T is

$$\Delta G = N_v \Delta H_f - T \Delta S \tag{B.4}$$

where ΔH_f is the enthalpy of formation of a vacancy in the lattice and ΔS is the entropy increase from the ground state. If we take the ground state to be unique, its entropy is zero (the third law of thermodynamics), and furthermore, if we ignore the vibrational entropy contribution, we have

$$\Delta S = S - 0 = k \ln \Gamma$$

At thermal equilibrium at temperature T, the chemical potential of a vacancy or of an atom in the lattice is the same, so we have

$$\frac{\partial \Delta G}{\partial N_v} = 0$$

$$\Delta H_f - T \frac{\partial \Delta S}{\partial \Delta N_v} = 0$$

Since

$$\frac{\partial \Delta S}{\partial N_v} = k \ln \frac{N - N_v}{N_v}$$

we obtain

$$\frac{N_v}{N - N_v} = \exp\left(-\frac{\Delta H_f}{kT}\right) \tag{B.5}$$

Since $N \gg N_v$, this equation shows that the equilibrium vacancy concentration obeys the Boltzmann distribution.

In Chapter 4, Section 4.9.2, we showed that in Al, the value of $\Delta H_f = 0.76$ eV. Therefore, near the melting point of 660°C, we have

$$\frac{N_v}{N} \cong 10^{-4}$$

which is the experimentally measured vacancy concentration. Since the concentration decreases with temperature, we can quench a sample from the temperature near its melting point to a lower temperature to produce a supersaturation of vacancies in the sample. The rate of re-establishing equilibrium can be measured by resistivity change in the sample and enables us to determine the activation energy of motion of vacancies.

Appendix C Derivation of Huntington's electron wind force

In the following, we present the assumptions and step-by-step derivation in Huntington's model of electron wind force.

(1) Considerations are hemi-classical. Each electron is treated as a group of waves or Bloch waves with an average wave vector k and group velocity of

$$\overline{V} = \frac{1}{\hbar} \frac{\partial E(\overline{k})}{\partial \overline{k}}$$

where the function $E(\overline{k})$ should be found from the electron band theory (dispersion law).

For free electrons, $E(\overline{k}) = \hbar^2 k^2 / 2m^*$, and for electrons at the bottom of the conduction band, $E(\overline{k}) = E_{min} + \hbar^2 k^2 / 2m_0$, where $m^* = \hbar^2 (\partial^2 E / \partial k^2)^{-1}$ is the effective electron mass. Note that $\partial E / \partial \overline{k}$ means the gradient in k-space, e.g. a vector with components of $\partial E / \partial k_x$, $\partial E / \partial k_y$, $\partial E / \partial k_z$.

For Bloch waves, according to Bloch's theorem, recall that each quantum state of an independent electron in the periodic potential $U(\overline{r} + \overline{R}) = U(\overline{r})$ and $\overline{R} = n_1 \overline{a_1} + n_2 \overline{a_2} + n_3 \overline{a_3}$ can be described by the product of a planar wave and periodic function $\Psi_{\hbar \overline{k}}(\overline{r}) = e^{i \overline{k} \overline{r}} W_{\hbar \overline{k}}(\overline{r})$, where $W_{\hbar \overline{k}}(\overline{r} + \overline{R}) = W_{\hbar \overline{k}}(\overline{r})$ and n is the band index.

(2) $(1/\hbar)[(\partial E / \partial \overline{k'}) - (\partial E / \partial \overline{k})] = \overline{V'} - \overline{V}$ is the change of electron's group velocity as a result of scattering.

(3) $-\frac{m_0}{\hbar}[(\partial E / \partial k'_x) - (\partial E / \partial k_x)] = -(p'_x - p_x)$ is momentum along the x-axis, transfer to defect during mentioned individual scattering.

(4) $f(\overline{k})$ is a probability that the quantum state $\overline{k}$ is occupied by some electron. The quantum cell in the k-space with a "k-volume" is given by $\Omega = (2\overline{\mu}/L_x) \cdot (2\overline{\mu}/L_y) \cdot (2\overline{\mu}/L_z) = 8\pi^3 / V$, where V is the real total volume). In equilibrium, we have $f_0 = 1/(e^{E - \mu / kT} + 1)$ (Fermi–Dirac distribution).

(5) $1 - f(\overline{k'})$ is a probability that the quantum state $\overline{k'}$ was free or unoccupied before scattering, so that the Pauli principle (exclusion principle) does not forbid the $\overline{k} - \overline{k'}$ transition.

(6) $W_d(\overline{k} \rightarrow \overline{k'})$ is a probability of this transition per unit time. It means that the product $W_d dt$ is a probability of transition during dt, if $dt \ll \tau_d$.

(7) According to the Pauli principle, each quantum cell in k-space (with $\Omega = (8\pi^3 / V)$) may contain up to two electrons with opposite spins, so the k-volume per electron is $\Omega/2 = 4\pi^3 / V$.

(8) Now we consider the unit volume $V = 1 \text{ m}^3$.

(9) The number of possible electron states in the "elementary" k-volume $d_k^3 = dk_x dk_y dk_z$ is $d^3 k/(\Omega/2) = d^3 k/4\pi^3$. The elementary k-volume is physically small.

(10) The momentum, M_x along the x-axis, transferring from electrons to the defects in the unit volume $V = 1 \text{ m}^3$ per unit time is given as

$$-\iint \frac{d^3 k}{4\pi^3} \frac{d^3 k'}{4\pi^3} (p'_x - p_x) f(\overline{k})(1 - f(\overline{k'})) W_d(\overline{k}, \overline{k'}).$$

or

$$\frac{dM_x}{dt} = -\left(\frac{1}{4\pi^3}\right)^2 \iint \frac{m_0}{\hbar}\left(\frac{\partial E}{\partial k'_x} - \frac{\partial E}{\partial k_x}\right) f(\overline{k})(1 - f(\overline{k'})) W_d(\overline{k}, \overline{k'}) d^3 k' d^3 k$$

(11) We shall represent the last equation by two integrals:

$$\frac{dM_x}{dt} = I_1 + I_2$$

where

$$I_1 = -\left(\frac{1}{4\pi^3}\right)^2 \iint \frac{m_0}{\hbar} \frac{\partial E}{\partial k'_x} f(\overline{k})(1 - f(\overline{k'})) W_d(\overline{k}, \overline{k'}) d^3 k' d^3 k$$

$$I_2 = -\left(\frac{1}{4\pi^3}\right)^2 \iint \frac{m_0}{\hbar} \frac{\partial E}{\partial k_x} f(\overline{k})(1 - f(\overline{k'})) W_d(\overline{k}, \overline{k'}) d^3 k' d^3 k$$

Since the integration is being made over all $\overline{k}$ and all $\overline{k'}$, we can interchange the variables in the first integral as:

$$I_1 = -\left(\frac{1}{4\pi^3}\right)^2 \iint \frac{m_0}{\hbar} \frac{\partial E}{\partial k_x} f(\overline{k'})(1 - f(\overline{k})) W_d(\overline{k'}, \overline{k}) d^3 k' d^3 k$$

Then, in I_1 and I_2, we have the same $\partial E/\partial k_x$; it means that we now have

$$\frac{dM_x}{dt} = (-I_2) - (-I_1)$$

$$= \left(\frac{1}{4\pi^3}\right)^2 \iint \frac{m_0}{\hbar} \frac{\partial E}{\partial k_x}$$

$$\times \left[f(\overline{k'})(1 - f(\overline{k})) W_d(\overline{k}, \overline{k}) - f(\overline{k'})(1 - f(\overline{k})) W_d(\overline{k'}, \overline{k}) \right] d^3 k' d^3 k$$

$$\text{(C.1)}$$

(12) To simplify the expression of the last equation, Huntington used the concept of relaxation time τ_d. This notion was first introduced for the analysis of the kinetic Boltzmann equation for gases. With a certain approximation, the rate of change of

the distribution function can be represented as

$$\frac{\partial f(t,\bar{k})}{\partial t} = \frac{1}{4\pi^3} \int \left\{ f(\bar{k})(1 - f(\bar{k}'))W_d(\bar{k},\bar{k}') - f(\bar{k}')(1 - f(\bar{k}))W_d(\bar{k}',\bar{k}) \right\} d^3k'$$
$$- \frac{f(t,\bar{k}) - f(\bar{k})}{\tau_d}$$

for the equilibrium distribution. For the stationary case, $\partial f/\partial t = 0$, so that

$$\frac{1}{4\pi^3} \int \left\{ f(\bar{k})(1 - f(\bar{k}'))W_d(\bar{k},\bar{k}') - f(\bar{k}')(1 - f(\bar{k}))W_d(\bar{k}',\bar{k}) \right\} d^3k'$$
$$= \frac{f(t,\bar{k}) - f(\bar{k})}{\tau_d} \tag{C.4}$$

On the right-hand side of the above equation, $f(\bar{k})(1-f(\bar{k}'))W_d(\bar{k},\bar{k}')$ is a probability per unit time of the $\bar{k} \to \bar{k}'$ transition, provided that the state $\bar{k}$ before transition was filled and the state $\bar{k}'$ was empty. The function $f(\bar{k}')(1 - f(\bar{k}))W_d(\bar{k}',\bar{k})$ is a probability per unit time of the inverse transition.

(13) By substituting Eq. (C.4) into Eq. (C.1), we have

$$\frac{dM_x}{dt} = \frac{1}{4\pi^3} \int d^3k \frac{m_0}{\hbar} \frac{\partial E(\bar{k})}{\partial k_x} \frac{f(\bar{k}) - f_0(\bar{k})}{\tau_d}$$

(14) Let the relaxation time be independent of $\bar{k}$ and $\tau_d = $ constant. Then

$$\frac{dM_x}{dt} = \frac{m_0}{\hbar\tau_d} \frac{1}{4\pi^3} \int d^3k \frac{\partial E(\bar{k})}{\partial k_x} f(\bar{k}) - \frac{m_0}{\hbar\tau_d} \frac{1}{4\pi^3} \int d^3k \frac{\partial E(\bar{k})}{\partial k_x} f_0(\bar{k})$$

(15) Evidently, the average vector velocity of electrons in equilibrium is zero:

$$\bar{V}_x = \frac{1}{\hbar} \frac{\overline{\partial E}}{\partial k_x}\bigg|_{eq} = \frac{1}{\hbar} \frac{\overline{\partial E}}{\partial k_y}\bigg|_{eq} = \frac{1}{\hbar} \frac{\overline{\partial E}}{\partial k_z}\bigg|_{eq} = 0$$

Therefore

$$\int \frac{\partial E}{\partial k_x} f_0(\bar{k}) d^3k = 0$$

Thus,

$$\frac{dM_x}{dt} = \frac{m_0}{\hbar\tau_d} \frac{1}{4\pi^3} \int \frac{\partial E}{\partial k_x} f(\bar{k}) d^3k \tag{C.5}$$

(16) To relate the momentum change to force, we have the current density given as

$$j_x = (-e)n\bar{V}_x = (-e) \int \frac{d^3k}{4\pi^3} f(\bar{k}) \cdot \frac{1}{\hbar} \frac{\partial E(\bar{k})}{\partial k_x} \tag{C.6}$$

where $n = (d^3k/4\pi^3)f(\bar{k})$ is the number of electrons per unit volume with $\bar{k}$ belonging to d^3k. Indeed, $d^3k/4\pi^3$ is the number of "single electron cells" in the "volume" of d^3k of k-space, and $f(\bar{k})$ is the "inhabitance" of the cell.

(17) Combining Eqs. (C.5) and (C.6), we obtain

$$\frac{dM_x}{dt} = -\frac{j_x m_0}{e\tau_d} \qquad (C.7)$$

It is a momentum change along the x-direction, transferred to defects (the diffusing atoms) per unit time per unit volume.

(18) Let N_d be the density of defects (number of defects per unit volume). Then, according to Newton's second law, the force at one defect, caused by electron wind, is

$$F_x = \frac{1}{N_d}\frac{dM_x}{dt} = -\frac{j_x m_0}{e\tau_d N_d} \qquad (C.8)$$

This force has a clear physical meaning assuming the condition that during the atomic jump the defect feels much more than one collision. The characteristic time of one successful jump is of the order of Debye time, $\tau_{Debye} \sim 10^{-13}$ s. So for Eq. (C.8) to be reasonable, it is necessary that the scattering frequency product, $\nu_{scatter}$, and Debye time should be much less than unity:

$$\nu_{scatter} \approx \frac{kT}{\varepsilon_p}\frac{V_F}{l}$$

where l is the mean-free path length of electron around defect, V_F/l is the frequency of "possible" collisions, and kT/ε_p is the fraction of electrons which are able to be scattered according to the Pauli principle.

$$l \approx \frac{1}{n\sigma},$$

where σ is the cross-section and is about 10^{-19} m^2 (according to Huntington's estimate)

$$n \sim 10^{29} \text{ m}^{-3}\left(n_{ex} \approx \frac{kT}{\varepsilon_p}n \approx 10^{27} \text{ m}^{-3}\right), \frac{kT}{\varepsilon_p} \approx 10^{-2}, V_F = \frac{\hbar k_F}{m_0} \approx 10^6 \text{ m/s}$$

Thus, $\nu_{scatter} \approx 10^{-1}10^6 n\sigma \approx 10^{-2}10^6 10^{29} 10^{-19} \approx 10^{14}\text{s}^{-1}$

So $\nu_{scatter}\tau_{Debye} \approx 10 \gg 1$

(19) Let us now transform Eq. (C.8) in terms of electric field: $j_x = \varepsilon_x/\rho$, where ρ is an average resistance of metal. According to the Drude–Lorentz–Sommerfeld model, the resistance ρ of a metal can be written as

$$\rho = \frac{|m^*|}{ne^2\tau}$$

where $m^* = \hbar^2/(\partial^2 E/\partial k^2)$ is the effective electron mass.

Huntington used the same expression for the resistance of defects, $\rho_d = (|m^*|/ne^2\tau_d)$, so that we have $\tau_d = (|m^*|/ne^2\rho_d)$.

Thus, from Eq. (C.8), we obtain

$$F_x = -\frac{\varepsilon_x}{\rho} \frac{m_0}{eN_d} \frac{ne^2\rho_d}{|m^*|} = -\left(\frac{m_0}{|m|} \frac{ZN}{N_d} \frac{\rho_d}{\rho}\right) e\varepsilon_x \qquad (C.9)$$

where N is the density of ions and Z is the valence number; $n = ZN$.

Thus, we have the effective charge

$$Q^* = -Z^*e, \text{ where } Z^* = \frac{m_0}{|m^*|} \frac{ZN}{N_d} \frac{\rho_d}{\rho} = Z\frac{m_0}{|m^*|} \frac{\rho_d/N_d}{\rho/N}$$

(20) Now, let us take into account the fact that τ_d, ρ_d, and F_x change from position to position. Obviously, they shall reach maximum in the saddle-point of diffusion.

Assume that $F(y) = F_m \sin^2(\pi y/d)$, where y is not the y-axis. Rather, it is a coordinate along the jumping path, which usually does not coincide with the x-axis. Work or change of potential barrier is

$$U_j = \int_0^{a_j/2} F(y)dy = F_m \cos\theta_j \int_0^{a/2} \sin^2\frac{\pi y}{a}dy = \frac{a_j F_m}{4}\cos\theta_j$$

After averaging on all possible jump directions, we have

$$J_x = C\frac{D}{kT}\frac{1}{2}F_m$$

The factor of $1/2$ is due to the integral of

$$\int_0^{a/2} \sin^2\frac{\pi y}{a}dy = \frac{1}{2}\frac{a}{2}$$

(21) Thus, we finally have the effective charge number as below,

$$Z_{eff}^* = \frac{1}{2}Z_{max}^* - Z = Z\left(\frac{1}{2}\frac{m_0}{|m^*|}\frac{(\rho_d^{max}/N_d)}{\rho/N} - 1\right)$$

Appendix D Elastic constants tables and conversions

The purpose of this section is to give a brief review of the formulae and relationships among the commonly used elastic constants and to provide tables of elastic constants used in the design of elastic materials as discussed in Chapter 6 and 14.

D.1 Formulae and definitions

D.1.1 Stiffness and compliance tensors

For the electronic materials of interest, the stress and strain can be related through a 6×6 matrix as shown below.

$$
\left\| \begin{matrix}
c_{11} & c_{12} & c_{13} & c_{14} & c_{15} & c_{16} \\
c_{21} & c_{22} & c_{23} & c_{24} & c_{25} & c_{26} \\
c_{31} & c_{32} & c_{33} & c_{34} & c_{35} & c_{36} \\
c_{41} & c_{42} & c_{43} & c_{44} & c_{45} & c_{46} \\
c_{51} & c_{52} & c_{53} & c_{54} & c_{55} & c_{56} \\
c_{61} & c_{62} & c_{63} & c_{64} & c_{65} & c_{66}
\end{matrix} \right\|
\qquad
\left\| \begin{matrix}
s_{11} & s_{12} & s_{13} & s_{14} & s_{15} & s_{16} \\
s_{21} & s_{22} & s_{23} & s_{24} & s_{25} & s_{26} \\
s_{31} & s_{32} & s_{33} & s_{34} & s_{35} & s_{36} \\
s_{41} & s_{42} & s_{43} & s_{44} & s_{45} & s_{46} \\
s_{51} & s_{52} & s_{53} & s_{54} & s_{55} & s_{56} \\
s_{61} & s_{62} & s_{63} & s_{64} & s_{65} & s_{66}
\end{matrix} \right\|
$$

In short-hand these tensors relate the stress and strain via Hooke's law:

$$\overset{\leftrightarrow}{\varepsilon} = \|s\| \overset{\leftrightarrow}{\sigma}$$

$$\overset{\leftrightarrow}{\sigma} = \|c\| \overset{\leftrightarrow}{\varepsilon}$$

where $\overset{\leftrightarrow}{\varepsilon}$ and $\overset{\leftrightarrow}{\sigma}$ are the strain and stress tensors, respectively. The s_{ij} are referred to as the elastic compliance and the c_{ij} are referred to as the elastic stiffness constant.

D.1.2 Cubic materials

Most of the semiconductors and metals of interest are cubic materials, i.e. the single crystal form is a cubic crystal. Cubic materials are characterized by three independent values: c_{11}, c_{12}, c_{44}. For cubic crystals: $c_{ij} = c_{ji}$ and

$$c_{11} = c_{22} = c_{33}; c_{12} = c_{13} = c_{23}; c_{44} = c_{55} = c_{66}; c_{45} = c_{46} = c_{56} = 0$$

$$s_{11} = (s_{11} + s_{12}) / (s_{11} - s_{12}) (s_{11} + 2s_{12})$$
$$c_{12} = -s_{12}/ (s_{11} - s_{12}) (s_{11} + 2s_{12})$$
$$c_{44} = 1/s_{44}$$

$$s_{11} = (c_{11} + c_{12}) / (c_{11} - c_{12}) (c_{11} + 2c_{12})$$
$$s_{12} = -c_{12}/ (c_{11} - c_{12}) (c_{11} + 2c_{12})$$
$$s_{44} = 1/c_{44}$$

$$1/ (s_{11} + 2s_{12}) = c_{11} + 2c_{12}$$
$$1/ (s_{11} - s_{12}) = c_{11} - c_{12}$$

D.1.3   Isotropic materials

Polycrystalline materials and amorphous materials are usually described as "isotropic." They are characterized by two independent parameters: c_{11} and c_{12} or s_{11} and s_{12}.

The relations between the c's and the s's are the same as for the cubic material with the additional relation:

$$c_{44} = (c_{11} - c_{12}) /2$$

$$s_{44} = 2 (s_{11} - s_{12})$$

D.1.4   Common elastic constants (isotropic materials)

Young's Modulus (Y)
$Y = (c_{11} - c_{12}) (c_{11} + 2c_{12}) / (c_{11} + c_{12})$
$Y = 1/s_{11}$

Bulk Modulus (K) ($-V \cdot dp/dV$)
$K = (c_{11} + 2c_{12}) /3$
$K = 1/3 (s_{11} + 2s_{12})$

Poisson's Ratio (v)
$v = c_{12}/ (c_{11} + c_{12})$
$v = -s_{12}Y = -s_{12}/s_{11}$

Linear Compressibility, (β) ($l \cdot dp/dl$)$^{-1}$
$\beta = 1/ (c_{11} + 2c_{12})$
$\beta = s_{11} + 2s_{12}$

Shear Modulus (μ)
$\mu = c_{44} = (c_{11} - c_{12}) /2$
$\mu = 1/s_{44} = 1/2 (s_{11} - s_{12})$
$\mu = Y/ [2 (1 + v)]$

D.1.5 c's in terms of Y, v, μ (isotropic materials)

$$c_{11} = Y(1-v)/(1+v)(1-2v)$$

$$c_{12} = Yv/(1+v)(1-2v)$$

$$c_{44} = Y/2(1+v) = \mu$$

D.1.6 Lamé constants (isotropic materials)

A commonly used notation involving the stiffness constants is the Lamé constants λ, μ for isotropic materials:

$$c_{12} = \lambda, \quad c_{44} = \mu, \quad c_{11} = 2\mu + \lambda$$

D.1.7 Young's modulus—single cubic crystals

In single crystals Young's modulus is not isotropic but depends on the crystallographic direction of the applied stress. For a cubic system:

$$1/Y_{l_1 l_2 l_3} = s_{11} - 2(s_{11} - s_{12} - s_{44}/2)\left(l_1^2 l_2^2 + l_2^2 l_3^2 + l_3^2 l_1^2\right)$$

where l_i represent the direction cosines referred to the <100> axes. For a [100] orientation $\left(l_1^2 l_2^2 + l_2^2 l_3^2 + l_3^2 l_1^2\right) = 0$ and $Y = 1/s_{11}$, the same value as for an isotropic material.

Direction	$l_1^2 l_2^2 + l_2^2 l_3^2 + l_3^2 l_1^2$	$Y_{l_1 l_2 l_3}$
<100>	0	$1/s_{11}$
<111>	1/3	$3/(s_{11} + 2s_{12} + s_{44})$
<110>	1/4	$4/(2s_{11} + 2s_{12} + s_{44})$

D.1.8 Elastic strain energy

The energy of a thin film, thickness h, with strain ε parallel to the film plane is given by Cahn [6] as

$$E_\varepsilon = B\varepsilon^2 h$$

For common electric materials, growth is usually along one of the principal crystalline directions: $<100>$, $<111>$, $<110>$. In this case, B is given exactly by

$$B = \frac{1}{2} (c_{11} + 2c_{12}) \cdot \left[3 - \frac{c_{11} + 2c_{12}}{c_{11} + 2 (2c_{44} - c_{11} + c_{12}) (l_1^2 l_2^2 + l_2^2 l_3^2 + l_3^2 l_1^2)} \right]$$

where l_1, l_2, l_3 are the direction cosines that relate the direction normal to the interface or cube axes.

	$l_1^2 l_2^2 + l_2^2 l_3^2 + l_3^2 l_1^2$	B
$<100>$	0	$\dfrac{(c_{11} + 2c_{12})(c_{11} - c_{12})}{c_{11}}$
$<111>$	1/3	$\dfrac{6(c_{11} + 2c_{12})c_{44}}{c_{11} + 2c_{12} + 4c_{44}}$
$<110>$	1/4	$\left(\dfrac{c_{11} + 2c_{12}}{2}\right)\left(\dfrac{c_{11} - c_{12} + 6c_{44}}{c_{11} + c_{12} + 2c_{44}}\right)$

Note that there is a difference between $Y(l_1, l_2, l_3)$ and $B(l_1, l_2, l_3)$. The former is the correct cubic crystal value the latter is the appropriate strain factor for a film under biaxial stress. B is occasionally referred to as "Young's modulus under biaxial stress."

In an "isotropic approximation" $B(100) = 2\mu \left(\frac{1+\nu}{1-\nu}\right)$, often used in strain energy calculations.

D.2 Tables of elastic constants

D.2.1 Group IV Semiconductors (units of 10^{11} dyne/cm^2 at 300 K)

	c_{11}	c_{12}	c_{44}
Diamond	107.6	12.5	57.6
Si	16.56	6.39	7.90
Ge	12.88	4.83	6.71

From Gray [7]

	c_{11}	c_{12}	c_{44}
Diamond	107.64	15.2	57.4
Si	16.577	6.393	7.962
Ge	12.40	4.13	6.83

From Böer [4]

D.2.2 **Group III-V Semiconductors (10^{11} dyne/cm^2 at 300 K)**

	c_{11}	c_{12}	c_{44}
AlAs	12.02	5.70	5.89
AlSb	8.77	4.34	4.976
GaP	14.050	6.203	7.033
GaAs	11.90	5.38	5.95
GaSb	8.834	4.023	4.322
InP	10.11	5.61	4.56
InAs	8.329	4.526	3.959
InSb	6.669	3.645	3.020

From Böer [4]

	c_{11}	c_{12}	c_{44}
$Al_xGa_{1-x}As$	$11.88 + 0.14x$	$5.38 + 0.32x$	$5.94 - 0.05x$

From Adachi [1]

D.2.3 **Common metals (10^{11} dyne/cm^2 at 300 K)**

	c_{11}	c_{12}	c_{44}
Al	10.82	6.13	2.85
Ag	12.40	9.34	4.61
Au	18.6	15.7	4.20
Cr	35.0	6.78	10.08
Cu	16.84	12.14	7.54
Ni	24.65	14.73	12.47
Mo	46	17.6	11.0
W	50.1	19.8	15.14

From Huntington [8]

D.2.4 **Common insulators (10^{11} dyne/cm^2)**

	Y	μ	ν
Fused quartz[a]	7.26	3.10	0.17
Vitreous silica[b]	7.29	3.13	0.17
Silicon nitride[c]	3.00	—	0.22

[a] Huntington [8]
[b] Bansal and Doremus [2]
[c] Battelle [3]

D.3   Useful combinations for the common semiconductors (10^{11} dyne/cm^2)

	Y<100>	Y<111>	Y<011>
GaAs	8.53	14.12	12.13
GaP	10.34	16.69	14.47
Si	13.02	18.75	16.89
Ge	10.37	15.51	13.80

From Brantley [5]

D.4   Conversion factors

To convert from dyne/cm^2

To:	Multiply by:
Atmospheres	9.87×10^{-7}
Bars	1×10^{-6}
Pounds/sq in	1.45×10^{-5}
Newtons/m^2	0.1
Pascal ($= 1$ N/m^2)	0.1
Mm of Hg	7.5×10^{-4}

$$1 \text{ dyne/cm}^2 = 1 \text{ erg/cm}^3 = 6.24 \times 10^{11} \text{ eV/cm}^3$$

$$1 \text{ N/m}^2 = 1 \text{ J/m}^3 = 6.24 \times 10^{18} \text{ eV/m}^3$$

References

[1] S. Adachi, "GaAs, AlAs, and Al$_x$Ga$_{1-x}$As material parameters for use in research and device applications," *J. Appl. Phys.* **58**, R1 (1985).

[2] N. P. Bansal and R. H. Doremus, *Handbook of Glass Properties* (Academic Press, Orlando, 1986).

[3] Battelle-Columbus Laboratories, Engineering Property Data on Selected Ceramics, Vol. 1, Nitrides, Metals and Ceramics Information Center, Battelle's Columbus Laboratories (1976). Internal Report-M.C.I.C.-HB-07 Vol I.

[4] K. W. Böer, *Survey of Semiconductor Physics* (Von Nostrand Reinhold, 1990).

[5] W. A. Brantley, "Calculated elastic constants for stress problems associated with semiconductor devices," *J. Appl. Phys.* **44** (1973), 534.

[6] J. W. Cahn, "On spinodal decomposition in cubic crystals," *Acta. Met.* **10** (1962), 179.

[7] D. E. Gray (Coord. Ed.), *American Institute of Physics Handbook* (McGrawHill Book Company, 1972).

[8] H. B. Huntington, in *Solid State Physics* Vol. 7, eds F. Seitz and D. Turnbull (Academic Press, New York, 1958), 213.

[9] C. Kittel, *Introduction to Solid State Physics*, 2nd edn (John Wiley & Sons, New York, 1953).

[10] J. F. Nye, *Physical Properties of Crystals* (Oxford, Clarendon Press, London, 1957).

[11] G. Simmons and H. Wing, *Single Crystal Elastic Constants and Calculated Aggregate Properties* (MIT Press, Cambridge, 1971).

Appendix E   Terrace size distribution in Si MBE

In Chapter 7 (Section 7.5) we discussed the step periodicity in Si MBE. In this appendix we show that the standard deviation of the terrace size distribution reduces upon deposition of a small amount of material, under the assumption that the material grows via step growth. In this simplified argument we further assume that growth is entirely one way (i.e. attachment of all atoms occurs at the up-step only). This is a critical assumption which may not always apply, although the successful growth of step-mediated structures illustrates that the assumption is applicable in some cases.

Consider a terrace-size distribution, representing an irregular terrace-size array.

The average terrace width $\bar{l}$ is given by

$$\bar{l} = \frac{1}{N} \sum_{i=1}^{N} l_i \tag{E.1}$$

where we will exclude the first and last step for convenience. In general, the number of elements N is so large that there is no diffusivity with this assumption.

The standard deviation SD of the terrace-size distribution is given by

$$SD = \frac{1}{N} \sum_{i=1}^{N} \left(l_i - \bar{l} \right)^2 \tag{E.2}$$

and is a measure of the uniformity of terrace sizes. If SD approaches zero, we have an infinitely sharp terrace-size distribution and a perfectly periodic step spacing.

Note that SD can also be written as

$$SD = \frac{1}{N} \sum_{i=1}^{N} \left(l_i^2 - 2l_i\bar{l} + \bar{l}^2 \right) = \left(\frac{1}{N} \sum_{i=1}^{N} \left(l_i^2 \right) \right) - \bar{l}^2 \tag{E.3}$$

since

$$\frac{1}{N} \sum_{i=1}^{N} 2l_i\bar{l} = \left(\frac{2\bar{l}}{N} \sum_{i=1}^{N} l_i \right) = 2\bar{l}^2 \tag{E.4}$$

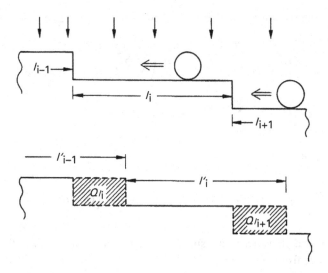

Fig. E.1 Representation of the surface cross-section during step-mediated growth, where a fraction Q of a monolayer of material leads to an increase in terrace length.

Assume that we add a fraction of a monolayer of material, Q, which diffuses to the nearest step riser (Fig. E.1). Then the terrace l_i acquires a new size l'_i given by

$$l'_i = l_i - Ql_i + Ql_{i+1}$$

or

$$l'_i = (1 - Q)\, l_i - + Ql_{i+1} \tag{E.4}$$

The original terrace becomes smaller by a fraction proportional to the original terrace size and larger by a fraction proportional to the adjacent terrace size. The new average terrace width $\bar{l}'$ becomes

$$\bar{l}' = \frac{1}{N} \sum_{i=1}^{N} [(1 - Q)\, l_i + Ql_{i+1}] = \bar{l} \tag{E.5}$$

where we have taken $l_1 = l_{N+1}$ for convenience. This result shows that the average terrace width of the new distribution is the same as that of the original distribution. This is intuitively sensible. The average terrace width is determined by the miscut angle and the lattice constant, $a/\tan\theta$, where θ is the miscut angle. Adding epitaxial material can never change this geometric relation, so the average terrace width remains the same. This is equivalent to saying that homoepitaxy neither increases nor decreases the number of steps per length.

Now consider the standard deviation of the new terrace size distribution,

$$SD' = \frac{1}{N} \sum_{i=1}^{N} (l'_i - \bar{l}')^2 = \left(\frac{1}{N} \sum_{i=1}^{N} l'^2_i \right) - \bar{l}'^2 \tag{E.6}$$

$$SD' = \frac{1}{N} \sum_{i=1}^{N} [l_i (1 - Q) + l_{i+1} Q]^2 - \bar{l}'^2 \tag{E.7}$$

Expanding the squared term,

$$SD' = \frac{1}{N} \sum_{i=1}^{N} \left[l_i^2 - 2Q l_i^2 + Q^2 l_i^2 + Q^2 l_{i+1}^2 + 2l_i l_{i+1} \left(Q - Q^2 \right) \right] - \bar{l}'^2 \tag{E.8}$$

The sum of the first and last terms are simply the standard deviation SD of the original distribution $(\bar{l}' = \bar{l})$ from Eq. (E.3), so

$$SD' = SD - \frac{2}{N} \sum_{i=1}^{N} l_i^2 \left(Q - Q^2 \right)^2 + \frac{2}{N} \sum_{i=1}^{N} l_i l_{i+1} \left(Q - Q^2 \right)$$

$$SD' = SD - \frac{2 \left(Q - Q^2 \right)}{N} \left(\sum_{i=1}^{N} \left(l_i^2 - l_i l_{i+1} \right) \right) \tag{E.9}$$

Our goal is to show that $SD' < SD$, which follows if the quantity following the negative sign is positive. The factor $2 \left(Q - Q^2 \right) / N$ is certainly positive since $0 < Q < 1$. For the second factor, consider the quantity:

$$\sum_{i=1}^{N} (l_i - l_{i+1})^2 = 2 \sum_{i=1}^{N} \left(l_i^2 - l_i l_{i+1} \right) \tag{E.10}$$

since

$$\sum_{i=1}^{N} l_i^2 = \sum_{i=1}^{N} l_{i+1}^2 \qquad l_1 = l_{N+1}$$

Since the left-hand side of Eq. (E.10) is positive, the right-hand side is also positive, thus

$$SD' < SD \tag{E.11}$$

This is the main result of our proof; the standard deviation decreases upon deposition in (preferentially) step-mediated growth. It is clear that one can continue to deposit material and each time the resulting standard deviation will decrease and approach zero. This corresponds to approaching the perfect periodic step distribution.

The step distribution can never become perfect due to at least two effects. The first is the *statistical nature of the growth process*. The number of deposited atoms M in any one

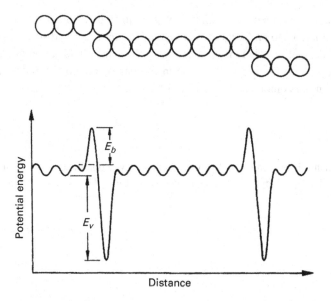

Fig. E.2 Schematic of a stepped terrace (side view) and the potential energy of an adsorbed atom.

interaction is assumed to be proportional to the terrace size. Since this is a "Poisson-like" process, there will be an uncertainty in this number of the order of $\sqrt{M}$, giving rise to imperfection in growth. The second reason the step distribution cannot become perfect is because of *thermal fluctuations*. A step is not perfectly sharp but has a roughness due to thermal fluctuations. This imperfect step, originating from the thermal processes, also yields an imperfect template resulting in a lack of perfect periodicity.

A more complete derivation (Gossmann *et al.* [1]) shows that after S interactions of deposited material Q, the standard deviation SD^S in the terrace width relative to the initial standard deviation is

$$SD^S / SD = \frac{1}{(2\pi\Theta)^{1/4}} \tag{E.12}$$

where $\Theta = SQ$. Note that the approach to periodicity is slow because the driving force for reorganization depends on the adjacent terrace size differences which decrease as growth proceeds.

One more word about terrace size distributions and the approach to periodicity: the derivation given assumes that atoms only stick at an up-step and do not stick at a down-step. This assumption is investigated more thoroughly in the work of Gossmann and others [1]; its ultimate justification must be related to the atomic structure at a step.

The atomic parameters which determine this situation can be viewed as a potential energy diagram (Fig. E.2). E_b is the energy barrier, which may or may not be present, and E_v is the potential well associated with the enhanced binding energy. Some limiting cases are immediately apparent: if $E_b = 0$ and $E_v \to \infty$, then it is no more probable for an atom to stick at an up-step than at a down-step, and periodicity is not achieved. If

$E_b \to \infty$ and $E_v \to \infty$, then atoms stick only at an up-step. It is important to recognize that if $E_b = 0$, there is no preferential sticking and the terrace size distribution remains the same upon deposition. There is no decrease of the standard deviation in this case. If sticking is only partially preferential, there is an approach to periodicity, but the deviation is considerably more complicated than that given here.

References

[1] H.-J. Gossmann, F. W. Sinden and L. C. Feldman, "Evolution of terrace size distributions during thin-film growth by step-mediated epitaxy," *J. Appl. Phys.* **67** (1990), 745.

Appendix F   Interdiffusion coefficient

In Chapter 5, we derived the interdiffusion coefficient by using Fick's first law where the driving force is the concentration gradient. In this Appendix, we derive the coefficient by using the chemical potential gradient as the driving force.

The marker velocity can be given as

$$v = (D_B - D_A)\frac{\partial X_B}{\partial x} = D_B\frac{\partial X_B}{\partial x} + D_A\frac{\partial X_A}{\partial x}$$

By substituting v into the equation of J_B, we have

$$J_B = j_B + C_B \left(D_B\frac{\partial X_B}{\partial x} + D_A\frac{\partial X_A}{\partial x} \right) = j_B - X_B(j_B + j_A)$$

In the analysis above, we have presented flux in terms of concentration gradient and we obtain a pair of equations of marker velocity v and interdiffusion coefficient, so that we can calculate the intrinsic diffusion coefficient of D_A and D_B. All the diffusion coefficients indicate that atomic diffusion goes with concentration gradient, i.e. from high to low concentration. However, in spinodal decomposition, it is against concentration gradient, the diffusion coefficient becomes negative in the field of concentration. Recall that diffusion should be driven by chemical potential gradient. Below, we shall present atomic flux in terms of chemical potential gradient. Recall that

$$j = C <v> = CMF = CM\left(-\frac{\partial \mu}{\partial x} \right)$$

where μ is the chemical potential in the alloy and M is mobility. We shall use the chemical potential gradient instead of concentration gradient as the driving force of interdiffusion. We take

$$j_B = -C_B M_B \frac{\partial \mu_B}{\partial x} = -C X_B M_B \frac{\partial \mu_B}{\partial x}$$

$$j_A = -C_A M_A \frac{\partial \mu_A}{\partial x} = -C X_A M_A \frac{\partial \mu_A}{\partial x}$$

By substituting j_B and j_A into the J_B equation, we have

$$J_B = -CX_BM_B\frac{\partial \mu_B}{\partial x} + CX_B\left[X_BM_B\frac{\partial \mu_B}{\partial x} + (1-X_B)M_A\frac{\partial \mu_A}{\partial x}\right]$$

$$= -C\left[X_BM_B\frac{\partial \mu_B}{\partial x} - X_B^2M_B\frac{\partial \mu_B}{\partial x} - (1-X_B)M_A\frac{\partial \mu_A}{\partial x}\right]$$

$$= -C\left\{X_B(1-X_B)\left[M_B\frac{\partial \mu_B}{\partial x} - M_A\frac{\partial \mu_A}{\partial x}\right]\right\}$$

From the Gibb–Duhem equation, we have

$$X_Ad\mu_A + X_Bd\mu_B = 0$$

$$(1-X_B)d\mu_A + X_Bd\mu_B = 0$$

Thus, we have

$$(1-X_B)M_A\frac{d\mu_A}{dx} + X_BM_A\frac{d\mu_B}{dx} = 0 \tag{F.1}$$

$$(1-X_B)M_B\frac{d\mu_A}{dx} + X_BM_B\frac{d\mu_B}{dx} = 0 \tag{F.2}$$

Now if we add Eq. (F.1) and subtract Eq. (F.2) in the J_B bracket, in other words we just add a zero and subtract a zero from the bracket, we have

$$M_B\frac{\partial \mu_B}{\partial x} + (1-X_B)M_A\frac{\partial \mu_A}{\partial x} + X_BM_A\frac{\partial \mu_B}{\partial x} - M_A\frac{\partial \mu_A}{\partial x}$$

$$- (1-X_B)M_B\frac{\partial \mu_B}{\partial x} - X_BM_B\frac{\partial \mu_B}{\partial x}$$

$$= M_B\frac{\partial \mu_B}{\partial x} + M_A\frac{\partial \mu_A}{\partial x} - X_BM_A\frac{\partial \mu_A}{\partial x} + X_BM_A\frac{\partial \mu_B}{\partial x} - M_A\frac{\partial \mu_A}{\partial x}$$

$$- M_B\frac{\partial \mu_A}{\partial x} + X_BM_B\frac{\partial \mu_A}{\partial x} - X_BM_B\frac{\partial \mu_B}{\partial x}$$

$$= M_B\left(\frac{\partial \mu_B}{\partial x} - \frac{\partial \mu_A}{\partial x}\right) + X_B(M_A - M_B)\left(\frac{\partial \mu_B}{\partial x} - \frac{\partial \mu_A}{\partial x}\right)$$

$$= [(1-X_B)M_B + X_BM_A]\left(\frac{\partial \mu_B}{\partial x} - \frac{\partial \mu_A}{\partial x}\right)$$

Thus, we have

$$J_B = -C[X_B(1-X_B)][(1-X_B)M_B + X_BM_A]\left(\frac{\partial \mu_B}{\partial x} - \frac{\partial \mu_A}{\partial x}\right)$$

$$= -CX_BM\frac{\partial}{\partial y}(\mu_B - \mu_A)$$

where $M = (1-X_B)[(1-X_B)M_B + X_BM_A] = X_A[X_AM_B + X_BM_A]$

By virtue of the definition of chemical potential,

$$dG = \mu_A dC_A + \mu_B dC_B = (\mu_B - \mu_A)dC_B$$

Or we have

$$\frac{dG}{dC_B} = \mu_B - \mu_A$$

Thus,

$$J_B = -CX_B M \frac{\partial}{\partial x}\left(\frac{\partial G}{\partial C_B}\right) = -C_B M \left(\frac{\partial^2 G}{\partial C_B^2}\frac{\partial C_B}{\partial y}\right) = -C_B M G''\frac{\partial C_B}{\partial x}$$

Or we have $-J_B/(\partial C_B/\partial x) = C_B M G'' = \overline{D}$, which is the interdiffusion coefficient. Recall that

$$J_B = -\overline{D}\frac{\partial C_B}{\partial x}$$

where $\overline{D} = X_A D_B + X_B D_A$ is the interdiffusion coefficient. Also as shown above, $\overline{D} = C_B M G''$, so the interdiffusion coefficient takes the sign of G''. Note that G'' is negative within spinodal region, so $\overline{D}$ is negative, indicating that the diffusion is against the concentration gradient, i.e. an uphill diffusion. Outside the spinodal region, G'' is positive and the interdiffusion coefficient is positive as in Section 5.3.2.1.

Appendix G Table of physical properties

G.1 Physical properties of elements used in electronic materials

Element (Z)	Melting point [°C]	Density [g/cm³]	Electrical resistance [$10^{-6}\Omega$cm]	[°C]	Thermal conductivity [cal/cm sec°C]	[°C]	Coefficient of thermal expansion [10^{-6}/°C]	[°C]
Aluminium A1 (13)	659.7	2.7	2.6	0	0.48	18	23.8	0–100
Bismuth Bi (83)	271.3	9.8	119.0	18	0.019	18	14.0	–
Chromium Cr (24)	1785	7.19	12.8	20	0.16	20	6	25
Copper Cu (29)	1083	8.96	1.7	20	0.98	20	16.6	25
Cobalt Co (27)	1495	8.9	6.3	20	0.16	18	12	25
Germanium Ge (32)	936.0	5.4	60*[Ωcm]	20	0.14	25	5.3	20
Gold Au (79)	1063.0	19.3	2.4	20	0.7	18	14.2	–
Indium In (49)	156.4	7.3	8.4	0	0.057	–	33.0	20
Iron Fe (26)	1536	7.86	10.0	20	0.18	20	12	25
Lead Pb (82)	327.4	11.3	22.0	20	0.083	18	29.5	0–100
Iridium Ir (77)	2454	22.5	5.3	20	0.17	20	6	25
Molybdenum Mo (42)	2620±10	10.2	5.7	20	0.34	20	5.1	25–100
Nickel Ni (28)	1455.0	8.9	6.8	20	0.14	18	13.0	50
Polladium Pd (46)	1552	12	10.7	20	0.17	20	–	25
Platinum Pt (78)	1773.5	21.4	10.0	20	0.17	20	9.0	40
Silver Ag (47)	961.0	10.5	1.5	0	0.97	18	18.7	20
Silicon Si (14)	1420.0	2.3	4000*[Ωcm]	20	0.35	–	2.6	40
Tantalum Ta (73)	2996	16.6	12.3	20	0.13	20	6.5	25
Titanium Ti (22)	1668	4.51	41.6	20	–	20	8.5	25
Tin Sn (50)	231.9	7.3	11.5	20	0.15	18	26.7	18–100
Tungsten W (74)	3370.0	19.3	5.5	20	0.476	17	4.5	–
Vanadium V (23)	1990	6.1	25	20	–	20	8	25
Zinc Zn (30)	419.4	7.1	5.7	0	0.269	0	26.3	0–100

*Depends on purity

Periodic Table of the elements listing symbol, atomic number, atomic mass, and structure at 20°C.

1	2	3	4	5	6	7	8	9	10	11	12	13	14	15	16	17	18
H 1 1.008																	He 2 4.00
Li 3 6.94 bcc	Be 4 9.01 hex											B 5 10.8 rhom	C 6 12.01 dia	N 7 14.01	O 8 16.0	F 9 19.0	Ne 10 20.17
Na 11 22.99 bcc	Mg 12 24.31 hex											Al 13 26.98 fcc	Si 14 28.09 dia	P 15 30.97 orth	S 16 32.06 orth	Cl 17 35.43	Ar 18 39.95
K 19 39.1 bcc	Ca 20 40.1 fcc	Sc 21 44.96 hex	Ti 22 47.88 hex	V 23 50.94 bcc	Cr 24 52.0 bcc	Mn 25 54.94 cubic	Fe 26 55.85 bcc	Co 27 58.93 hex	Ni 28 58.73 fcc	Cu 29 63.55 fcc	Zn 30 65.39 hex	Ga 31 69.72 orth	Ge 32 72.64 dia	As 33 74.92 rhom	Se 34 78.99 hex	Br 35 79.9	Kr 36 83.8
Rb 37 85.47 bcc	Sr 38 87.62 fcc	Y 39 88.91 hex	Zr 40 91.24 hex	Nb 41 92.91 bcc	Mo 42 95.89 bcc	Tc 43 – hex	Ru 44 101.0 hex	Rh 45 101.0 fcc	Pd 46 106.4 fcc	Ag 47 107.9 fcc	Cd 48 112.4 hex	In 49 114.8 tetr	Sn(α) 50 118.7 dia	Sb 51 121.8 rhom	Te 52 127.6 hex	I 53 126.9	Xe 54 131.3
Cs 55 132.9 bcc	Ba 56 137.3 bcc	La 57 138.9 hex	Hf 72 178.5 hex	Ta 73 180.9 bcc	W 74 183.8 bcc	Re 75 186.2 hex	Os 76 190.3 hex	Ir 77 192.2 fcc	Pt 78 195.1 fcc	Au 79 197.0 fcc	Hg 80 200.6 rhom	Ti 81 204.4 hex	Pb 82 207.2 fcc	Bi 83 209 rhom	Po 84 (210) hex	At 85 (210)	Rn 86 (222)
Fr 87	Ra 88	Ac 89 fcc															

Ce 58 140.1 fcc	Pr 59 140.9 hex	Nd 60 144.2 hex	Pm 61	Sm 62 150.4	Eu 63 152.0 bcc	Gd 64 157.3 hcp	Tb 65 158.9 hcp	Dy 66 162.5 hcp	Ho 67 164.9 hcp	Er 68 167.3 hcp	Tm 69 168.9 hcp	Yb 70 173.0 fcc	Lu 71 175.0 hcp
Th 90 fcc	Pa 91 tetr	U 92 238.0	Np 93	Pu 94	Am 95 hex	Cm 96	Bk 97	Cf 98	Es 99	Fm 100	Md 101	No 102	Lr 103

Legend:
Si — Symbol
14 — Atomic No.
28.09 — Atomic Mass
dia — Structure

G.3 Physical constants, conversions, and useful combinations

Physical constants

Avogadro constant	$N_A = 6.022 \times 10^{23}$ particles/mole
Boltzmann constant	$k = 8.617 \times 10^{-5}$ eV/K $= 1.38 \times 10^{-23}$ J/K
Elementary charge	$e = 1.602 \times 10^{-19}$ coulomb
Planck constant	$h = 4.136 \times 10^{-15}$ eV $\cdot$ s
	$= 6.626 \times 10^{-34}$ joule $\cdot$ s
Speed of light	$c = 2.998 \times 10^{10}$ cm/s
Permittivity (free space)	$\epsilon_0 = 8.85 \times 10^{-14}$ farad/cm
Electron mass	$m = 9.1095 \times 10^{-31}$ kg
Coulomb constant	$k_c = 8.988 \times 10^9$ newton-m^2/(coulomb)^2
Atomic mass unit	$u = 1.6606 \times 10^{-27}$ kg
Acceleration of gravity	$g = 980$ dyn/gm

Useful combinations

Thermal energy (300 K)	$kT = 0.0258$ eV $\approx (1/40)$ eV
Photon energy	$E = 1.24$ eV at $\lambda = 1$ μm
Permittivity (Si)	$\epsilon = \epsilon_r \epsilon_0 = 1.05 \times 10^{-12}$ farad/cm

Conversions

1 nm $= 10^{-9}$ m $= 10$ Å $= 10^{-7}$ cm

1 eV $= 1.602 \times 10^{-19}$ joule $= 1.602 \times 10^{-12}$ erg

1 eV/particle $= 23.06$ kcal/mol

1 newton $= 0.102$ kg$_{force}$ $= 1$ coulomb $\cdot$ volt/meter

10^6 newton/m^2 $= 146$ psi $= 10^7$ dyn/cm^2

1 μm $= 10^{-4}$ cm

0.001 inch $= 1$ mil $= 25.4$ μm

1 bar $= 10^6$ dyn/cm^2 $= 10^5$ N/m^2

1 weber/m^2 $= 10^4$ gauss $= 1$ tesla

1 pascal $= 1$ N/m^2 $= 7.5 \times 10^{-3}$ torr

1 erg $= 10^{-7}$ joule $= 1$ dyn-cm

1 joule $= 1$ newton $\cdot$ meter $= 1$ watt $\cdot$ second

1 calorie $= 4.184$ joules

G.4 Properties of Si, Ge, GaAs and SiO_2 at 300 K

Properties	Si	Ge	GaAs	SiO_2
Atoms/cm^3, molecules/cm$^3 \times 10^{22}$	5.0	4.42	4.42	2.27
Structure	diamond	diamond	zincblende	amorphous
Lattice constant (nm)	0.543	0.565	0.565	—
Density (g/cm^3)	2.33	5.32	5.32	2.27
Relative dielectric constant, ϵ	11.9	16.0	13.1	3.9
Permittivity, $\epsilon = \epsilon_r \epsilon_0$(farad/cm) $\times 10^{-12}$	1.05	1.42	1.16	0.34
Expansion coefficient $\times (10^{-6}/K)$	2.6	5.8	6.86	0.5
Specific heat (joule/g K)	0.7	0.31	0.35	1.0
Thermal conductivity (watt/cm K)	1.48	0.6	0.46	0.014
Thermal diffusivity (cm^2/s)	0.9	0.36	0.44	0.006
Energy gap (eV) (300 K)	1.12	0.67	1.424	~9
Debye temperature (K)	645	347	360	—
Young's modulus Y(100) (10^{10} N/m^2)	13.0	10.3	8.55	—
Shear modulus μ (10^{10} N/m^2)	5.1	4.04	3.26	—
Bulk modulus K(10^{10} N/m^2)	9.8	7.52	7.55	—
Poisson's ratio ν	.28	.27	0.31	—

Elastic moduli: GaAs: S. Blakemore, *J. Appl. Phys.* **53**, R123 (1982)
Si, Ge: W. E. Beadle *et al.*, eds., *Quick Reference Manual For Silicon Integrated Circuit Technology* (Wiley, New York, 1985).

Index

AC electromigration 266, 302
Activation energy 61–62, 73
Activity coefficient 108
Adatoms 143
Aluminum
   diffusion coefficient 74–75
   effective charge number 228, 250
   electromigration 224–226, 257
   thermal expansion coefficient 387
   vacancy concentration 80
   void formation 317
   grain boundary diffusion 196
Amorphous alloy 178, 213
Amorphous thin film 152
Anharmonicity 134
Anisotropic conductor 261, 264
Atomic density 53
Atomic volume 17, 53
Au-Al compound formation 172
Avogadro's number 18
Avrami's equation 359

Biaxial stress 124
Binding energy 31
Black's equation 356
Blech and Herring model 244
Blech structure 238
Boltzmann's distribution function 22
Boltzmann–Matano analysis 102
Bravais lattices 52
Built-in potential 60
Bulk diffusion couples 95
Bulk modulus 324, 373
Burgers vector 136

Capillary effect 38
Change-over thickness 184
Chemical potential 64, 311
Chemical vapor deposition 15
Clusters
   coalescence 163
   ripening 164
Cohesive energy 36
Collision frequency 21

Compliance 372
Compound formation 186
Compression 119
Conjugate force and flux 220
Constant volume process 337
Continuity equation 66
Contact angle 40
Coordination number of nearest
   neighbor 66
Copper
   effective charge number 249
   electromigration 279
   thermal expansion coefficient 387
   tin 319
Correlation factor 66
Creep
   Nabarro–Herring creep 313
   Coble creep 317
   zero creep 41
Critical disc of nucleation 157
Critical thickness 186
Crystal systems 52
Crystallographic axes 51
Current crowding 271, 346
Current density gradient force 259

Darken's analysis 98
DC electromigration 245–249, 302
Debye frequency 79
Debye temperature 79
Defect concentration 79
Deformation potential 229
Deposition parameters 26
Desorption frequency 143
Diffusion
   adatoms 143
   lattice 60, 86
   grain boundary 192
   surface 146
Diffusion barrier 281
Diffusion coefficient
   interdiffusion 101, 107, 384
   intrinsic 100
   tracer 108

Diffusion equation 71–73
Diffusion-induced grain boundary migration
    (DIGM) 207
Diffusional creep 41, 223, 313
Diffusional flux 6
Disilicide 175, 177
Dislocation
   core energy 135–137
Divergence 70, 338, 341
Drift velocity 16, 241
Driving force
   electromigration 220, 226, 246, 367
   stress-migration (creep) 226, 311
   thermomigration 220, 299

Effective charge number 226, 246,
    250–251
Einstein frequency 78
Einstein mode of vibration 249
Elastic modulus 121
Elastic limit 123
Elastic strain energy 123
Electric field 219, 228, 241, 246
Electrical potential 241
Electromigration
   alternative current 266, 302
   Al interconnects 276
   back-stress 251, 254
   critical length 251
   Cu interconnects 279
   direct current 245–249, 302
   flip-chip solder joints 341
   in metals 223–225, 237
   failure 270, 276, 279
Electron effective mass 247, 249
Electron wind force 246
Energy
   activation 61, 73
   binding 31
   bond 31
   dislocation 135
   elastic 123
   evaporation 35
   grain boundary 193
   interatomic potential 32, 36
   kinetic 22
   pair potential 31
   solid surface 46
   strain 123
   sublimation 35
   surface 35, 37
Enthalpy 65, 362
Entropy 65, 362, 365
Entropy generation 216
Epitaxy
   homo-epitaxial growth 149
Equilibrium binary phase diagram 170

Error function 97
Eutectic structure 293

Fick's first law of diffusion 64
Fick's second law of diffusion 66
Flip-chip technology
   solder joint 341
First law of thermodynamics 362
First-phase formation 174, 178
Fisher's analysis 197
Flux equation 15
Formation energy
   silicides 179
Free electron model of conductivity 237
Frequency
   attempt 62
   exchange 62
   atomic vibration 76
   surface vibration 143
Friction coefficient 241

Gas law
   ideal gas law 17
Gauss theorem 68, 70
Gibbs free energy function 362
Gibbs–Thomson potential 114
Gibbs–Thomson equation 164
Gold
   interatomic potential energy 36
   latent heat 35
   surface energy 36
Grain boundary
   diffusivity 194, 202
   energy 207
   electromigration 264
   migration 207
   small-angle 206
   tilt-type 193
Grain boundary diffusion
   comparison with bulk 194
   Fisher's analysis 197
   small-angle 206
   Whipple's analysis 202
Grain boundary migration
   diffusion-induced 207
Grain growth
   abnormal 281
   bamboo-type 276
Growth
   layered compound 84
   two-layered compounds 185
Growth kinetics
   diffusion control 180
   first phase 178
   interfacial reaction control 180
Growth modes 153

Ham's model of growth 109
Heat
  crystallization 35
  evaporation 35
  sublimation 35
  melting 35
  latent 35
  internal 362
Heat capacity 363
Helmholtz free energy 311, 362
Heterogeneous nucleation 155
Hillock growth 320, 337
Homoepitaxy
  growth modes 153
  growth rate 149
Homologous temperature 244
Hooke's Law 119
Huntington's electron wind force 246, 367

Ideal gas law 17
Ideal solution 65
Interatomic distance 31
Interatomic potential energy 31, 36
Interconnects 4, 212, 243, 270–279, 317
Interdiffusion 95, 170
Interdiffusion coefficient 107, 384
Interfacial reaction coefficient 183
Intermetallic compounds 70
Internal energy 362
Interstitial diffusion 61, 75
Intrinsic diffusion coefficient 107–108
Irreversible processes 212
Isotropic materials 373

Johnson–Mehl–Avrami theory 358
Jump distance 63
Jump frequency 61

Kidson's analysis 89
Kinetics
  growth 68, 108, 149, 17, 346
  nucleation 155
  precipitation 109, 113
Kink site 142
Kirkendall shift 98
Kirkendall (Frenkel) void 104

Lateral diffusion couples 174
Lattice constants 50
Lattice point 51
Lattice self diffusion 75
Lattice shift 104, 338
Lennard–Jones potential 33

Marker
  motion 98
  thin film reaction 186

Marker analysis 98, 186
Matano interface 103
Maxwell velocity distribution 24
Mean field consideration 112
Mean kinetic energy 19
Mean square velocity 26
Mean time to failure 355
Metals
  elastic constants 372
Metastable materials 213
Microstructure 276, 281
Miller indices 54
Miscut 141
Misfit dislocation
  elastic energy 135
Mobility
  atom 64
  electron 241
Modulus 121
Molar volume 107
Monolayer 189
MOSFET 2
Motion energy 62

Nabarro–Herring equation 313
Nano-twin 286
Nearest neighbors 66, 246
Ni-silicide
  lateral diffusion couple 174
  sequence of phase formation 174–175
Notation
  surface structure 51
Nucleation
  homogeneous 155
  heterogeneous 159
  surface disc 155
  void 104

Ohm's law 242
Overhang structure 258

Packaging technology 212
Pancake-type void 346
Patterned surfaces 159
Phase transformation 358
Peltier effect 235
Penetration depth 200
Pipe diffusion 206
$P–n$ junction 60
Poisson's ratio 121
Potential energy 39
Pre-exponential factor 81
Pressure
  equilibrium vapor 20
Pulsed direct current 245

Random walk 61
Reconstruction 54

Reliability
  science 336
  failure 11, 270, 341, 350
  physical analysis 341
  statistical analysis 350
Residence time 143
RHEED 155
Ripening 113
Root mean square velocity 26
Rutherford backscattering spectrometry 175

Second law of thermodynamics 362
Sectioning 200
Seebeck effect 233
Self-aligned silicide 5
Shear modulus 123
Shear strain 122
Shear stress 123
Silicide
  epitaxial 190
  formation 172
  formation energy 179
  transition metal silicides 179
Silicon
  activation energy of sublimation 145
  diffusion coefficient 76
  dopants 60
  elastic constants 375
  homoepitaxy 149
  self diffusion 75
  thermal expansion coefficient 387
  vapor pressure 146
  surface energy 159
Silver
  bulk diffusion 194
  grain boundary diffusion 194, 202
Single-phase formation 170
Small-angle grain boundaries 206
SnPb solder 291, 295
Solder joints 291
Steps 142, 147
Step-mediated growth 149
Step spacing 141
Step velocity 149
Sticking coefficient 144
Stirling's approximation 365
Stoney's equation 127
Strain
  energy 123
  shear 121
Stress
  backstress in electromigration 251–254
  biaxial 124
  chemical potential 311
  compressive 118
  effect of stress on electromigration 285
  in-plane 129

  relaxation 332
  migration 309
  thermal 131–134
  intrinsic stress in thin films 134
Strip 226
Supersaturation 146, 159
Surface
  atom density 35
  curvature 143
  desorption 143
  diffusion 146
  diffusivity 143
  energy 30
  flux 150
  kinetic processes 141
  reconstruction 54
  relaxation 54
  step 154
  step nucleation 156
Surface energy
  mechanical approach 46
  surface energy systematic 44
  thermodynamic approach 46
Surface step 142
Surface structure 51
Surface tension 36, 45
Synchrotron radiation
  micro X-ray diffraction 328

Tensile stress 120, 126, 317
Tensor 372
Terraces 142
Theoretical strength 119
Thermal-electric effect 232
Thermal expansion 133–134
Thermodynamics
  energy functions 362
Thermomigration
  driving force 299
  Pb-free solder joints 303
Thickness
  changeover 184
  critical 186
Thin-film couple 171
Thin-film marker 186–187
Thin-film reactions
  diffusion control 89
  first phase 178
  reaction control 180
Thomson effect 233
Tin
  cry 319
  hillock 320
  pest 319
  whisker 319
Tracer diffusion coefficient 108
Transition metal silicides 179

Triple point 196
TTT diagram 360

Under-bump-metallization 7

Vacancy
    concentration 79–80, 366
    diffusivity 315
    formation energy 79
    motion energy 79
Vapor pressure 14–21, 145
Velocity
    drift 16, 64, 241
    gas particles 19
    mean 25
    root mean square 26
    sound 19
Velocity distribution functions 22, 24
Vibrational frequency 76
Void formation 197, 272, 317, 346

Volume
    collision 21
    constant volume process 337
    non-constant volume process 337

Wear-out mechanism of electromigration 277
Weibull distribution 353, 391
Whipple's analysis 202
Whisker
    broken oxide model 332
    stress generation 323
    stress relaxation 332

X-ray diffraction
    glancing incidence 176

Young's modulus 121

Zero creep 41
Zener's growth model 87

Printed in the United States
by Baker & Taylor Publisher Services